The Origin and Early Evolution of Life

The Origin and Early Evolution of Life: Prebiotic Chemistry of Biomolecules

Special Issue Editor

Michele Fiore

MDPI • Basel • Beijing • Wuhan • Barcelona • Belgrade

MDPI

Special Issue Editor
Michele Fiore
University of Lyon
France

Editorial Office
MDPI
St. Alban-Anlage 66
4052 Basel, Switzerland

This is a reprint of articles from the Special Issue published online in the open access journal *Life* (ISSN 2075-1729) from 2018 to 2019 (available at: https://www.mdpi.com/journal/life/special_issues/ Prebiotic_Chemistry)

For citation purposes, cite each article independently as indicated on the article page online and as indicated below:

LastName, A.A.; LastName, B.B.; LastName, C.C. Article Title. *Journal Name* **Year**, *Article Number, Page Range.*

ISBN 978-3-03921-606-2 (Pbk)
ISBN 978-3-03921-607-9 (PDF)

Cover image courtesy of Michele Fiore.

Contents

About the Special Issue Editor

Michele Fiore, Associate Professor of organic chemistry at the University of Lyon, Claude Bernard, Lyon 1, conducts research in the field of systems chemistry, prebiotic chemistry, and on the preparation of synthetic protocells to identify abiotic models that enable understanding or a possible explanation for the origin of life. Michele Fiore is a Doctorate of the University of Naples "Federico II", where he completed his thesis on the chemical and biological characterization of phytotoxins produced by phytopathogenic fungi under the direction of Prof. A. Evidente. Michele Fiore has worked at the USDA, ARS, NPURU (United States Department of Agriculture, Agricultural Research Service, Natural Product Utilization Research Unit) under the direction of Dr. S. O. Duke and Dr. A. M. Rimando. He was Senior Postdoc at the University of Ferrara (Prof. A. Dondoni) and Grenoble Rhone-Alpes (Prof. O. Renaudet) where he studies the synthesis of bioactive molecules against degenerative and proliferative diseases and the formulation of vaccine prototypes against cancer. In the last five years, he has collaborated with Prof. P. Strazewski of the SysChem laboratory at the Institut de Chimie et Biochimie Moléculaires et Supramoléculaires of the University of Lyon on the formulation of evolutionary chemical systems. On the 4th of July 2019, he received the "Habilitation à Diriger des Recherches" from University of Lyon for his thesis entitled "Prebiotic Synthesis of Phospholipids and Membranogenic Compounds: Use and Application in Systems Chemistry. At present, Michele Fiore is Editor for the Royal Chemical Society.

Editorial

The Origin and Early Evolution of Life: Prebiotic Chemistry

Michele Fiore

Université de Lyon, Claude Bernard Lyon 1, Institut de Chimie et Biochimie Moléculaires et Supramoléculaires, Batiment Lederer, Bureau 11.002, 1 Rue Victor Grignard, F–69622 Villeurbanne CEDEX, France; michele.fiore@univ-lyon1.fr; Tel.: +33-(0)472-448-080

Received: 4 September 2019; Accepted: 9 September 2019; Published: 12 September 2019

Microfossil evidence indicates that cellular life on Earth emerged during the Paleoarchean era be-tween 3.6 and 3.2 thousand million years ago (Gya) [1]. But what is really what we call life? How, where, and when did life arise on our planet? These questions have remained most-fascinating over the last hundred years. The German biologist Carl Richard Woese emphasized the urgency of conducting in-depth studies in search of what in the early days of the formation of the universe and then of our planet, gave rise to what is called Life and he wrote *"Biology today is no more fully understood in principle than physics was a century or so ago. In both cases the guiding vision has (or had) reached its end, and in both, a new, deeper, more invigorating representation of reality is (or was) called for."* [2] From the beginning of the last century, and in accord with what David Deamer highlighted "Life can emerge where physics and chemistry intersect" and for this reason the study of the origin of Life intersect not only the organic and inorganic chemistry but also biology, astrophysics, geochemistry, geophysics, planetology, earth science, bioinformatics, complexity theory, mathematics and philosophy from the equation. From an evolutionary chemical point of view, is possible to presume that life emerged from a mixture of inanimate matter: Organic and inorganic compounds. Such compounds reacted under favorable conditions, forming molecules that are commonly called "biotic" and that, thanks to a kind of self-organization, gave rise to the first biopolymers and to proto-metabolisms. The geology and the chemistry of Earth before the advent of life was completely different from what we know today. At that time, sunlight, volcanic heat, and hydrothermal sites were the main energy sources that could drive the synthesis of many molecules, including nucleosides, peptides, sugars and amphiphilic compounds. The atmosphere was mostly nitrogen (N_2), as today, with a substantial amount of carbon dioxide (CO_2) and much smaller amounts of carbon monoxide, ammonia, and methane (CO, NH_3, CH_4). It is also likely that water, present in locally limited amounts, contained hydrogen cyanide (HCN), formaldehyde (HCHO) and formamide ($HCONH_2$). Intriguingly, those molecules are found in the interstellar space together with many other that can be considered as building blocks for the assembling of biomolecules such as water (H_2O), formic acid (HCOOH), methanol (CH_3OH) cyanamide (NH_2CN), acetic acid (CH_3COOH), acetamide (CH_3CONH_2), ethylene glycol ($HOCH_2CH_2OH$) and glycine [3,4]. Prebiotic chemistry experiences showed that the chemical combinations of different building blocks can give rise to the formations of different classes of biotic molecules such as 2′,3′-cyclic pyrimidine nucleotides, various–amino acids and glycerol phosphate [5–11]. The plausible scenarios for the assembling of these building blocks thus of such complex biomolecules are depicted as two: Hydrothermal vents and hydrothermal pools. Hydrothermal vents are systems whose heat source is the underlying magma or hot water generated by convection currents due to high thermal gradients [12]. The alternatives to hydrothermal vents are hydrothermal fields known also as hydrothermal pools. Recently, Damer and Deamer pointed out that fluctuating hydrothermal pools (FHPs) could be considered as plausibly prebiotic reactors for the synthesis of several key molecules for the development of life, including lipids, nucleic acids and peptides [13]. This short résumé is to say that the seventeen papers published in this special issue perfectly matches with the aim of the study of the origin of Life from a system chemistry and prebiotic chemistry perspective. We expect that this collection of original articles and reviews will provide the reader with an updated view of some important aspects of prebiotic chemistry thought. We hope that in the further investigations on the origin of Life will bring scientist to combine prebiotic chemistry and system chemistry in order to develop new strategies for the best understanding of how life emerged on planet based on the use of protocells models that can encapsulate sort of primitive metabolisms [14,15].

Acknowledgments: Michele Fiore wish to warmly thank all the contributors of the special issue of LIFE (ISSN 2075-1729): "The Origin and Early Evolution of Life: Prebiotic Chemistry". My daily work is dedicated to the memory of my beloved daughter Océane (2015–2017).

Conflicts of Interest: The author declares no conflict of interest.

References

1. Wacey, D.; Kilburn, M.R.; Saunders, M.; Cliff, J.; Brasier, M.D. Microfossils of sulphur-metabolizing cells in 3.4-billion-year-old rocks of Western Australia. *Nat. Geosci.* **2011**, *4*, 698–702.
2. Woese, C.R. A new biology for a new century. *Microbiol. Mol. Biol. Rev.* **2004**, *68*, 173–186. [CrossRef] [PubMed]
3. Cleaves, H.J., II. Prebiotic Chemistry: Geochemical Context and Reaction Screening. *Life* **2013**, *3*, 331–345. [CrossRef] [PubMed]
4. Zahne, K.; Schaefer, L.; Fegley, B. Earth's Earliest Atmospheres. *Cold Spring Harbor Perps Biol.* **2010**, *2*, a004895.
5. Patel, B.H.; Percivalle, C.; Ritson, D.J.; Duffy, C.D.; Sutherland, J.D. Common origins of RNA, protein and lipid precursors in a cyanosulfidic protometabolism. *Nat. Chem.* **2015**, *7*, 301–307. [CrossRef] [PubMed]
6. Saladino, R.; Crestini, C.; Pino, S.; Costanzo, G.; Di Mauro, E. Formamide and the origin of life. *Phys. Life Rev.* **2012**, *9*, 84–104. [CrossRef] [PubMed]
7. Fiore, M.; Strazewski, P. Bringing Prebiotic Nucleosides and Nucleotides Down to Earth. *Angew. Chem. Int. Ed.* **2016**, *55*, 13930–13933. [CrossRef] [PubMed]
8. Fiore, M.; Strazewski, P. Prebiotic Lipidic Amphiphiles and Condensing Agents on the Early Earth. *Life* **2016**, *6*, 17. [CrossRef] [PubMed]
9. Fiore, M. The synthesis of mono-alkyl phosphates and their derivatives: An overview of their nature, preparation and use, including synthesis under plausible prebiotic conditions. *Org. Biomol. Chem.* **2018**, *16*, 3068–3086. [CrossRef] [PubMed]
10. Fayolle, D.; Altamura, E.; D'Onofrio, A.; Madanamothoo, W.J.; Fenet, B.; Mavelli, F.; Buchet, R.; Stano, P.; Fiore, M.; Strazewski, P. Crude phosphorylation mixtures containing racemic lipid amphiphiles self-assemble to give stable primitive compartments. *Sci. Rep.* **2017**, *7*, 18106. [CrossRef] [PubMed]
11. Fiore, M.; Madanamoothoo, W.; Berlioz-Barbier, A.; Manniti, O.; Girard-Egrot, A.; Buchet, R.; Strazewski, P. Giant vesicles from rehydrated crude phosphorylation mixtures containing mono-alkyl phosphoethanolamine and its analogues. *Org. Biomol. Chem.* **2017**, *15*, 4231–4238. [CrossRef] [PubMed]
12. Miller, S.L.; Bada, J.L. Submarine hot springs and the origin of life. *Nature* **1988**, *334*, 609–611. [CrossRef] [PubMed]
13. Damer, B.; Deamer, D. Coupled Phases and Combinatorial Selection in Fluctuating Hydrothermal Pools: A Scenario to Guide Experimental Approaches to the Origin of Cellular Life. *Life* **2015**, *5*, 872–887. [CrossRef] [PubMed]
14. Fiore, M.O.; Maniti, O.; Girard-Egrot, A.; Monnard, P.-A.; Strazewski, P. Glass Microsphere-Supported Giant Vesicles as Tools for Observation of Self-reproduction of Lipid Boundaries. *Angew. Chem. Int. Ed.* **2018**, *57*, 282–286. [CrossRef] [PubMed]
15. Lopez, A.; Fiore, M. Investigating prebiotic protocells for a comprehensive understanding of the origins of life: A prebiotic systems chemistry perspective. *Life* **2019**, *9*, 49. [CrossRef] [PubMed]

life

MDPI

Review

The Beginning of Systems Chemistry

Peter Strazewski

Institut de Chimie et Biochimie Moléculaires et Supramoléculaires (Unité Mixte de Recherche 5246), Université de Lyon, Claude Bernard Lyon 1, 43 bvd du 11 Novembre 1918, 69622 Villeurbanne Cedex, France; strazewski@univ-lyon1.fr; Tel.: +33-472-448-234

Received: 2 January 2019; Accepted: 17 January 2019; Published: 24 January 2019

Abstract: Systems Chemistry has its roots in the research on the autocatalytic self-replication of biological macromolecules, first of all of synthetic deoxyribonucleic acids. A personal tour through the early works of the founder of Systems Chemistry, and of his first followers, recalls what's most important in this new era of chemistry: the growth and evolution of compartmented macromolecular populations, when provided with "food" and "fuel" and disposed of "waste".

Keywords: population growth; replication; growth order; Darwinian evolution; selection

---

Dedicated to Günter von Kiedrowski, the Founder of Systems Chemistry, on the Occasion of His Retirement

Leslie Eleazer Orgel (1927–2007) was the prophet of Systems Chemistry, his pupil Günter von Kiedrowski is the founder and name inventor of Systems Chemistry, and Eörs Szathmáry is the mastermind of the first theoretical concepts in Systems Chemistry. I am an active witness of Günter's and Eörs' first steps in laying the grounds for Systems Chemistry one year before the first workshop on Systems Chemistry took place in Venice, 2005 [1]. So let me give a very short, very personal and subjective view on how Systems Chemistry started. Ever since, the field has evolved in wide steps, but the first questions still remain generally unanswered.

Orgel's immense work in prebiotic chemistry and on enzyme-free template-directed nucleic acid chain elongation had a profound influence on the founder of Systems Chemistry (Figure 1).

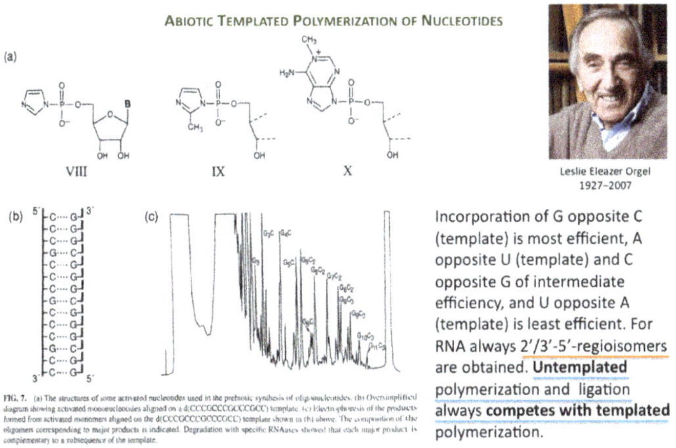

ABIOTIC TEMPLATED POLYMERIZATION OF NUCLEOTIDES

Leslie Eleazer Orgel
1927–2007

Incorporation of G opposite C (template) is most efficient, A opposite U (template) and C opposite G of intermediate efficiency, and U opposite A (template) is least efficient. For RNA always 2'/3'-5'-regioisomers are obtained. Untemplated polymerization and ligation always competes with templated polymerization.

Figure 1. Template-directed enzyme-free RNA chain elongation versus untemplated polymerization and ligation. Parts (a,b,c) taken from [2] and reproduced with permission from Taylor & Francis © 2004.

Life **2019**, 9, 11

The first success in understanding autocatalytic molecular replicators was pioneered by Günter's experiments on the enzyme-free autocatalytic chemical fuel-driven ligation of synthetically end-capped DNA fragments A* + B, in particular, the discovery by minute HPLC analysis of the growth rate of these ligated templates T (Figure 2).

ABIOTIC TEMPLATED LIGATION OF OLIGONUCLEOTIDES

Angew. Chem. Int. Ed. Engl. 25 (1986) No. 10

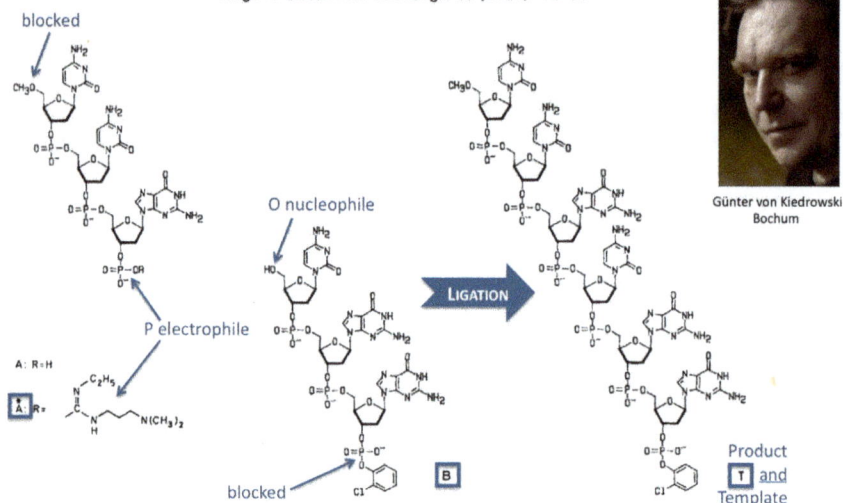

Figure 2. Abiotic templated ligation of oligonucleotides. Structure formula taken from [3] and reproduced with permission from Wiley-VCH Verlag GmbH & Co. KGaA © 1986.

The formulation of an experimentally derived "square-root law" from the fitting of the obtained peak intensities has proven to be a robust concept and general molecular property of self-replicating and cross-replicating macromolecules [4,5] that are in principle able to carry over sequence information through multiple rounds of ligation (Figure 3).

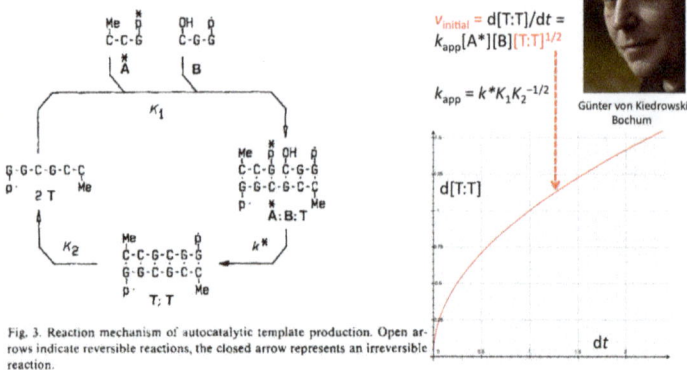

Autocatalysis kinetics of oligonucleotide ligation follows the "square-root law":

$$v_{initial} = d[T{:}T]/dt = k_{app}[A^*][B][T{:}T]^{1/2}$$

$$k_{app} = k^* K_1 K_2^{-1/2}$$

Fig. 3. Reaction mechanism of autocatalytic template production. Open arrows indicate reversible reactions, the closed arrow represents an irreversible reaction.

Figure 3. Autocatalytic production of ligation product T from oligodeoxynucleotides A* and B follows a square-root law. Reaction scheme taken from [3] and reproduced with permission from Wiley-VCH Verlag GmbH & Co. KGaA © 1986.

Ten years after, Reza Ghadiri and coworkers showed that the square-root law also applies to the kinetics of autocatalytic ligation of synthetically activated peptide fragments, one being electrophilic at its C-terminus (thioester), the other nucleophilic at its N-terminus (Cys thiol) through template-directed native chemical ligation (Figure 4).

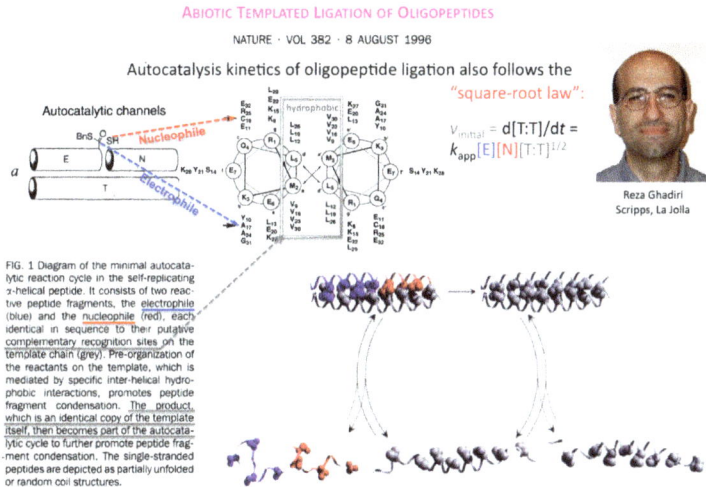

Figure 4. Autocatalytic production of ligation product T from oligopeptides E and N follows the square-root law. Figures and text taken from [6] and reproduced with permission from https://www.nature.com/ © 1996.

The concentration or density of autocatalytic or cross-catalytic molecular—as opposed to supramolecular—replicators in well mixed homogeneous milieus thus grows sub-exponentially with time t (Figure 5). For any doubling template population {[T:T] + [T]} = x, at any apparent growth rate constant k, the resulting parabolic growth order $0 < p < 1$ describes a growth dynamics where each generation produces on the average fewer descendants per parent than the previous generation (see also right graph in Figure 3). This contrasts exponential and hyperbolic growth orders ($p \geq 1$) where in each generation, on average, the same number or even more descendants are produced per parent than in the previous generation.

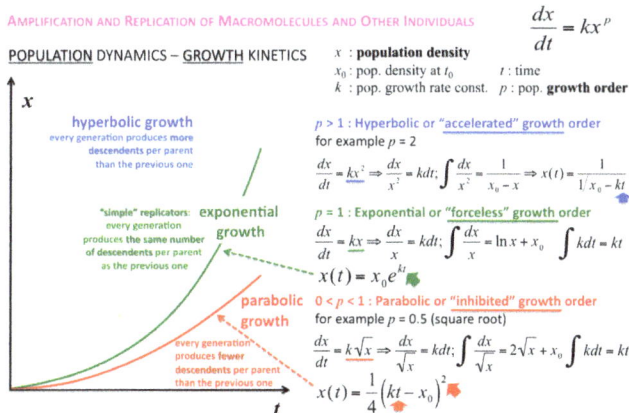

Figure 5. Parabolic (inhibited), exponential (forceless, simple) and hyperbolic (accelerated) growth orders (regimes). Exemplary integrations apply to 1→2 stoichiometric growth (doublings) only.

The corresponding growth regimes are termed "inhibited", "forceless" ("simple") and "accelerated", respectively; they apply to all stoichiometries (doubling, tripling and so forth), and explicitly include any selection of the fittest fertile individuals from changes in the environment and the degradation or death rates over time. For example, the human population, domesticated animals and plants—like pigs, cows, chicken, wheat, rice, maize, potatoes, tomatoes, grapes and oranges—globally spread in the accelerated growth regime, owing to increasingly optimised life qualities such as food, fertiliser, health, genetic manipulation, safe transportation and peace. Persisting populations of wild animals and plants, also cloned bacteria and in vitro selected macromolecules (cf. PCR), spread in the forceless growth regime, unless the animals or plants belong to endangered species, the resources are diminishing or the waste is undisposed of for some reason. The inhibited growth regime for the doubling of well-mixed and resourceful autocatalytic and cross-catalytic macromolecules has its roots in a general self-capturing phenomenon termed "strand inhibition". Without external "help", usually from enzymes, the unfolding of T:T double-strands (T:T:T triple-strands and so forth, if applicable) is difficult for intrinsic molecular reasons, which is hardly the case for bacterial populations, plants and animals. It is as if grown-up children could not become fully reproductive because, during much of their fertile time, the siblings would prefer to stay together on the playground rather than to go out and mate. Hence, in spite of plentiful resources, fully suppressed side reactions—no degradation or chain elongation instead of replication—and negligible waste product concentrations, viz. under ideal initial conditions, the growth order of the vast majority of macromolecular replicators remains parabolic. The second phenomenal coup out of Günter's kitchen was to show SPREAD, that is, that the exponential regime can be achieved enzyme-free through the surface-promoted replication and exponential amplification of DNA analogues [7]. The immobilisation of the template strand allows for sequential enzyme-free ligation. The copy is released, and reimmobilised at another part of the solid support to become a template for the next cycle of steps. Irreversible immobilisation of template molecules is thus a means to overcome strand inhibition. In other words, once the grown-up children happen to be out of the playground, don't let them go back.

Before that demonstration, and soon after Günter's first pioneering discovery, Eörs' and colleagues' early insight was to realise that this general strand inhibition was a problem for competing parabolic replicators, and how generally it could be solved [8,9]. In the absence of efficient T:T double-strand unfolders, different macromolecular replicators, bearing markedly different sequences and lengths for example, that are competing for the same resources, can all slowly thrive in the parabolic growth regime, but will virtually never outcompete one another in a well-mixed milieu where food is plentiful and their waste is properly disposed of (Figure 6).

In such a situation, Darwinian evolution, being defined as evolution through natural selection, as opposed to evolution through genetic drift, migration, mutations, etc., cannot commence. All abiotically produced parabolic replicators will coexist and spread at different rates. In other words, no speciation at the well-mixed macromolecular level is possible. The idea how to solve the problem originates from the notion of group selection. Rather than being well-mixed, compartmented parabolic replicators are in a different population dynamic situation, since selective forces do not affect them directly but address the fitness of whole systems (Figure 7). Eörs calls it the "stochastic (error) corrector" model [10]. This is the most fundamental reason for why life needs to be cellular—other important reasons being confinement, protection, concentration, import-export control, and so forth. My naïve human equivalent: as long as the grown-up children insist on playing instead of mating, those clans that furnish the best housing conditions can maintain their collective fertility potential longer than other clans, who may be at risk of dying without progeny.

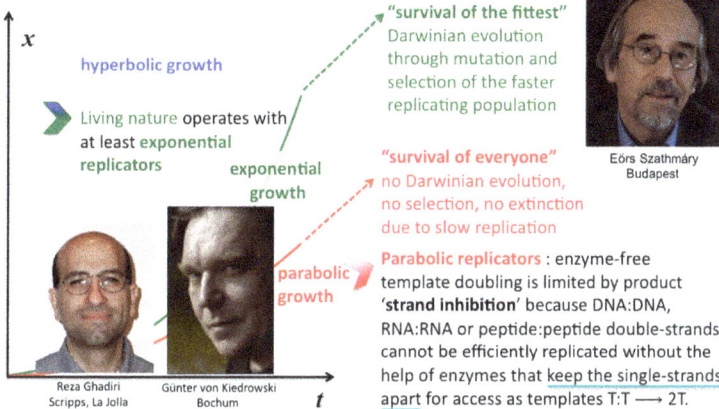

Figure 6. Survival of everyone. Competing but different parabolic replicators (different *k* but same *p*, cf. Figure 5) cannot outcompete one another in a well-mixed milieu.

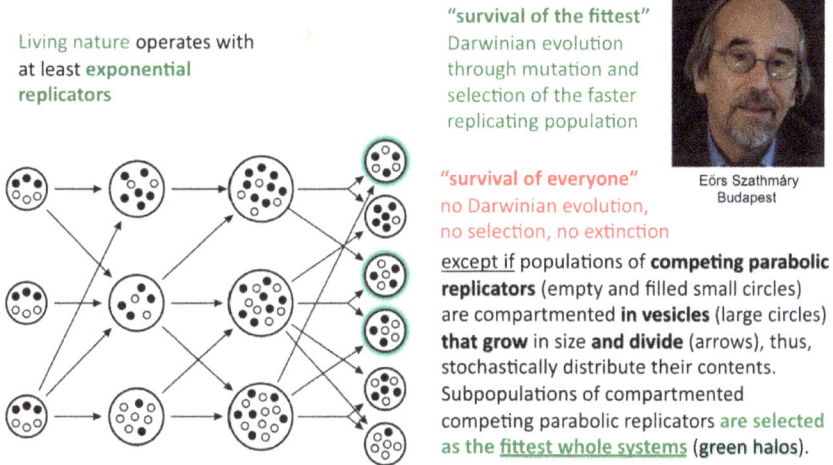

Figure 7. Survival of the fittest whole systems. Once different parabolic replicators are randomly distributed over periodically growing and randomly dividing compartments, the fittest compartments can outcompete less fit compartments, thus, whole populations specify despite the absence of efficient T:T double-strand unfolders.

Of course, once exponential replicators self-evolve inside selected compartments, the hosting populations are predisposed for their spreading rates to "shoot off exponentially", if sufficiently fed and disposed of waste products. Such populations can outcompete without hesitation the throng of selected parabolic compartments, now compete with one another in the exponential regime, and spread by Darwinian evolution as we know it from biological cells, organisms and populations. Just how exactly can the integrity of parabolic replicators be maintained long enough throughout their spreading? How can parabolic replicators self-evolve at all? These questions did escort Systems Chemistry right from the start; Eörs exposed yet another fundamental problem that needs to be solved

(Figure 8). Manfred Eigen realised long before the founding works of Systems Chemistry that any error propagation sets limits to the amount of information that can be soundly and recurrently inherited through many generations [11]. Solutions to the problem of the self-evolution of the replication fidelity of parabolic, exponential and hyperbolic replicators have been proposed ever since, and are manifold [12], still under vivid debate, and out of the scope of this article.

Problem no. 2 : Replication fidelity limits maximum polymer length

match vs. mismatch

$$Q \frac{dx_f}{dt} > \frac{dx_m}{dt}$$

Q : replication fidelity of a given sequence N
x : population density f : functional m : mutant
t : time

Replication fidelity Q of any replicator is a function of the sequence length N and the mutation probability of every of its digits $0 < \mu_b < 1$:

$$Q = \left(1 - \mu_b\right)^N \cong \frac{1}{e^{N\mu_b}}$$

μ_b : mutation probability (mutation rate) of digit b

Manfred Eigen
MPI Göttingen

Eliminating Q from the above expressions sets a maximum length of any functional and replicating sequence N_f to successfully persist in a mixture of mutant replicators N_m that all compete for the same resources :

$$N < \frac{1}{\mu_b} \ln \frac{dx_f}{dx_m}$$ **ERROR THRESHOLD**

The lower the mutation rate (the closer μ_b to 0), the longer the maximal sequence length N.

Figure 8. Minimal mutability needed for the robust spreading of useful information. Taken from [12].

What can we learn from the pioneering works? The *pudels kern* of Systems Chemistry always was, and still is, the growth and evolution of molecular populations, when provided with "food" and "fuel", and when disposed of "waste". Formidable work has been published in the decades that followed this pioneering phase, but there remains much chemistry to be discovered where chemical systems are developed that can "inherit", i.e., transmit through replication a large amount of highly diverse information (open-ended evolution), that remain robust and dynamically stable over many rounds of replication in the presence of competing replicators and parasites, and that are also sufficiently diverse to be useful for the whole system—therefore, most likely localised in, and carried over from covalent macromolecules—but nevertheless subtly mutable, evolvable, and self-evolvable. This is the essence of Systems Chemistry (to be continued elsewhere).

Acknowledgments: This work summarises results of a collaborative effort undertaken within the COST Actions D27 "Prebiotic Chemistry and Early Evolution", CM0703 "Systems Chemistry", and CM1304 "Emergence and Evolution of Complex Chemical Systems". This article is based on parts of the author's presentation "Population growth and encoding principles in self-evolving chemical systems" held on 2 August 2018 at the Gordon Research Conference on *Systems Chemistry: From Concepts to Conception* in Newry, ME, USA. The financial support from the *Volkswagen Foundation* for the project "Molecular Life" (Az 92850) is gratefully acknowledged.

Conflicts of Interest: The authors declare no conflict of interest.

References

1. Stankiewicz, J.; Eckhardt, L.H. Meeting Reviews: Chembiogenesis 2005 and Systems Chemistry Workshop. *Angew. Chem. Int. Ed.* **2006**, *45*, 342–344. [CrossRef]
2. Orgel, L.E. Prebiotic Chemistry an the Origin of the RNA World. *Crit. Rev. Biochem. Mol. Biol.* **2004**, *39*, 99–123. [PubMed]
3. Von Kiedrowski, G. A Self-Replicating Hexanucleotide. *Angew. Chem. Int. Ed.* **1986**, *25*, 932–935. [CrossRef]
4. Sievers, D.; von Kiedrowski, G. Self-Replication of Complementary Nucleotide-Based Oligomers. *Nature* **1994**, *369*, 221–224. [CrossRef] [PubMed]

5. Sievers, D.; von Kiedrowski, G. Self-Replication of Hexanucleotide Analogues: Autocatalysis Versus Cross-Catalysis. *Chem. Eur. J.* **1998**, *4*, 629–641. [CrossRef]

6. Lee, D.H.; Granja, J.R.; Martinez, J.A.; Severin, K.; Ghadiri, R. A Self-Replicating Peptide. *Nature* **1996**, *382*, 525–528. [CrossRef] [PubMed]

7. Luther, A.; Brandsch, R.; von Kiedrowski, G. Surface-Promoted Replication and Exponential Amplification of DNA Analogues. *Nature* **1998**, *396*, 245–248. [CrossRef] [PubMed]

8. Szathmáry, E.; Demeter, L. Group Selection of Early Replicators and the Origin of Life. *J. Theor. Biol.* **1987**, *128*, 463–486. [CrossRef]

9. Szathmáry, E.; Gladkih, I. Sub-Exponential Growth and Coexistence of Non-Enzymatically Replicating Templates. *J. Theor. Biol.* **1989**, *138*, 55–58. [CrossRef]

10. Szathmáry, E.; Maynard-Smith, J. From Replicators to Reproducers: The First Major Transitions Leading to Life. *J. Theor. Biol.* **1997**, *187*, 555–571. [CrossRef] [PubMed]

11. Eigen, M. Selforganization of Matter and the Evolution of Biological Macromolecules. *Naturwissenschaften* **1971**, *58*, 465–523. [CrossRef] [PubMed]

12. Szilágyi, A.; Zachar, I.; Scheuring, I.; Kun, Á.; Könnyű, B.; Czárán, T. Ecology and Evolution in the RNA World Dynamics and Stability of Prebiotic Replicator Systems. *Life* **2017**, *7*, 48. [CrossRef] [PubMed]

Review

Chemomimesis and Molecular Darwinism in Action: From Abiotic Generation of Nucleobases to Nucleosides and RNA

Raffaele Saladino [1], Judit E. Šponer [2,*], Jiří Šponer [2], Giovanna Costanzo [3], Samanta Pino [1] and Ernesto Di Mauro [1,*]

[1] Biological and Ecological Department, University of Tuscia, 01100 Viterbo, Italy; saladino@unitus.it (R.S.); samantapino78@libero.it (S.P.)
[2] Institute of Biophysics of the Czech Academy of Sciences, Královopolská 135, 61265 Brno, Czech Republic; sponer@ncbr.muni.cz
[3] Institute of Molecular Biology and Pathology, CNR, 00185 Rome, Italy; giovanna.costanzo@uniroma1.it
* Correspondence: judit@ncbr.muni.cz (J.E.Š.); ernesto.dimauro@uniroma1.it (E.D.M.)

Received: 23 May 2018; Accepted: 19 June 2018; Published: 20 June 2018

Abstract: Molecular Darwinian evolution is an intrinsic property of reacting pools of molecules resulting in the adaptation of the system to changing conditions. It has no a priori aim. From the point of view of the origin of life, Darwinian selection behavior, when spontaneously emerging in the ensembles of molecules composing prebiotic pools, initiates subsequent evolution of increasingly complex and *innovative* chemical information. On the *conservation* side, it is a posteriori observed that numerous biological processes are based on prebiotically promptly made compounds, as proposed by the concept of Chemomimesis. Molecular Darwinian evolution and Chemomimesis are principles acting in balanced cooperation in the frame of Systems Chemistry. The one-pot synthesis of nucleosides in radical chemistry conditions is possibly a telling example of the operation of these principles. Other indications of similar cases of molecular evolution can be found among biogenic processes.

Keywords: origin of life; systems chemistry; Chemomimesis; Molecular Darwinism

1. Introduction

In the absence of life, the components of biogenic processes were necessarily generated in abiotic reactions [1–5]. The conditions under which these syntheses occurred and may still occur are multiform and, as such, are widespread in the Universe. Hence the observations in different interstellar spaces and in different lifeless celestial bodies of molecules which, on our Planet, are starting points and/or are part of biological systems [6–10].

The chemical composition and complexity of the pools of potentially biogenic compounds differ, necessarily depending on a large number of parameters. Many of these parameters are still poorly characterized or are possibly unknown. Nevertheless, it is increasingly clear that prebiotic syntheses occur under a variety of energy sources, of different mixtures of simple starting compounds, of catalysts, and of physico-chemical conditions. Is it possible to identify some of the principles guiding their evolution towards Life?

2. The Principles Underlying Progress towards Further Complexity

Darwinian selection has no aims; it does not work for a purpose. It only has consequences, the major of which being the adaptive variation, otherwise called "evolution", of the system following the modification of the conditions in which the system has existed thus far. The process of "adaptive variation" implies "adaptation" to the new conditions. The word "adaptation" describes the qualitative/quantitative modification of the components of the system as a consequence of the process started by the variation of the initial conditions.

When dealing with a population of molecules generated in a synthetic system endowed with biogenic potential, the variation of the conditions of the system depends upon external and internal factors. Internal factors essentially consist of the singly independent and/or of the multiple interacting reactivity of the molecules present. In the absence of special quenching factors, all the molecular populations produced in prebiotic synthetic pools are bound to evolve up to a given point, adapting themselves to the environment that their synthesis has contributed to establish, till exhaustion of the intrinsic reactivity of the system. As discussed below, energy aspects are paramount. What could be hinted at by recent findings on prebiotic synthetic pools about prebiotic evolutionary processes?

3. Principles for Systems Chemistry

One way of considering the progress of first-generation prebiotic pools towards biogenic processes is to consider them at the light of Systems Chemistry. In Systems Chemistry [11–14] the focus does not a priori lie on individual chemical components, but rather on the overall ensemble of interacting molecules and on their emergent properties. Systems Chemistry would benefit from the definition of working principles. We propose to use the expression "Systems Chemistry" for the ensemble of considerations dictated by Molecular Darwinism and Chemomimesis.

Molecular Darwinism is a term first introduced, to the best of our knowledge, by J. S. Wicken [15], who critically considered it as a primordial selective process based on unwarranted assumptions. The term was later progressively used as a means to refer to genetic phenomena at the molecular level, like the principle underlying spontaneously occurring genetic variants as driving force of biological evolution by W. Arber [16]. The complexity level considered was high: local sequence changes, intragenomic reshuffling of DNA segments, acquisition of a segment of a foreign DNA, and the like. This complexity defines the purport of the term at a mature biological level. In what follows, we use this term in the meaning originally suggested by the Göttingen school of Molecular Darwinism, which extends the operation of Darwinian principles of random mutations and selection to chemical processes occurring at the abiotic level of complexity [17]. Molecular Darwinism is Chemical Evolution in Higgs' purport [18], with the additional attributes of intrinsic selection and competition processes which is the core essence of Darwinism.

Chemomimesis is a term introduced by A. Eschenmoser and E. Loewenthal [19] to indicate that chemical compounds and processes characterizing biological phenomena often have purely abiotic precedents: something is copied and used that already existed. In trying to understand the mechanisms characterizing the passages from the abiotic through the prebiotic to the biotic, Chemomimesis is a powerful concept [20,21] which, however, can only be applied according to a posteriori logics: a natural process becomes chemomimetic *after* the organisms which use it have come into being. The combination of Molecular Darwinism and Chemomimesis may be instrumental for a fact-based understanding of the abiotic-to-prebiotic-to-biotic paths.

4. One-Pot Initial Events under a Variety of Energy Sources: An Example

Pools of potentially prebiotic compounds are obtained in early-Earth conditions [22,23], in hydrothermal environments [24,25], and in irradiated and/or impacted Earth atmosphere [26,27]. The HCN and formamide (NH_2COH) chemistries are interrelated [28,29] and are the natural chemical frames into which a rich panel of prebiotic compounds have been obtained. The ubiquitous [30–33] compound formamide has in particular shown its worth [34–37], due to its peculiar physico-chemical properties [34,35], allowing its liquid state to have a 200 °C-wide interval, as well as its facile accumulation [38,39]. In Ref. [40], we suggested a prebiotic scenario, which assumed that liquid formamide could accumulate on some hot surface on the early Earth at temperatures around 180 °C as a thermal dissociation product of ammonium formate. This paper responds to the critical notes of

Bada et al. [41], who demonstrated that at room temperature formamide is highly hygroscopic, i.e., in these conditions it could not accumulate in a concentrated form. We have been pleased to learn that in a recent report, one of the authors of Ref. [41] has changed his mind and has experimentally demonstrated that far above the boiling point of water (in line with our proposal in Ref. [40]) formamide can be accumulated in concentrated form [42].

Noteworthily, formamide was shown to be the key intermediate in the Urey-Miller reactions [43]. It has been proposed that among all the chemical scenarios tested so far, formamide is a favored starting compound as far as the complexity of the resulting mixtures of products is concerned [34,35]. Formamide is also remarkable for its versatility: its synthetic capacities are evident in all the physico-chemical environments tested and under the effect of a great variety of catalysts [34,35,44–51]. As for the energy source triggering the formamide-based prebiotic syntheses, large panels of products were observed under heat, UV, ionizing and proton irradiation, as reviewed [9].

The idea that prebiotic pathways were straight, streamlined and fastidiously demanding (see Figure 1, in Ref. [37]) is contradicted by the promiscuous efficiency of formamide-based comprehensive syntheses. We suggest that a combination of external conditions, for example, exposing the system simultaneously to variable proton irradiation, UV irradiation and temperature conditions may lead to further complexification of the chemical composition, as the individual external factors may act in synergy. A physicist would suggest that combining variations in several external factors in a combinatorial way may create a "multidimensional response" in the resulting composition of the chemical system, driving the system through diverse "chemical trajectories" on the "chemical compositional landscape". Such scenarios are certainly not irrelevant in the context of the prebiotic Earth, considering the time and size scale available for the onset of chemical evolution.

Formamide-based syntheses carried out under proton irradiation yielded the structurally most complex set of compounds obtained in one-pot reactions [46], including the sugars ribose and 2′-deoxyribose, and the canonical nucleobases (cytosine, uracil, adenine, guanine, and thymine). Most notably in the prebiotic perspective, the four nucleosides uridine, cytidine, adenosine, and thymidine were also synthesized. In this latter study, proton irradiation of formamide was performed in highly controlled conditions using 170 MeV proton beams generated by accelerated Helium at the Phasotron facility, Joint Institute for Nuclear Research, Dubna, Russia. In these conditions the prevailing chemical scenario is a bona fide radical chemistry, implying participation of cyano radical species ($\bullet$CN).

Radical chemistry may occur in different environments and may be triggered by different causes, as reactions at very low temperatures, break-up of larger molecules, heat, electrical discharges, electrolysis or, in particular, ionizing radiations. One remarkable property of radical chemistry is that the presence of unpaired electrons makes free radicals highly reactive both towards neutral molecules as well as towards themselves. As a consequence, radical chemistry may be particularly relevant when dealing with dimerizations and polymerizations. If a reactant has been activated by, say, irradiation, its activated radical state could allow its further reactions to occur in conditions in which closed-shell molecules could not react (i.e., at lower temperatures). Hence, radical chemistry is the favored system for space and/atmospheric prebiotic chemistry studies and could allow reactions to proceed through intermediate stages which are not accessible to non-radical compounds. An example is provided by the above-mentioned one-pot condensation of formamide up to nucleosides [46], accompanied in the same reaction pool by the synthesis of other different types of molecules, from amino acids and carboxylic acids, to molecules as complex as C18 and C20-compounds, like stearic acid and arachidic acid. In addition to the interest of the production of chemical information-bearing molecules per se, the prebiotic relevance of radical chemistry-based scenarios is enhanced by the overall increased complexity of the pool of the afforded compounds. Radical chemistry-related synthesis of nucleic acid bases was also reported in other studies in different conditions [27,52,53].

As Benner noted, prebiotic chemistry without selection leads to tar formation [54,55]. As several literature examples on the oligomerization of HCN [56,57] demonstrate, this statement is especially true for radical chemistry. A possible way to overcome tar formation is binding to minerals [58]. As demonstrated in Refs. [27,46], formamide-based radical chemistry combined with catalysis by meteorites could provide a plausible solution for this problem.

5. From Complex Mixtures to Pre-Genetic Materials

A reproducible transmission of genotype is the consensual essence of "Life" [59,60]. It could only have started through the fertile interaction of pre-genetic materials with metabolism-wise energy control and membrane-based containment devices. The pools of molecules obtained in one-pot syntheses from formamide encompass compounds relevant for each of these three independent to-be-converged domains. Recent progress has been marked more for pre-genetics and for bio-vesicles [61–63] than for pre-metabolisms. Focusing on pre-genetics, new data and new scenarios are currently being proposed. Reports abound on the abiotic syntheses of nucleobases and of nucleosides, on the mechanisms for their possible phosphorylation, on their oligomerizations, and on the properties which endow them with possible selective evolutionary advantages, as we describe below.

6. The Nucleobases are the Right Ones since the Beginning. The Case of AICA and fAICA

Adenine and guanine are the pivotal compounds for genetics (the genotype) and for metabolism (the phenotype). Protein synthesis, which connects the two, is not conceivable without ATP or GTP. The imidazoles AICA (4-aminoimidazole-5-carboxamide) and AICAI (4-aminoimidazole-5-carboxamidine) are the relevant intermediates in the chemical synthesis of purines, as first described for the synthesis of adenine from a concentrated solution of ammonia and HCN [64]. AICA and fAICA (5-formamidoimidazole-4-carboxamide) are intermediates of the biosynthesis of inosine-5'-monophosphate (IMP), the main route to purine nucleotides in extant cells. The similarity between the intermediates of this metabolic process and the chemical route described by Oró even increases when considering the compounds obtained in the condensation of formamide into adenine and hypoxanthine (which is the stable version of guanine, because it lacks the labile NH_2 group in C-4 position of the purine ring) when reacted in the presence of a variety of catalysts and under different energy sources.

Similarly, in the frame of formamide chemistry, six of the eight carboxylic acids which are intermediates of the extant Krebs cycle have been detected under UV irradiation in the presence of titanium dioxide, highlighting the possibility of the total synthesis of a large part of the chemical machinery utilized by one of the cell's oldest metabolic pathways [65]. The robustness of this chemical pathway is further evidenced by the formation of Krebs cycle intermediates from formamide under a variety of prebiotic scenarios, including iron-sulfur minerals [66], borates [67], zirconium minerals [68], and meteorites [45].

These observations provide a clear indication of the operation of Chemomimesis for compounds which are central to both genetics and energy control, apparently starting from the very beginning.

7. Focusing on Nucleosides

Performing, in the same conditions as those used for their synthesis [46], proton irradiation on mixtures of preformed sugars and adenine in the presence of a chondrite meteorite allowed the analysis of the reaction leading to the formation of the β-glycosidic bond [69]. These conditions simulate the presumptive conditions in space or on an early Earth fluxed by slow protons from the solar wind, sketching a potentially prebiotic scenario. The reaction consists of the formation of the β-glycosidic bond between separately preformed sugar and nucleobase moieties (both of which can be prebiotically obtained in the same reacting pool, as described in Ref. [46]), thus providing a simple alternative to the complex pathways suggested for the prebiotic formation of nucleosides. These latter ones are based on the involvement of oxazoline chemistry [70] in the synthesis of pyrimidine nucleosides [71–73], and on the synthesis of purine nucleosides through the formamido-pyrimidines (FPy) chemistry [74]. The point on these approaches was recently made [75].

The possibility of studying the formation of nucleosides in one-pot reactions makes possible the analysis of the factors that might have played a role in the condensation of nucleotides into polymers, eventually leading to the evolution of extant nucleic acids. The analysis of stereoselectivity, regioselectivity, and the possibility of (poly)glycosylation of the nucleosides formed in this reaction set was, in particular, made possible [69].

The relevance of this detailed information resides in the fact that extant RNA is built based on a structure consisting of phosphodiester bonds formed along a sequence of strictly stereo- and regioselective nucleosides. DNA has conserved these selectivities. Thus, a selection was exerted on the pool of sugars potentially formed in the synthetic first ur-reactions, eventually leading to the phenotype of the polymeric molecule that resulted in being the most adaptable to self-reproduction and to codogenic roles. In the absence of any finalism, the selection was necessarily initially based on the most basic phenotypes: reciprocal structural affinity of the precursors, energetic compatibility in the polymerization process, survival capacity of the resulting polymer. Stabilization of the phosphorylated precursors may be acquired through several mechanisms, important among which is the cyclization of the phosphate moiety and the self-protection through polymerization [76–78] (see below). Survival of the polymer mostly depended on resilience towards hydrolysis and other degradative reactions, thus entailing its possible accumulation.

8. Regio- and Stereoselectivity of Nucleoside Formation is Conserved from the Beginning

In the radical chemistry-based proton irradiation-powered one-pot reaction between adenine and 2-deoxyribose (for a summary of the mechanism, see Figure 1), the formation of mono- and poly-glycosylated nucleosides was observed, affording: α-D-2′-deoxy-ribofuranosyl adenine, β-D-2′-deoxy-ribofuranosyl adenine, α-D-2′-deoxy-ribopyranosyl adenine, and β-D-2′-deoxy-ribopyranosyl adenine. Poly-glycosylated N^6-2′-deoxy-ribofuranosyl- and N^6-2′-deoxy-ribopyranosyl-2′-deoxyadenosine isomers were detected, and higher molecular weight poly-glycosylated derivatives, corresponding to the addition of up to six sugar moieties, were also observed [69].

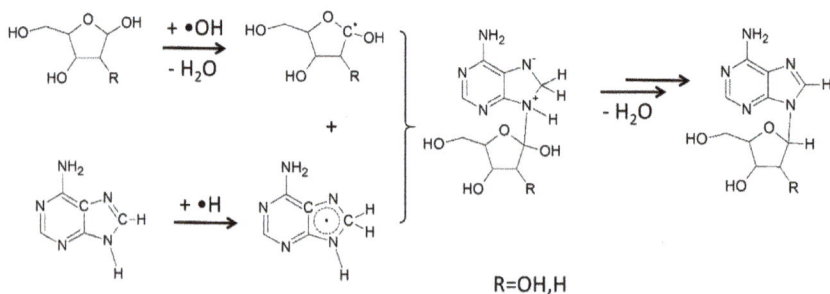

Figure 1. Proposed mechanism of the proton irradiation induced N-glycosidation between adenine and ribose [69].

The reaction of adenine with ribose afforded α-D-ribofuranosyl adenine, β-D-ribofuranosyl adenine, α-D-ribopyranosyl adenine, and β-D-2′-deoxy-ribopyranosyl adenine. Furanosides are the anomeric form present in extant nucleic acids. 2′-Deoxyribonucleosides formed more efficiently than ribonucleosides, and the β-isomer prevailed over the α-isomer.

As for the nucleobase regioselectivity of the glycosylation, the reaction selectively afforded N9 isomers, which are the isomers that molecular evolution has selected for the formation of nucleic acids. A mechanistic explanation was given for the absence of glycosylation on N1 and N7 of adenine [69].

These observations point to the fact that at least one reaction system exists [69] which allows the one-pot synthesis of the right components right from the beginning. Here, Darwinism worked on the evolution of new functions and Chemomimesis maintained the chemical structures.

9. Chemomimetic RNA

Artificial nucleic acids may exist in a large number of chemical alternatives [79–82]. The exploration of all the alternative possibilities is limited only by the ingenuity of the chemist. On the other hand, biological RNA and DNA are universal, unique, and very conserved. All the existing biological variants are epigenetic modifications of an evolutionarily unaltered chemical blueprint. If the initial pool had a nucleotide composition similar to the one that we have just described, we could a posteriori reason that RNA evolved to be composed of N9 isomers and of furanosides just because these forms were present as major species already in Darwin's *"warm little pond"*. This is largely an example of Chemomimesis. Furthermore, polymers built as RNA and DNA are built (on that very backbone and using those very furanosides to which N9 isomers are bound) have the balanced properties of (i) stability, (ii) possibility of replication, and (iii) codogenicity, which makes them the best fit to fulfill the multiple roles that genetics needs from them. These properties were acquired, certified, and maintained through Darwinian Molecular evolution.

10. Exploring the Environment for the First Effective Phosphorylation Agents

The reasons why nature chose phosphate as the link to hold and maneuver genetic information have been conclusively reviewed [83]. The sources of phosphate for the prebiotic phosphorylation of nucleosides have long been debated [84–87], and the topic has been surrounded by a reasoned skepticism about the possibility of ever knowing them [88,89]. A. M. Schoffsthal reported in four studies between 1976 and 1988 the phosphorylation of nucleosides in the presence of formamide [90–93] from soluble phosphates and, lastly, from hydroxylapatite. Extending these studies [94,95], it was shown that nucleosides could be phosphorylated in the presence of many different phosphate minerals, provided the presence of a dissolving agent and of high temperature ($\geq$400 K). Formamide efficiently fulfills this latter role, as does (less efficiently, and/or requiring longer times) water. The time scale of the mineral world may well be different from that of biology. Especially hydroxylapatite was shown to be a good source of phosphate for nucleoside phosphorylation. In this reaction, phosphorylation occurs at every possible position of the sugar moiety, at 2', or at 3', or at 5' [94]. With time, the open forms (2'-, or 3'-, or 5'-XMPs) are degraded, while the more stable cyclic forms remain. These can be 2', 3' or 3', 5' cyclic XMPs. Chemically related solvent systems, based on urea as originally proposed in L. Orgel's studies [86,87], have also recently been shown to be effective [96]. In addition, phosphorylation under aqueous conditions may occur from diamidophosphate, a compound derived from trimetaphosphate [97], whose prebiotic plausibility is claimed ibidem.

In conclusion, the phosphorylation of nucleosides may occur from numerous sources of phosphates and under a variety of conditions [98]. Which source [99] and which condition, among the various possibilities, was actually frequented in the *warm little pond* depends on their coherence with the steps that followed on the evolutionary path. From the point of view of Chemomimesis, the logic is clear: phosphorylation spontaneously occurs if the mineral environment, the solvent and the temperature are the right ones. The RNA structure a posteriori tells us that the process has been chemomimetically adopted and copied over. From the point of view of Molecular Darwinism, it all depends on the phenotype considered.

11. Focusing on Differential Kinetic Stability

In terms of possibility to evolve, the key molecular phenotype of any compound, both pre-biological and biological, is *stability*. Stability may be of kinetic or of thermodynamic nature. Among the two, thermodynamic stability is more universal (as Clemens Richert once said, "one can never fool thermodynamics"), whereas kinetic stability may be easily tuned with catalysts/inhibitors. At lower levels of chemical complexity (synthesis of prebiotic building blocks [27,69]), thermodynamic stability plays a more decisive role, while kinetic stability dominates at the level of biological molecules.

Metabolic cycles in modern organisms, which are based on non-equilibrium chemistry, are kept alive due to kinetic barriers. Indeed, enhancement of the kinetic stability of nucleosides and nucleotides could drive Molecular Darwinism towards oligonucleotide sequences. The stability of the components of nucleic acids was analyzed in the 1960s and the 1970s under various physico-chemical conditions. Several differences were determined: (i) the rate of cleavage of the glycosidic bonds of free deoxynucleosides [100,101] is 10–50 times higher relative to that in single-stranded DNA [102]; (ii) the rate of hydrolysis of glycosidic bonds varies in the order deoxynucleosides > deoxynucleotides > DNA [103–105]; (iii) the depurination is 4-fold in single- versus double-stranded DNA (rate constant of single-stranded DNA = 4×10^{-9} s^{-1}, 70 °C, pH 7.4) [106]. As a trend, higher molecular complexity allows higher stability. The stability of the phosphoester bonds determined in early studies has been reviewed [107]. A systematic comparison of the stability of the phosphoester bonds in precursor monomers (both ribo- and 2'-deoxyribo-) and that in DNA or RNA [76,77], has shown that the stability of the phosphoester bonds strongly depends on the molecular structure in which it is embedded, as well as on the solvent environment. In particular, the 3' phosphoester bond (the fragile and active site of the RNA molecule) is more stable towards hydrolysis when incorporated in RNA than in monomers, both in water and at high concentrations of formamide [77]. The higher stability of the polymeric form establishes a basically important evolutionary advantage: longer survival. Interestingly, and expectedly, different RNA sequences have different stabilities [78], thus being endowed with different fitness. In this respect, exploration of sequence space corresponds to exploration of safer thermodynamical niches.

12. Focusing on Oligomerization

Attention was devoted in the 1970s to the cyclic forms of nucleotides as potential actors of abiotic polymerization [108,109] and as effectors of nucleic acids ligation and stability [110,111]. Studies on the possible origin of RNA self-polymerization from 3', 5' cyclic monophosphates were later resumed, showing that self-oligomerization may occur efficiently for 3', 5' cGMP [112–115], and less so for 3', 5' cAMP [116] and 3', 5' cCMP [117]. Thermodynamic arguments suggested the preferential accumulation of 3', 5' cyclic nucleotides over the 2', 3' isomers in a formamide-rich environment [118]. Among the four 3', 5' cyclic nucleotides, the markedly higher observed oligomerization efficiency of 3', 5' cGMP was explained by the unique self-assembling properties of this molecule [114]. It was shown that in this particular case a special stacked supramolecular architecture formed which provided optimum steric conditions for an anionic ring-opening living polymerization mechanism (see Figure 2). Thus, the favorable entropic factor ensured a kinetic fitness for the oligomerization reaction and for Molecular Darwinism to give rise to the emergence of oligoG sequences.

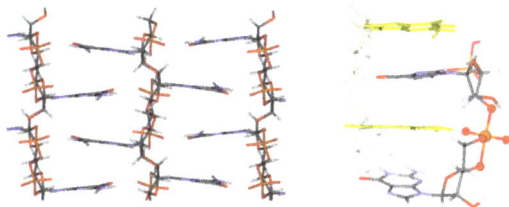

Figure 2. A ladder-like stacked supramolecular architecture provides optimum steric conditions for the oligomerization of 3′, 5′ cGMP. Left: Nucleobase stacking in the crystal structure of 3′, 5′ cGMP [119]. Right: Proposed structure of the trigonal bipyramidal intermediate of the chain-extension reaction from TPSS-D2/TVZP calculations [114]. The yellow nucleotides serve as mediators of the transphosphorylation reactions.

Non-enzymatic polymerization of RNA precursors occurs in several systems other than cyclic nucleotides. The efficient polymerization of highly activated precursor monomers (usually phosphorimidazolides) has been explored [120–122], the results providing important information on in vitro evolutionary behaviors of RNA populations and on their interactions with other systems, i.e., membranes [123]. The relevance of phosphorimidazolides or even of triphosphate nucleotides in early prebiotic scenarios has been, however, questioned [108,124], due to their elaborate synthesis and high energy content, resulting in their difficult accumulation and prebiotic availability. Shortly, from the point of view of Molecular Darwinism, activation by imidazole breaks the systematic trend outlined by the energetics of the so far demonstrated synthesis of nucleic acid building blocks, which always progresses towards increasing stability. A reaction system allowing copying of RNA sequences based on local and transient formation of phosphorimidazolides was described [125]. The polymerization of acyclic monophosphate nucleosides in acidic conditions has been reported [126,127]. From an energetic point of view this process is compatible with the concept of Molecular Darwinism: in this case, binding of a proton/positively charged cation to the phosphate moiety of the nucleotides [128] combined with the formation of a stacked and H-bonded supramolecular architecture ensures increased kinetic fitness to the transphosphorylation reactions that lead to oligonucleotide formation.

13. Chemomimesis of Cyclic Nucleotides

Cyclic nucleotides provide a striking example of functional plasticity in compounds that can be adapted to multiple uses while remaining in the genetic domain. The facile prebiotic phosphorylation of nucleosides [94], their higher stability as cyclic structures [94], the higher thermodynamic stability of the 3′, 5′ forms in formamide [118], all point to their presence in the *warm little pond* since the beginning, endowed with dedicated functions. As demonstrated [114], oligomerization of 3′, 5′ cyclic nucleotides offers a plausible way to overcome the water-paradox [4] that principally hampers oligonucleotide formation from acyclic nucleoside phosphate precursors in an aqueous environment.

Chemomimesis is considered to be a principle valid both for monomeric molecules and for processes. Nothing prevents the extension of its validity also to polymers. Given the spontaneous polymerizations reported for cyclic nucleotides, for preactivated phosphoimidazolides, and for monophosphates, it seems reasonable to assume that, one way or another, RNA prebiotically generated itself. Thus, extant biological RNA itself is a product of Chemomimesis. For RNA, mimesis is valid for the overall structure since the very beginning, while evolution pertains to the functions that it acquired along the way.

14. Concluding Remarks

Proto-life resulted from processes which were not programmed a priori, undergoing molecular adaptations dictated by the compositions of the prebiotic molecular pools, by the properties of the constituent molecules, by the environment, and by the history of the system. The principles at the basis of this selection processes are largely dictated by the quality/quantity of the compounds present, which results from both the first-run synthetic events and from their further-generation reactions, and is strongly influenced by the energetics of the system. A marked initial complexity is instrumental in determining the evolution of additionally complex chemical systems.

The rich and variegated pools of compounds, encompassing from nucleobases to nucleosides, generated by formamide chemistry in conditions allowing their further reactions and development of complexity, are a possible example of a Darwin's *warm little pond*. Up to what point does the initial reactivity generate pre-genetic complexity? Somehow, to our surprise, we observed that the one-carbon atom formamide system may go a long way, especially so in radical chemistry conditions.

In complex systems, selection functions based on the most adapted, not on the most abundant. "*Adapted to what?*" can be rephrased into: which are the relevant phenotypes for selection in the absence of an established and functioning biological apparatus?

Among the properties that may drive the system towards complexity, creating the selective conditions for further interactions and higher complexity, one should consider *thermodynamic stability*. Another important parameter is the *kinetics of the reactions* because it determines which reaction occurs preferentially before the final steady-state of the reacting pool is reached. Kinetics and thermodynamics are both influenced by *entropic factors*, which is the manifestation of the remarkable role played by structural effects in molecular evolution.

In conclusion, complex mixtures undergo processes subjected to Molecular Darwinism. This term is particularly useful for summarizing what happens at the borderline between prebiotic Chemistry and rudimental pre-Biology, possibly opening up the clarification of relevant mechanisms and interactions. From our privileged point of observation of living beings, we may a posteriori add Chemomimesis as an analytical tool to indirectly reconstruct these phenomena.

Funding: This work was supported by the Italian Space Agency (ASI) project: "Esobiologia e Ambienti Estremi" number 2014-026-R.O (CUP: F 92I14000030005). Financial support from the grant GAČR 17-05076S is greatly acknowledged.

Conflicts of Interest: The authors declare no conflict of interest.

References

1. Miller, S.L.; Urey, H.C. Organic compound synthesis on the primitive earth. *Science* **1959**, *130*, 245–251. [CrossRef] [PubMed]
2. Chyba, C.; Sagan, C. Endogenous production, exogenous delivery and impact-shock synthesis of organic molecules: An inventory for the origins of life. *Nature* **1992**, *355*, 125–132. [CrossRef] [PubMed]
3. Benner, S.A.; Ricardo, A.; Carrigan, M.A. Is there a common chemical model for life in the universe? *Curr. Opin. Chem. Biol.* **2004**, *8*, 672–689. [CrossRef] [PubMed]
4. Benner, S.A. Paradoxes in the origin of life. *Orig. Life Evol. Biosph.* **2014**, *44*, 339–343. [CrossRef] [PubMed]
5. Chyba, C.F.; Thomas, P.J.; Brookshaw, L.; Sagan, C. Cometary delivery of organic molecules to the early Earth. *Science* **1990**, *249*, 366–373. [CrossRef] [PubMed]
6. Schmitt-Kopplin, P.; Gabelica, Z.; Gougeon, R.D.; Fekete, A.; Kanawati, B.; Harir, M.; Gebefuegi, I.; Eckel, G.; Hertkorn, N. High molecular diversity of extraterrestrial organic matter in Murchison meteorite revealed 40 years after its fall. *Proc. Natl. Acad. Sci. USA* **2010**, *107*, 2763–2768. [CrossRef] [PubMed]
7. Burton, A.S.; Stern, J.C.; Elsila, J.E.; Glavin, D.P.; Dworkin, J.P. Understanding prebiotic chemistry through the analysis of extraterrestrial amino acids and nucleobases in meteorites. *Chem. Soc. Rev.* **2012**, *41*, 5459–5472. [CrossRef] [PubMed]
8. James, E.G.; Padelis, P.P. Molecular and atomic line surveys of galaxies. I. The dense, star-forming gas phase as a beacon. *Astrophys. J.* **2012**, *757*, 156. [CrossRef]

9.   Carota, E.; Botta, G.; Rotelli, L.; Di Mauro, E.; Saladino, R. Current advances in prebiotic chemistry under space conditions. *Curr. Org. Chem.* **2015**, *19*, 1963–1979. [CrossRef]

10.   Rubin, R.H.; Swenson, G.W., Jr.; Benson, R.C.; Tigelaar, H.L.; Flygare, W.H. Microwave detection of interstellar formamide. *Astrophys. J.* **1971**, *169*, L39. [CrossRef]

11.   Ashkenasy, G.; Hermans, T.M.; Otto, S.; Taylor, A.F. Systems chemistry. *Chem. Soc. Rev.* **2017**, *46*, 2543–2554. [CrossRef] [PubMed]

12.   Stankiewicz, J.; Eckardt, L.H. Chembiogenesis 2005 and systems chemistry workshop. *Angew. Chem. Int. Ed.* **2006**, *45*, 342–344. [CrossRef]

13.   Kindermann, M.; Stahl, I.; Reimold, M.; Pankau, W.M.; von Kiedrowski, G. Systems chemistry: Kinetic and computational analysis of a nearly exponential organic replicator. *Angew. Chem. Int. Ed.* **2005**, *44*, 6750–6755. [CrossRef] [PubMed]

14.   Sadownik, J.W.; Mattia, E.; Nowak, P.; Otto, S. Diversification of self-replicating molecules. *Nat. Chem.* **2016**, *8*, 264–269. [CrossRef] [PubMed]

15.   Wicken, J.S. An organismic critique of molecular Darwinism. *J. Theor. Biol.* **1985**, *117*, 545–561. [CrossRef]

16.   Arber, W. Molecular Darwinism: The contingency of spontaneous genetic variation. *Genome Biol. Evol.* **2011**, *3*, 1090–1092. [CrossRef] [PubMed]

17.   Küppers, B.-O. *Information and the Origin of Life*; MIT Press: Cambridge, MA, USA, 1990.

18.   Higgs, P.G. Chemical evolution and the evolutionary definition of life. *J. Mol. Evol.* **2017**, *84*, 225–235. [CrossRef] [PubMed]

19.   Eschenmoser, A.; Loewenthal, E. Chemistry of potentially prebiological natural products. *Chem. Soc. Rev.* **1992**, *21*, 1–16. [CrossRef]

20.   Menor-Salván, C.; Marín-Yaseli, M.R. A new route for the prebiotic synthesis of nucleobases and hydantoins in water/ice solutions involving the photochemistry of acetylene. *Chem. Eur. J.* **2013**, *19*, 6488–6497. [CrossRef] [PubMed]

21.   Pereto, J. Out of fuzzy chemistry: From prebiotic chemistry to metabolic networks. *Chem. Soc. Rev.* **2012**, *41*, 5394–5403. [CrossRef] [PubMed]

22.   Airapetian, V.S.; Glocer, A.; Gronoff, G.; Hebrard, E.; Danchi, W. Prebiotic chemistry and atmospheric warming of early Earth by an active young sun. *Nat. Geosci.* **2016**, *9*, 452–455. [CrossRef]

23.   Saladino, R.; Botta, G.; Bizzarri, B.M.; Di Mauro, E.; Garcia Ruiz, J.M. A global scale scenario for prebiotic chemistry: Silica-based self-assembled mineral structures and formamide. *Biochemistry* **2016**, *55*, 2806–2811. [CrossRef] [PubMed]

24.   Mulkidjanian, A.Y.; Bychkov, A.Y.; Dibrova, D.V.; Galperin, M.Y.; Koonin, E.V. Origin of first cells at terrestrial, anoxic geothermal fields. *Proc. Natl. Acad. Sci. USA* **2012**, *109*, E821–E830. [CrossRef] [PubMed]

25.   Djokic, T.; van Kranendonk, M.J.; Campbell, K.A.; Walter, M.R.; Ward, C.R. Earliest signs of life on land preserved in ca. 3.5 Ga hot spring deposits. *Nat. Commun.* **2017**, *8*, 15263. [CrossRef] [PubMed]

26.   Ferus, M.; Pietrucci, F.; Saitta, A.M.; Knížek, A.; Kubelík, P.; Ivanek, O.; Shestivska, V.; Civiš, S. Formation of nucleobases in a Miller–Urey reducing atmosphere. *Proc. Natl. Acad. Sci. USA* **2017**, *114*, 4306–4311. [CrossRef] [PubMed]

27.   Ferus, M.; Nesvorný, D.; Šponer, J.; Kubelík, P.; Michalčíková, R.; Shestivská, V.; Šponer, J.E.; Civiš, S. High-energy chemistry of formamide: A unified mechanism of nucleobase formation. *Proc. Natl. Acad. Sci. USA* **2015**, *112*, 657–662. [CrossRef] [PubMed]

28.   Saladino, R.; Crestini, C.; Ciciriello, F.; Costanzo, G.; Di Mauro, E. Formamide chemistry and the origin of informational polymers. *Chem. Biodivers.* **2007**, *4*, 694–720. [CrossRef] [PubMed]

29.   Kua, J.; Thrush, K.L. HCN, formamidic acid, and formamide in aqueous solution: A free energy map. *J. Phys. Chem. B* **2016**, *120*, 8175–8185. [CrossRef] [PubMed]

30.   Adande, G.R.; Woolf, N.J.; Ziurys, L.M. Observations of interstellar formamide: Availability of a prebiotic precursor in the galactic habitable zone. *Astrobiology* **2013**, *13*, 439–453. [CrossRef] [PubMed]

31.   López-Sepulcre, A.; Jaber, A.A.; Mendoza, E.; Lefloch, B.; Ceccarelli, C.; Vastel, C.; Bachiller, R.; Cernicharo, J.; Codella, C.; Kahane, C.; et al. Shedding light on the formation of the pre-biotic molecule formamide with ASAI. *Mon. Notices R. Astron. Soc.* **2015**, *449*, 2438–2458. [CrossRef]

32.   Biver, N.; Bockelée-Morvan, D.; Debout, V.; Crovisier, J.; Boissier, J.; Lis, D.C.; Dello Russo, N.; Moreno, R.; Colom, P.; Paubert, G.; et al. Complex organic molecules in comets C/2012 F6 (Lemmon) and C/2013 R1 (Lovejoy): Detection of ethylene glycol and formamide. *Astron. Astrophys.* **2014**, *566*, L5. [CrossRef]

33. Kröcher, O.; Elsener, M.; Jacob, E. A model gas study of ammonium formate, methanamide and guanidinium formate as alternative ammonia precursor compounds for the selective catalytic reduction of nitrogen oxides in diesel exhaust gas. *Appl. Catal. B* **2009**, *88*, 66–82. [CrossRef]

34. Saladino, R.; Botta, G.; Pino, S.; Costanzo, G.; Di Mauro, E. Genetics first or metabolism first? The formamide clue. *Chem. Soc. Rev.* **2012**, *41*, 5526–5565. [CrossRef] [PubMed]

35. Saladino, R.; Crestini, C.; Pino, S.; Costanzo, G.; Di Mauro, E. Formamide and the origin of life. *Phys. Life Rev.* **2012**, *9*, 84–104. [CrossRef] [PubMed]

36. Saladino, R.; Crestini, C.; Costanzo, G.; Negri, R.; Di Mauro, E. A possible prebiotic synthesis of purine, adenine, cytosine, and 4(3H)-pyrimidone from formamide: Implications for the origin of life. *Bioorg. Med. Chem.* **2001**, *9*, 1249–1253. [CrossRef]

37. Saladino, R.; Šponer, J.E.; Šponer, J.; Di Mauro, E. Rewarming the primordial soup: Revisitations and rediscoveries in prebiotic chemistry. *ChemBioChem* **2018**, *19*, 22–25. [CrossRef] [PubMed]

38. Niether, D.; Afanasenkau, D.; Dhont, J.K.G.; Wiegand, S. Accumulation of formamide in hydrothermal pores to form prebiotic nucleobases. *Proc. Natl. Acad. Sci. USA* **2016**, *113*, 4272–4277. [CrossRef] [PubMed]

39. Niether, D.; Wiegand, S. Heuristic approach to understanding the accumulation process in hydrothermal pores. *Entropy* **2017**, *19*, 33. [CrossRef]

40. Šponer, J.E.; Šponer, J.; Nováková, O.; Brabec, V.; Šedo, O.; Zdráhal, Z.; Costanzo, G.; Pino, S.; Saladino, R.; Di Mauro, E. Emergence of the first catalytic oligonucleotides in a formamide-based origin scenario. *Chem. Eur. J.* **2016**, *22*, 3572–3586. [CrossRef] [PubMed]

41. Bada, J.L.; Chalmers, J.H.; Cleaves, H.J. Is formamide a geochemically plausible prebiotic solvent? *Phys. Chem. Chem. Phys.* **2016**, *18*, 20085–20090. [CrossRef] [PubMed]

42. Adam, Z.R.; Hongo, Y.; Cleaves, H.J.; Yi, R.; Fahrenbach, A.C.; Yoda, I.; Aono, M. Estimating the capacity for production of formamide by radioactive minerals on the prebiotic Earth. *Sci. Rep.* **2018**, *8*, 265. [CrossRef] [PubMed]

43. Saitta, A.M.; Saija, F. Miller experiments in atomistic computer simulations. *Proc. Natl. Acad. Sci. USA* **2014**, *111*, 13768–13773. [CrossRef] [PubMed]

44. Saladino, R.; Crestini, C.; Cossetti, C.; Di Mauro, E.; Deamer, D. Catalytic effects of Murchison material: Prebiotic synthesis and degradation of RNA precursors. *Orig. Life Evol. Biosph.* **2011**, *41*, 437–451. [CrossRef] [PubMed]

45. Saladino, R.; Botta, G.; Delfino, M.; Di Mauro, E. Meteorites as catalysts for prebiotic chemistry. *Chem. Eur. J.* **2013**, *19*, 16916–16922. [CrossRef] [PubMed]

46. Saladino, R.; Carota, E.; Botta, G.; Kapralov, M.; Timoshenko, G.N.; Rozanov, A.Y.; Krasavin, E.; Di Mauro, E. Meteorite-catalyzed syntheses of nucleosides and of other prebiotic compounds from formamide under proton irradiation. *Proc. Natl. Acad. Sci. USA* **2015**, *112*, E2746–E2755. [CrossRef] [PubMed]

47. Saladino, R.; Carota, E.; Botta, G.; Kapralov, M.; Timoshenko, G.N.; Rozanov, A.; Krasavin, E.; Di Mauro, E. First evidence on the role of heavy ion irradiation of meteorites and formamide in the origin of biomolecules. *Orig. Life Evol. Biosph.* **2016**, *46*, 515–521. [CrossRef] [PubMed]

48. Botta, L.; Saladino, R.; Bizzarri, B.M.; Cobucci-Ponzano, B.; Iacono, R.; Avino, R.; Caliro, S.; Carandente, A.; Lorenzini, F.; Tortora, A.; et al. Formamide-based prebiotic chemistry in the Phlegrean fields. *Adv. Space Res.* **2017**. [CrossRef]

49. Rotelli, L.; Trigo-Rodríguez, J.M.; Moyano-Cambero, C.E.; Carota, E.; Botta, L.; Di Mauro, E.; Saladino, R. The key role of meteorites in the formation of relevant prebiotic molecules in a formamide/water environment. *Sci. Rep.* **2016**, *6*, 38888. [CrossRef] [PubMed]

50. Ferus, M.; Michalčíková, R.; Shestivská, V.; Šponer, J.; Šponer, J.E.; Civiš, S. High-energy chemistry of formamide: A simpler way for nucleobase formation. *J. Phys. Chem. A* **2014**, *118*, 719–736. [CrossRef] [PubMed]

51. Bizzarri, B.M.; Botta, L.; Perez-Valverde, M.I.; Saladino, R.; Di Mauro, E.; Garcia Ruiz, J.M. Silica metal-oxide vesicles catalyze comprehensive prebiotic chemistry. *Chem. Eur. J.* **2018**. [CrossRef]

52. Ferus, M.; Kubelík, P.; Knížek, A.; Pastorek, A.; Sutherland, J.; Civiš, S. High energy radical chemistry formation of HCN-rich atmospheres on early Earth. *Sci. Rep.* **2017**, *7*, 6275. [CrossRef] [PubMed]

53. Nguyen, H.T.; Jeilani, Y.A.; Hung, H.M.; Nguyen, M.T. Radical pathways for the prebiotic formation of pyrimidine bases from formamide. *J. Phys. Chem. A* **2015**, *119*, 8871–8883. [CrossRef] [PubMed]

54. Benner, S.A.; Kim, H.-J.; Carrigan, M.A. Asphalt, water, and the prebiotic synthesis of ribose, ribonucleosides, and RNA. *Acc. Chem. Res.* **2012**, *45*, 2025–2034. [CrossRef] [PubMed]

55. Benner, S.A.; Kim, H.-J. The case for a martian origin for Earth life. In *Instruments, Methods, and Missions for Astrobiology XVII*; Proceedings of SPIE 9606; SPIE Press: San Diego, CA, USA, 2015. [CrossRef]

56. He, C.; Lin, G.; Upton, K.T.; Imanaka, H.; Smith, M.A. Structural investigation of HCN polymer isotopomers by solution-state multidimensional NMR. *J. Phys. Chem. A* **2012**, *116*, 4751–4759. [CrossRef] [PubMed]

57. Minard, R.D.; Hatcher, P.G.; Gourley, R.C.; Matthews, C.N. Structural investigations of hydrogen cyanide polymers: New insights using TMAH thermochemolysis/GC-MS. *Orig. Life Evol. Biosph.* **1998**, *28*, 461–473. [CrossRef] [PubMed]

58. Ricardo, A.; Carrigan, M.A.; Olcott, A.N.; Benner, S.A. Borate minerals stabilize ribose. *Science* **2004**, *303*, 196. [CrossRef] [PubMed]

59. Joyce, G.F. Foreword. In *Origins of Life: The Central Concepts*; Deamer, D.W., Ed.; Jones and Bartlett: Boston, MA, USA, 1994.

60. Trifonov, E.N. Vocabulary of definitions of life suggests a definition. *J. Biomol. Struct. Dyn.* **2011**, *29*, 259–266. [CrossRef] [PubMed]

61. Lane, N.; Martin, W.F. The origin of membrane bioenergetics. *Cell* **2012**, *151*, 1406–1416. [CrossRef] [PubMed]

62. Adamala, K.; Szostak, J.W. Competition between model protocells driven by an encapsulated catalyst. *Nat. Chem.* **2013**, *5*, 495–501. [CrossRef] [PubMed]

63. Zhu, T.F.; Szostak, J.W. Exploding vesicles. *J. Syst. Chem.* **2011**, *2*, 4. [CrossRef]

64. Oró, J.; Kimball, A.P. Synthesis of purines under possible primitive earth conditions. I. Adenine from hydrogen cyanide. *Arch. Biochem. Biophys.* **1961**, *94*, 217–227. [CrossRef]

65. Saladino, R.; Brucato, J.R.; De Sio, A.; Botta, G.; Pace, E.; Gambicorti, L. Photochemical synthesis of citric acid cycle intermediates based on titanium dioxide. *Astrobiology* **2011**, *11*, 815–824. [CrossRef] [PubMed]

66. Saladino, R.; Neri, V.; Crestini, C.; Costanzo, G.; Graciotti, M.; Di Mauro, E. Synthesis and degradation of nucleic acid components by formamide and iron sulfur minerals. *J. Am. Chem. Soc.* **2008**, *130*, 15512–15518. [CrossRef] [PubMed]

67. Saladino, R.; Barontini, M.; Cossetti, C.; Di Mauro, E.; Crestini, C. The effects of borate minerals on the synthesis of nucleic acid bases, amino acids and biogenic carboxylic acids from formamide. *Orig. Life Evol. Biosph.* **2011**, *41*, 317–330. [CrossRef] [PubMed]

68. Saladino, R.; Neri, V.; Crestini, C.; Costanzo, G.; Graciotti, M.; Di Mauro, E. The role of the formamide/zirconia system in the synthesis of nucleobases and biogenic carboxylic acid derivatives. *J. Mol. Evol.* **2010**, *71*, 100–110. [CrossRef] [PubMed]

69. Saladino, R.; Bizzarri, B.M.; Botta, L.; Šponer, J.; Šponer, J.E.; Georgelin, T.; Jaber, M.; Rigaud, B.; Kapralov, M.; Timoshenko, G.N.; et al. Proton irradiation: A key to the challenge of N-glycosidic bond formation in a prebiotic context. *Sci. Rep.* **2017**, *7*, 14709. [CrossRef] [PubMed]

70. Sanchez, R.A.; Orgel, L.E. Studies in prebiotic synthesis. V. Synthesis and photoanomerization of pyrimidine nucleosides. *J. Mol. Biol.* **1970**, *47*, 531–543. [CrossRef]

71. Powner, M.W.; Gerland, B.; Sutherland, J.D. Synthesis of activated pyrimidine ribonucleotides in prebiotically plausible conditions. *Nature* **2009**, *459*, 239–242. [CrossRef] [PubMed]

72. Powner, M.W.; Sutherland, J.D. Phosphate-mediated interconversion of ribo- and arabino-configured prebiotic nucleotide intermediates. *Angew. Chem. Int. Ed.* **2010**, *49*, 4641–4643. [CrossRef] [PubMed]

73. Xu, J.; Tsanakopoulou, M.; Magnani, C.J.; Szabla, R.; Šponer, J.E.; Šponer, J.; Góra, R.W.; Sutherland, J.D. A prebiotically plausible synthesis of pyrimidine β-ribonucleosides and their phosphate derivatives involving photoanomerization. *Nat. Chem.* **2017**, *9*, 303–309. [CrossRef] [PubMed]

74. Becker, S.; Thoma, I.; Deutsch, A.; Gehrke, T.; Mayer, P.; Zipse, H.; Carell, T. A high-yielding, strictly regioselective prebiotic purine nucleoside formation pathway. *Science* **2016**, *352*, 833–836. [CrossRef] [PubMed]

75. Fiore, M.; Strazewski, P. Bringing prebiotic nucleosides and nucleotides down to Earth. *Angew. Chem. Int. Ed.* **2016**, *55*, 13930–13933. [CrossRef] [PubMed]

76. Saladino, R.; Crestini, C.; Busiello, V.; Ciciriello, F.; Costanzo, G.; Di Mauro, E. Differential stability of 3'- and 5'-phosphoester bonds in deoxy monomers and oligomers. *J. Biol. Chem.* **2005**, *280*, 35658–35669. [CrossRef] [PubMed]

77. Saladino, R.; Crestini, C.; Ciciriello, F.; Di Mauro, E.; Costanzo, G. Origin of informational polymers: Differential stability of phosphoester bonds in ribomonomers and ribooligomers. *J. Biol. Chem.* **2006**, *281*, 5790–5796. [CrossRef] [PubMed]

78. Ciciriello, F.; Costanzo, G.; Pino, S.; Crestini, C.; Saladino, R.; Di Mauro, E. Molecular complexity favors the evolution of ribopolymers. *Biochemistry* **2008**, *47*, 2732–2742. [CrossRef] [PubMed]

79. Georgiadis, M.M.; Singh, I.; Kellett, W.F.; Hoshika, S.; Benner, S.A.; Richards, N.G.J. Structural basis for a six nucleotide genetic alphabet. *J. Am. Chem. Soc.* **2015**, *137*, 6947–6955. [CrossRef] [PubMed]

80. Benner, S.A. Understanding nucleic acids using synthetic chemistry. *Acc. Chem. Res.* **2004**, *37*, 784–797. [CrossRef] [PubMed]

81. Nielsen, P.E. Peptide nucleic acids and the origin of life. *Chem. Biodivers.* **2007**, *4*, 1996–2002. [CrossRef] [PubMed]

82. Eschenmoser, A. Towards a chemical etiology of nucleic acid structure. *Orig. Life Evol. Biosph.* **1997**, *27*, 535–553. [CrossRef] [PubMed]

83. Westheimer, F.H. Why nature chose phosphates. *Science* **1987**, *235*, 1173–1178. [CrossRef] [PubMed]

84. Beck, A.; Lohrmann, R.; Orgel, L.E. Phosphorylation with inorganic phosphates at moderate temperatures. *Science* **1967**, *157*, 952. [CrossRef] [PubMed]

85. Lohrmann, R.; Orgel, L.E. Prebiotic synthesis: Phosphorylation in aqueous solution. *Science* **1968**, *161*, 64–66. [CrossRef] [PubMed]

86. Osterberg, R.; Orgel, L.E.; Lohrmann, R. Further studies of urea-catalyzed phosphorylation reactions. *J. Mol. Evol.* **1973**, *2*, 231–234. [CrossRef] [PubMed]

87. Lohrmann, R.; Orgel, L.E. Urea-inorganic phosphate mixtures as prebiotic phosphorylating agents. *Science* **1971**, *171*, 490–494. [CrossRef] [PubMed]

88. Schwartz, A.W. Prebiotic phosphorus chemistry reconsidered. *Orig. Life Evol. Biosph.* **1997**, *27*, 505–512. [CrossRef] [PubMed]

89. Yamagata, Y.; Watanabe, H.; Saitoh, M.; Namba, T. Volcanic production of polyphosphates and its relevance to prebiotic evolution. *Nature* **1991**, *352*, 516–519. [CrossRef] [PubMed]

90. Schoffstall, A.M. Prebiotic phosphorylation of nucleosides in formamide. *Orig. Life* **1976**, *7*, 399–412. [CrossRef] [PubMed]

91. Schoffstall, A.M.; Barto, R.; Ramos, D. Nucleoside and deoxynucleoside phosphorylation in formamide solutions. *Orig. Life Evol. Biosph.* **1982**, *12*, 143–151. [CrossRef]

92. Schoffstall, A.M.; Laing, E. Phosphorylation mechanisms in chemical evolution. *Orig. Life Evol. Biosph.* **1985**, *15*, 141–150. [CrossRef]

93. Schoffstall, A.M.; Mahone, S.M. Formate ester formation in amide solutions. *Orig. Life Evol. Biosph.* **1988**, *18*, 389–396. [CrossRef] [PubMed]

94. Costanzo, G.; Saladino, R.; Crestini, C.; Ciciriello, F.; Di Mauro, E. Nucleoside phosphorylation by phosphate minerals. *J. Biol. Chem.* **2007**, *282*, 16729–16735. [CrossRef] [PubMed]

95. Saladino, R.; Crestini, C.; Ciciriello, F.; Pino, S.; Costanzo, G.; Di Mauro, E. From formamide to RNA: The roles of formamide and water in the evolution of chemical information. *Res. Microbiol.* **2009**, *160*, 441–448. [CrossRef] [PubMed]

96. Burcar, B.; Pasek, M.; Gull, M.; Cafferty, B.J.; Velasco, F.; Hud, N.V.; Menor-Salván, C. Darwin's warm little pond: A one-pot reaction for prebiotic phosphorylation and the mobilization of phosphate from minerals in a urea-based solvent. *Angew. Chem. Int. Ed.* **2016**, *55*, 13249–13253. [CrossRef] [PubMed]

97. Gibard, C.; Bhowmik, S.; Karki, M.; Kim, E.-K.; Krishnamurthy, R. Phosphorylation, oligomerization and self-assembly in water under potential prebiotic conditions. *Nat. Chem.* **2017**, *10*, 212–217. [CrossRef] [PubMed]

98. Pasek, M.A.; Gull, M.; Herschy, B. Phosphorylation on the early Earth. *Chem. Geol.* **2017**, *475*, 149–170. [CrossRef]

99. Hazen, R.M. Paleomineralogy of the Hadean Eon: A preliminary species list. *Am. J. Sci.* **2013**, *313*, 807–843. [CrossRef]

100. Shapiro, R.; Kang, S. Uncatalyzed hydrolysis of deoxyuridine, thymidine, and 5-bromodeoxyuridine. *Biochemistry* **1969**, *8*, 1806–1810. [CrossRef] [PubMed]

101. Garrett, E.R.; Mehta, P.J. Solvolysis of adenine nucleosides. II. Effects of sugars and adenine substituents on alkaline solvolyses. *J. Am. Chem. Soc.* **1972**, *94*, 8542–8547. [CrossRef] [PubMed]

102. Lindahl, T.; Karlstrom, O. Heat-induced depyrimidination of deoxyribonucleic acid in neutral solution. *Biochemistry* **1973**, *12*, 5151–5154. [CrossRef] [PubMed]

103. Shapiro, H.S.; Chargaff, E. Studies on the nucleoside arrangement in deoxyribonucleic acids. I. The relationship between the production of pyrimidine nucleoside 3′,5′-diphosphates and specific features of nucleotide sequence. *Biochim. Biophys. Acta* **1957**, *26*, 596–608. [CrossRef]

104. Venner, H. Research on nucleic acids. XII. Stability of the N-glycoside bond of nucleotides. *Hoppe Seylers Z. Physiol. Chem.* **1966**, *344*, 189–196. [CrossRef] [PubMed]

105. Shapiro, R.; Danzig, M. Acidic hydrolysis of pyrimidine deoxyribonucleotides. *Biochim. Biophys. Acta* **1973**, *319*, 5–10. [CrossRef]

106. Lindahl, T.; Nyberg, B. Rate of depurination of native deoxyribonucleic acid. *Biochemistry* **1972**, *11*, 3610–3618. [CrossRef] [PubMed]

107. Kochetov, N.K.; Budowski, E.L. (Eds.) *Organic Chemistry of Nucleic Acids*; Plenum Press: London, UK; New York, NY, USA, 1982; pp. 477–532.

108. Verlander, M.S.; Lohrmann, R.; Orgel, L.E. Catalysts for self-polymerization of adenosine cyclic 2′,3′-phosphate. *J. Mol. Evol.* **1973**, *2*, 303–316. [CrossRef] [PubMed]

109. Verlander, M.S.; Orgel, L.E. Analysis of high molecular weight material from the polymerization of adenosine cyclic 2′,3′-phosphate. *J. Mol. Evol.* **1974**, *3*, 115–120. [CrossRef] [PubMed]

110. Usher, D.A.; McHale, A.H. Nonenzymic joining of oligoadenylates on a polyuridylic acid template. *Science* **1976**, *192*, 53–54. [CrossRef] [PubMed]

111. Usher, D.A.; McHale, A.H. Hydrolytic stability of helical RNA—Selective advantage for natural 3′,5′-bond. *Proc. Natl. Acad. Sci. USA* **1976**, *73*, 1149–1153. [CrossRef] [PubMed]

112. Costanzo, G.; Pino, S.; Ciciriello, F.; Di Mauro, E. Generation of long RNA chains in water. *J. Biol. Chem.* **2009**, *284*, 33206–33216. [CrossRef] [PubMed]

113. Costanzo, G.; Saladino, R.; Botta, G.; Giorgi, A.; Scipioni, A.; Pino, S.; Di Mauro, E. Generation of RNA molecules by a base-catalysed click-like reaction. *ChemBioChem* **2012**, *13*, 999–1008. [CrossRef] [PubMed]

114. Šponer, J.E.; Šponer, J.; Giorgi, A.; Di Mauro, E.; Pino, S.; Costanzo, G. Untemplated nonenzymatic polymerization of 3′,5′ cGMP: A plausible route to 3′,5′-linked oligonucleotides in primordia. *J. Phys. Chem. B* **2015**, *119*, 2979–2989. [CrossRef] [PubMed]

115. Morasch, M.; Mast, C.B.; Langer, J.K.; Schilcher, P.; Braun, D. Dry polymerization of 3′,5′-cyclic GMP to long strands of RNA. *ChemBioChem* **2014**, *15*, 879–883. [CrossRef] [PubMed]

116. Costanzo, G.; Pino, S.; Timperio, A.M.; Šponer, J.E.; Šponer, J.; Nováková, O.; Šedo, O.; Zdráhal, Z.; Di Mauro, E. Non-enzymatic oligomerization of 3′,5′ cyclic AMP. *PLoS ONE* **2016**, *11*, e0165723. [CrossRef] [PubMed]

117. Costanzo, G.; Giorgi, A.; Scipioni, A.; Timperio, A.M.; Mancone, C.; Tripodi, M.; Kapralov, M.; Krasavin, E.; Kruse, H.; Šponer, J.; et al. Nonenzymatic oligomerization of 3′,5′-cyclic CMP induced by proton and UV irradiation hints at a nonfastidious origin of RNA. *ChemBioChem* **2017**, *18*, 1535–1543. [CrossRef] [PubMed]

118. Cassone, G.; Šponer, J.; Saija, F.; Di Mauro, E.; Saitta, A.M.; Šponer, J.E. Stability of 2′,3′ and 3′,5′ cyclic nucleotides in formamide and in water: A theoretical insight into the factors controlling the accumulation of nucleic acid building blocks in the prebiotic pool. *Phys. Chem. Chem. Phys.* **2017**, *19*, 1817–1825. [CrossRef] [PubMed]

119. Chwang, A.K.; Sundaralingam, M. The crystal and molecular structure of guanosine 3′,5′-cyclic monophosphate (cyclic GMP) sodium tetrahydrate. *Acta Crystallogr. B* **1974**, *30*, 1233–1240. [CrossRef]

120. Kervio, E.; Sosson, M.; Richert, C. The effect of leaving groups on binding and reactivity in enzyme-free copying of DNA and RNA. *Nucleic Acids Res.* **2016**, *44*, 5504–5514. [CrossRef] [PubMed]

121. Lohrmann, R. Formation of nucleoside 5′-phosphoramidates under potentially prebiological conditions. *J. Mol. Evol.* **1977**, *10*, 137–154. [CrossRef] [PubMed]

122. Kawamura, K.; Ferris, J.P. Kinetic and mechanistic analysis of dinucleotide and oligonucleotide formation from the 5′-phosphorimidazolide of adenosine on Na⁺-montmorillonite. *J. Am. Chem. Soc.* **1994**, *116*, 7564–7572. [CrossRef] [PubMed]

123. Mansy, S.S.; Schrum, J.P.; Krishnamurthy, M.; Tobe, S.; Treco, D.A.; Szostak, J.W. Template-directed synthesis of a genetic polymer in a model protocell. *Nature* **2008**, *454*, 122–125. [CrossRef] [PubMed]

124. Orgel, L.E. Prebiotic chemistry and the origin of the RNA world. *Crit. Rev. Biochem. Mol. Biol.* **2004**, *39*, 99–123. [PubMed]

125. Jauker, M.; Griesser, H.; Richert, C. Copying of RNA sequences without pre-activation. *Angew. Chem. Int. Ed.* **2015**, *54*, 14559–14563. [CrossRef] [PubMed]

126. Da Silva, L.; Maurel, M.C.; Deamer, D. Salt-promoted synthesis of RNA-like molecules in simulated hydrothermal conditions. *J. Mol. Evol.* **2015**, *80*, 86–97. [CrossRef] [PubMed]

127. DeGuzman, V.; Vercoutere, W.; Shenasa, H.; Deamer, D. Generation of oligonucleotides under hydrothermal conditions by non-enzymatic polymerization. *J. Mol. Evol.* **2014**, *78*, 251–262. [CrossRef] [PubMed]

128. Šponer, J.E.; Šponer, J.; Di Mauro, E. Four ways to oligonucleotides without phosphoimidazolides. *J. Mol. Evol.* **2015**, *82*, 5–10. [CrossRef] [PubMed]

Article

The Reaction of Aminonitriles with Aminothiols: A Way to Thiol-Containing Peptides and Nitrogen Heterocycles in the Primitive Earth Ocean

Ibrahim Shalayel, Seydou Coulibaly, Kieu Dung Ly, Anne Milet and Yannick Vallée *

Univ. Grenoble Alpes, CNRS, Département de Chimie Moléculaire, Campus, F-38058 Grenoble, France; ibrahim.shalayel@univ-grenoble-alpes.fr (I.S.); bouahcoul@gmail.com (S.C.); lykieudung1005@yahoo.com.vn (K.D.L.); anne.milet@univ-grenoble-alpes.fr (A.M.)
* Correspondence: yannick.vallee@univ-grenoble-alpes.fr

Received: 28 September 2018; Accepted: 18 October 2018; Published: 19 October 2018

Abstract: The Strecker reaction of aldehydes with ammonia and hydrogen cyanide first leads to α-aminonitriles, which are then hydrolyzed to α-amino acids. However, before reacting with water, these aminonitriles can be trapped by aminothiols, such as cysteine or homocysteine, to give 5- or 6-membered ring heterocycles, which in turn are hydrolyzed to dipeptides. We propose that this two-step process enabled the formation of thiol-containing dipeptides in the primitive ocean. These small peptides are able to promote the formation of other peptide bonds and of heterocyclic molecules. Theoretical calculations support our experimental results. They predict that α-aminonitriles should be more reactive than other nitriles, and that imidazoles should be formed from transiently formed amidinonitriles. Overall, this set of reactions delineates a possible early stage of the development of organic chemistry, hence of life, on Earth dominated by nitriles and thiol-rich peptides (TRP).

Keywords: origin of life; prebiotic chemistry; thiol-rich peptides; cysteine; aminonitriles; imidazoles

1. Introduction

In ribosomes, peptide bonds are formed by the reaction of the amine group of an amino acid with an ester function. For non-ribosomal peptides, the amide formation involves the reaction of an amine on a thioester [1]. In both cases, mixed phosphoric carboxylic anhydrides are transiently formed. Esters, thioesters, and anhydrides are activated forms of the carboxylic acid function. Their intermediacy is mandatory and no significant C-N bond formation would occur directly from the reaction of an acid function with an amine [2]. What is true in today's biology, was also true four billion years ago, when life was beginning its development in the terrestrial ocean. Activated derivatives had to be involved in the formation of prebiotic polymers. As a consequence, if acids were involved at some stage, a strong energy source was necessary. Nowadays, it is furnished by the cleavage of the triphosphate group of adenosine triphosphate [3].

Many simple aldehydes were probably present in the primitive ocean [4] and are plausible precursors for α-amino acids. Reacting with ammonia and hydrogen cyanide, they would have first given α-aminonitriles, which, upon hydrolysis, would have delivered amino acids (Figure 1) [5]. However, even though it is exothermic, the reaction of nitriles with water is a slow process [6]; so slow that, once formed in the ocean, aminonitriles would have had ample time to react with species more nucleophilic than water.

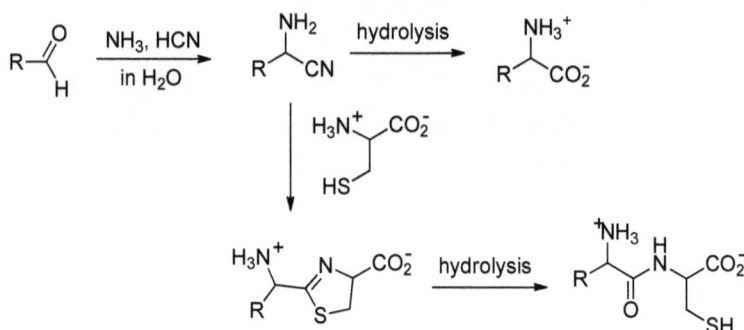

Figure 1. Strecker reaction followed by condensation of the obtained aminonitrile with cysteine.

Nitriles are known to react with aminothiols to give thiazolines, which in turn can be hydrolyzed to mercaptoamides [7]. Starting from α-aminonitriles and cysteine, the expected products of this two-step process are dipeptides (Figure 1). In the early ocean, this could have been an efficient and selective process to thiol-containing dipeptides [8].

Compared to any activation process starting from acids, this nitrile scenario has the advantage of not necessitating any strong energy source. The acid does not need to be activated as it is delivered directly in an activated form by the Strecker reaction.

HCN has long been given an important role in prebiotic molecular evolution [9]. As it is largely distributed in space, having been observed in various regions, for instance, near carbon stars [10] and in a proto-planetary nebula [11], as well as in comets [12,13], it is highly possible that HCN was present on the early Earth. Furthermore, it has been postulated that it could have been formed when numerous asteroids struck our planet during the Late Heavy Bombardment [14]. It might have been produced photochemically in the atmosphere [15,16]. It was ejected from volcanoes [17] and submarine hydrothermal vents [18].

Hydrogen sulfide is another important small molecule in our hypothesis. It would have been necessary for the formation of cysteine. It has often been detected in space [19], *inter alia* in star forming regions [20], and in cold clouds [21], as well as in comets [22]. Furthermore, it is abundantly ejected from volcanoes [23,24], so there is no doubt that it was effectively present on the primitive Earth. Its presence permitted the synthesis of cysteine and homocysteine [25]. Homocysteine would have been obtained by a Strecker reaction starting from the addition of the product H_2S onto acrolein ($HSCH_2CH_2CHO$). In a similar way, cysteine would have been synthesized from $HSCH_2CHO$, itself possibly obtained from glycolaldehyde.

2. Experimental Section

Products (thiazolines, dipeptides …) were identified in reaction mixtures by NMR spectroscopy (^1H and ^{13}C) and mass spectrometry. No attempt at purifying them was made (except for **11** and **12**).

NMR monitored reactions were run in D_2O solutions, in NMR tubes. NMR apparatus: Bruker Avance III 400 or 500. Classically, NMR experiments were run at concentrations of 5×10^{-3} to 5×10^{-2} mol/L.

For the mass experiment, H_2O was used as the solvent. High-resolution mass spectra were recorded on a Waters G2-S Q-TOF mass spectrometer or on a LTQ Orbitrap XL (Thermo Scientific) spectrometer. Low resolution ESI analysis was performed on an Amazon speed (Brucker Daltonics) IonTrap spectrometer.

(*R*)-2-((*S*)-1-amino-3-(methylthio)propyl)-4,5-dihydrothiazole-4-carboxamide (**11**)

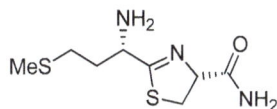

Met-CN (168 mg, 1.29 mmol) was dissolved in 15 mL H_2O. Cys-NH$_2$.TFA (280 mg; 1.29 mmol) was added. The pH of the solution was adjusted to 8 by adding Na_2CO_3. The solution was stirred at 45 °C for 2.5 h. The aqueous phase was extracted three times with ethyl acetate. The organic layer was dried over Na_2SO_4, filtered, and concentrated under vacuum. After purification by silica gel chromatography (1–10% MeOH/DCM), the thiazoline **11** was obtained as an orange oil (16% yield).

HRMS (ESI) for $C_8H_{16}ON_3S_2$: calc. m/z = 234.0735, Found m/z = 234.0740 [M + H]$^+$. **^{1}H-NMR** (D$_2$O, 400 MHz) (δ, ppm): 5.08 (1H, t, *J* = 8.98 Hz, CH), 3.95 (1H, t, *J* = 6.57 Hz, CH), 3.65 (1H, t, *J* = 10.82 Hz, CH$_2$), 3.46 (1H, dd, *J* = 11.30; 8.20, CH$_2$), 2.56 (2H, t, *J* = 7.08 Hz, CH$_2$), 2.07 (3H, s, CH$_3$), 1.97 (2H, sep, *J* = 7.0, CH$_2$). **^{13}C-NMR** (D$_2$O, 100 MHz) (δ, ppm): 182.99, 176.14, 77.00, 52.93, 34.89, 34.43, 29.18, 14.12.

(4*R*)-2-(1-amino-2-methylpropyl)-4,5-dihydrothiazole-4-carboxamide (**12**)

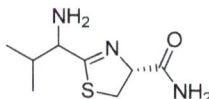

Val-CN.HCl (35 mg, 0.26 mmol) was dissolved in 5 mL H_2O. Cys-NH$_2$.TFA (57 mg; 0.26 mmol) was added. The solution was adjusted to pH = 7 by adding Na_2CO_3. The solution was stirred at 45 °C for 24 h. The aqueous phase was extracted three times with ethyl acetate. The organic layer was dried over Na_2SO_4, filtered, and concentrated under vacuum. After purification by silica gel chromatography (1–10% MeOH/DCM), the thiazoline **12** was obtained as a yellow oil (30% yield).

HRMS (ESI) for $C_8H_{16}ON_3S$: calc. m/z = 202.1014, Found m/z = 202.1016 [M + H]$^+$. **^{1}H-NMR** (D$_2$O, 500 MHz) (δ, ppm): 4.40 (1H, m, CH), 3.75 (1H, dd, 5.68; 2.59 Hz, CH), 2.70–2.99 (2H, m, CH$_2$), 2.12 (1H, sep, *J* = 6.60, CH), 0.85–0.94 (6H, m, CH$_3$). **^{13}C-NMR** (D$_2$O, 125 MHz, both isomers were observed) (δ, ppm): 174.02–173.74, 169.70–169.43, 58.66–58.38, 55.65–55.63, 30.01–29.94, 25.30–25.02, 17.79–17.62, 16.85–16.65.

Theoretical calculations were carried out using the Gaussian09, Revision D.01 software. All the geometries were optimized using the B3LYP functional in conjunction with the 6-31g(d,p) basis set and the water solvent effects were described by using the polarizable continuum model (PCM), namely IEFPCM (integral equation formalism PCM) [26]. These optimizations were followed by a frequency calculation at the same level to ensure that the geometry was indeed a real minimum, i.e., all the second derivatives were positive.

3. Results

We first studied the reaction of aminoacetonitrile (GlyCN **1a**, the nitrile derivative of glycine) with cysteine. Reactions were conducted in D$_2$O solutions and followed by NMR spectroscopy. Representative ^{1}H NMR spectra are shown in Figure 2. Formation of the expected thiazoline ring **2a** was evidenced by the apparition of signals at ca. 5 ppm (a triplet-like dd), and from 3.4 to 3.7 ppm (2 dd). After some time, new signals grew, including a triplet at 4.4 ppm and a thin doublet-like signal at ca. 2.9 ppm, both characteristic of Gly-Cys **3a**. We repeated this experiment many times, generally at a concentration of 5×10^{-3} to 5×10^{-2} mol/L, for practical NMR measurements. However, we also tested it at $3\ 10^{-4}$ mol/L, a concentration at which **2a** and **3a** were also obtained.

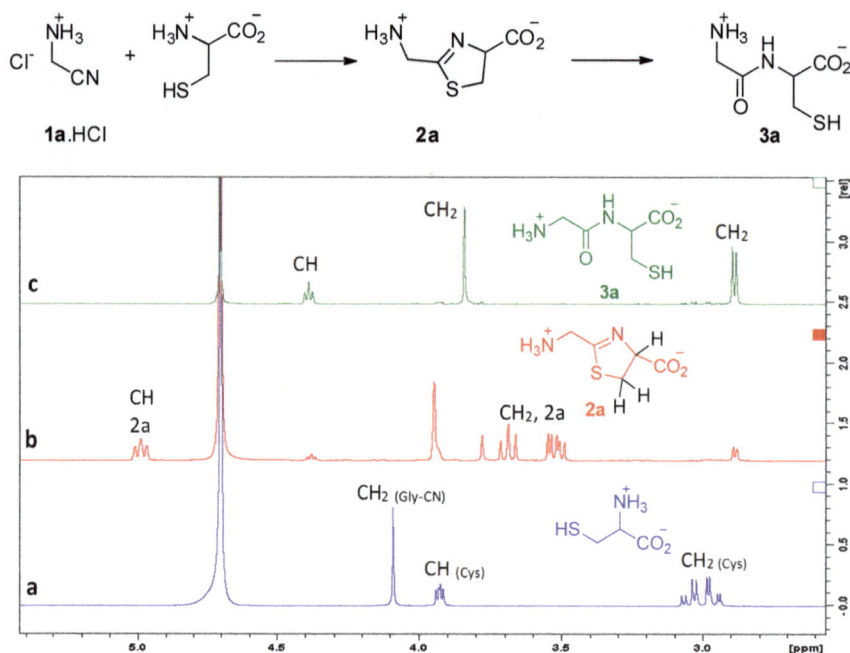

Figure 2. Reaction of aminoacetonitrile with cysteine, (**a**) mixture of starting materials, (**b**) mostly **2a**, (**c**) GlyCys **3a**. Conditions: room temperature, pH 6.5, concentration 10^{-2} mol/L.

We have studied the influence of the pH on these reactions. The results are summarized in Figure 3. The ring formation is quicker under basic conditions. We believe that under such conditions, the thiol function is deprotonated, giving the more nucleophilic thiolate species. Under an acidic condition, the nucleophilic species is probably the thiol itself. The hydrolysis step is quicker under acidic conditions. This probably implies that the thiazoline ring is activated through protonation of the double bonded nitrogen atom before H_2O addition.

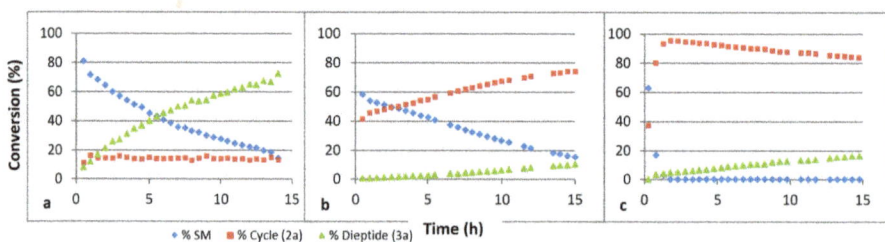

Figure 3. Evolution of a mixture of GlyCN and cysteine in D_2O at 45 °C followed by ^1H NMR, at various pH's. (**a**) pH 4, (**b**) pH 6, (**c**) pH 8. SM: starting materials. Concentration 4×10^{-2} mol/L.

We have also tested these reactions at 24 °C and 70 °C. Not surprisingly the process is quicker at a higher temperature, but also goes well at room temperature.

The conditions in the ocean four billion years ago are not precisely known. However, water was probably still hotter than now [27] and the presence of large amounts of CO_2 in the atmosphere might imply that it was slightly acidic (nowadays, ocean's pH is 8.1) [28]. Taking these considerations into account, we chose a temperature of 45 °C and a pH of ca. 5.5–6.5 as standard conditions.

Life **2018**, *8*, 47

Under such conditions, we observed no reaction between aminoacetonitrile and any other proteinogenic amino acid that we tested (glycine, alanine, serine, methionine, aspartic acid, histidine, and lysine). It is worth noting that serine did not react. It appears that its alcohol function is not nucleophilic enough to attack the CN triple bond. Hence, the presence of a thiol function is mandatory. Indeed, homocysteine (Hcy) did react with a reaction rate similar to that observed with cysteine. In this case, the intermediate is the six-membered ring **4a**, and the final product is Gly-Hcy **5a** (Figure 4).

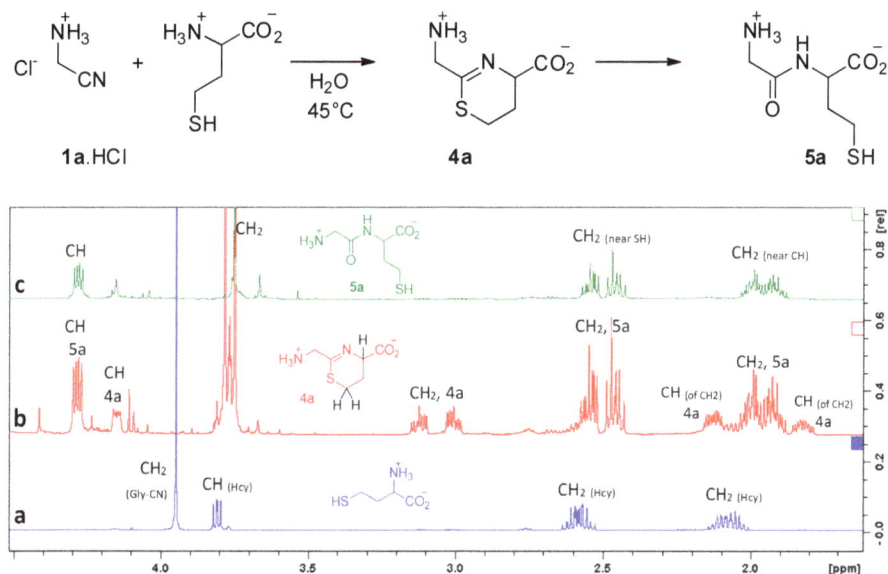

Figure 4. Reaction of homocysteine with GlyCN at 45 °C, pH = 6.5, 10^{-2} mol/L. ^{1}H NMR's show: (**a**) starting mixture, (**b**) reaction mixture after 6 h (**4a**/**5a** = 3/7), (**c**) after 24 h.

Some other representative results are presented in Figure 5. They show that the acid function of cysteine can be replaced by a primary or secondary amide. When Cys-Gly was used, the tripeptide Gly-Cys-Gly **9** was obtained with a very good conversion. *N*-Acetyl aminoacetonitrile **1b**, which can be considered as a model for any other *N*-acyl acetonitrile, including cyano-terminated peptides, also reacted with a good rate (Figure 6a). In contrast, the reaction was slower when aminoacetonitrile was replaced by β-aminopropionitrile **1c** (Figure 7). In these two last examples, the hydrolysis step was quick. No reaction was observed with the γ-nitrile of glutamic acid **1d** [29]. The selectivity in favor of α-aminonitriles was also exemplified when the bis-nitrile derived from aspartic acid **1e** [30] was used. In this case, only the α-aminonitrile reacted, giving the corresponding thiazoline **2e,** which was stable under these conditions (Figure 6c,d).

Finally, we also tested the reactivity of penicilamine, a sterically hindered aminothiol. In this case, the reaction was very slow, probably because of the bulkiness of the gem-dimethyl substituents. Furthermore, the only detected product was the final dipeptide **10** (Figure 6b). This might be due to the electron donating property of the methyl groups, making the nitrogen atom of the intermediate thiazoline ring more basic. Protonation of this nitrogen atom would thus be easier, hence the hydrolysis step quicker.

In order to explain the observed selectivity, we calculated the level of the π^* orbital of a series of nitriles (Table 1).

The lowest calculated orbital was that of the protonated form of aminoacetonitrile. Such a level would explain its greater reactivity compared to other nitriles. Noticeably, the non-protonated form of aminoacetonitrile is predicted to be much less reactive and so is probably not involved in the reaction mechanism. Also, the π^* orbital of β-propionitrile is higher (it is less reactive) and the simplest γ-aminonitrile is predicted to be even less reactive (no reaction from glutamic nitrile). In contrast, α-substitution of aminoacetonitrile, as in α-propionitrile, should not significantly alter its reactivity.

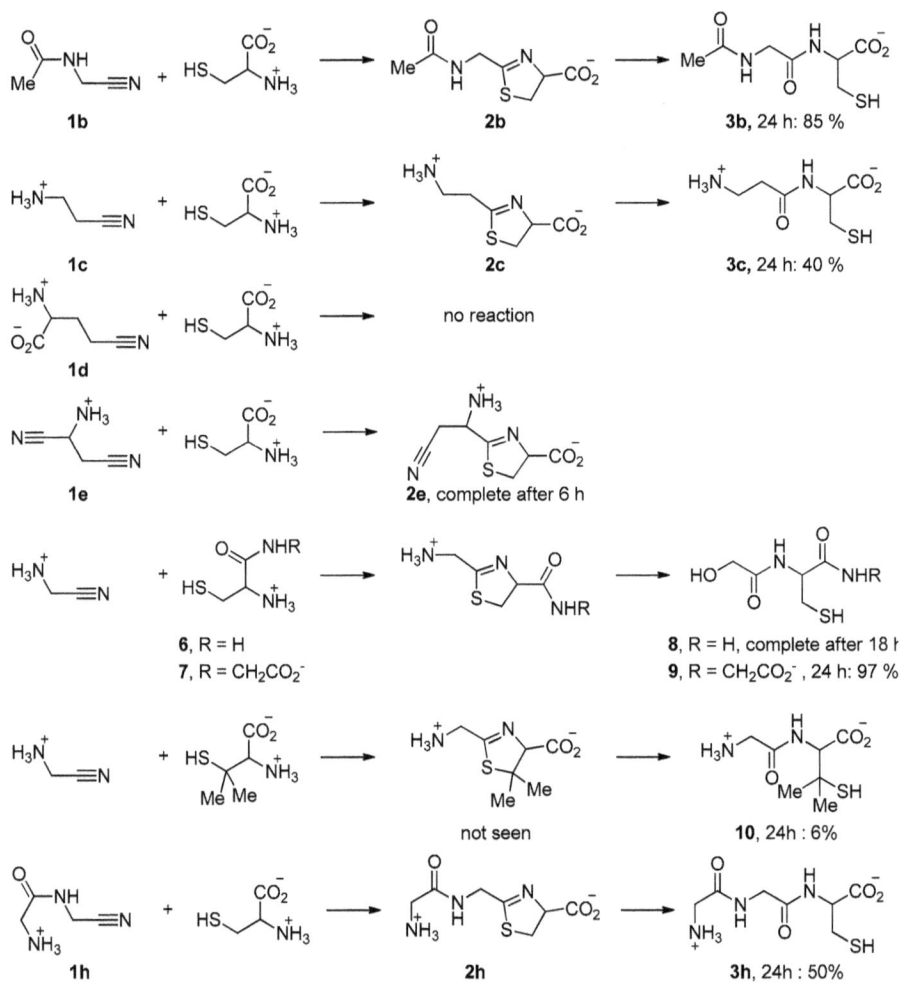

Figure 5. Some reactions of aminothiols with nitriles (45 °C, pH 5–6). The solvent was D_2O. Reactions were monitored by ^{1}H NMR.

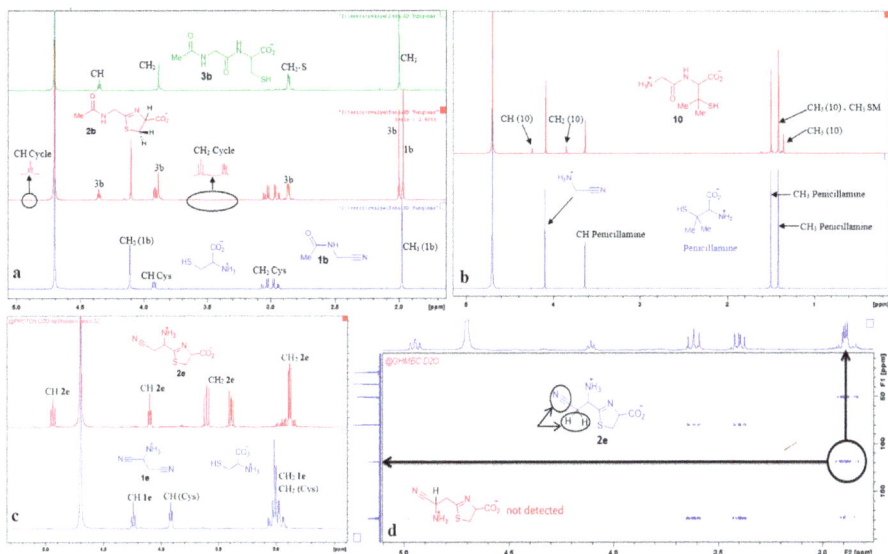

Figure 6. NMR spectra recorded during representative aminothiol + aminonitrile reactions. (**a**) Reaction of cysteine with *N*-acetyl aminoacetonitrile; (**b**) reaction of penicillamine with GlyCN; (**c**) reaction of aspartic acid bis-nitrile with cysteine; (**d**) 2D experiment demonstrating the regioselectivity of this last reaction towards α-nitrile.

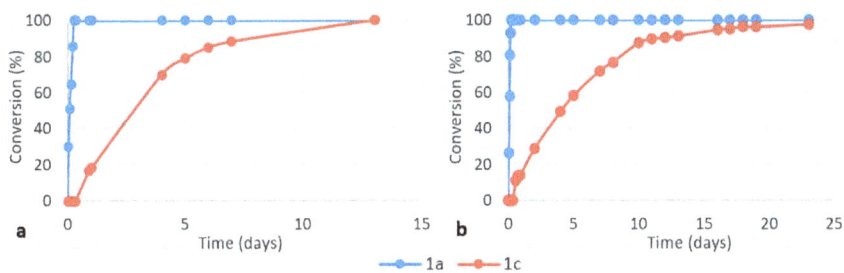

Figure 7. Consumption of aminoacetonitrile **1a** and β-aminopropionitrile **1c** in competition reactions with (**a**) cysteine and (**b**) homocysteine (ratio **1a/1c**/amino acid 1/1/2, pH ca. 6, 45 °C).

Table 1. Calculated level of the π* orbital of various nitriles.

Nitrile		π* Value	Reaction Rate
$H_3N(+)CH_2CN$ **1a** protonated		−0.03632	quick
H_2NCH_2CN **1a**		0.01698	No reaction?
$H_3N(+)CH(CH_3)CN$		−0.03010	quick
$H_3CCONHCH_2CN$ **1b**		0.00504	slower
$H_3N(+)CH_2CONHCH_2CN$ **1h**		0.00353	quick
$H_3N(+)CH_2CONHCH_2CONHCH_2CN$		−0.00547	quick
$H_3N(+)CH_2CH_2CN$ **1c**		0.00616	slower
$H_3N(+)CH_2CH_2CH_2CN$		0.01556	No reaction
H_3CCN		0.03491	No reaction
Aspartic acid bis-nitrile **1e**	αCN	−0.03566	Reacts at αCN
	βCN	−0.00500	

We studied this substitution effect using the nitriles derived from two other amino acids (Figure 8). L-Methionine nitrile **1f** [31] was prepared from N-protected L-methionine in a three-step process. Valine nitrile **1g**, was prepared as a racemic mixture using a Strecker reaction from the corresponding aldehyde [32].

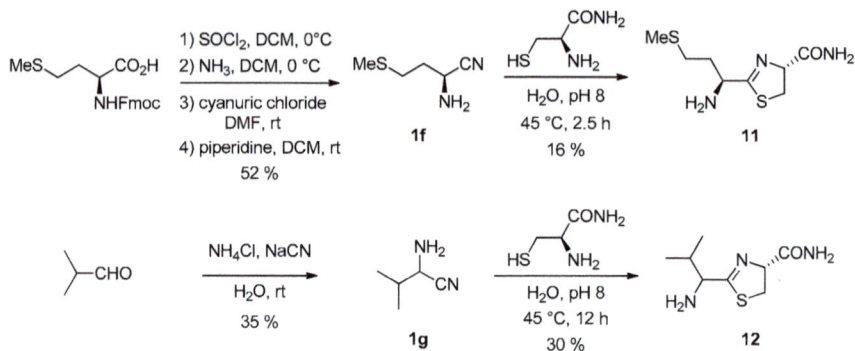

Figure 8. Synthesis of MeCN **1f** and ValCN **1g** and their reaction with cysteine amide.

Their reaction with cysteine amide [33] was studied (Figure 8). In these cases, we were able to isolate the intermediate cycles in pure form (as a 1/1 mixture of diastereoisomers from racemic ValCN). The deceptive isolated yields, despite the slightly basic conditions we used, which should have slowed the hydrolysis step, were probably due to important hydrolysis during column chromatography on silica. In addition, we found that the rate of hydrolysis in water of the valine-derived thiazoline **12** was much slower than the one of the methionine derivative **11** (and of the simplest Glycine derivative **2a**). This is probably due to the presence of the bulky isopropyl group in **12**.

On the basis of our experiments, we propose that AA-Cys and AA-Hcy dipeptides were over-represented in the primitive ocean (compared to non-thiol-containing dipeptides).

These dipeptides are thiols and as such, could be major players in a "thioester world" [34]. Indeed, when we mixed Gly-Cys **3a** (obtained from a 1 to 1 mixture of GlyCN and cysteine) with an excess of GlyCN in D_2O solution at 45 °C (Figure 9), a peak was observed at 194.16 ppm in the ^{13}C NMR of the reaction mixture (Figure 10). Such a chemical shift is characteristic of the thioester function. We believe that it belongs to compound **14**. We also noticed the formation of glycine amide **15**. These products would both derive from the first formed C=N double-bonded addition product **13**. The thioester was partly hydrolyzed to give glycine, but we were also able to characterize, among the reaction products, the amidonitrile Gly-GlyCN **1h** (^{13}C NMR: 27.58, 40.34, 116.74, 167.58 ppm), meaning that the thioester reacted with the non-protonated amino group of GlyCN (which is possible because of the low pKa of GlyCN: 5.55 [35]). For instance, in an experiment in which we used globally 4 eq. of GlyCN (relative to cysteine), after two days at 45 °C, the observed GlyOH/GlyNH$_2$/GlyGlyCN ratio was found to be 21/37/42. This demonstrates that Gly-Cys is able to promote the formation of a peptide bond from a nitrile. Similar results were obtained for Gly-Hcy. In addition, as our theoretical calculations predicted (Table 1), when Gly-GlyCN **1h** was mixed with cysteine, Gly-Gly-Cys **3h** [36] was readily formed (Figure 5), demonstrating that not only dipeptides, but also tripeptides, could have been formed by this process in the primitive ocean.

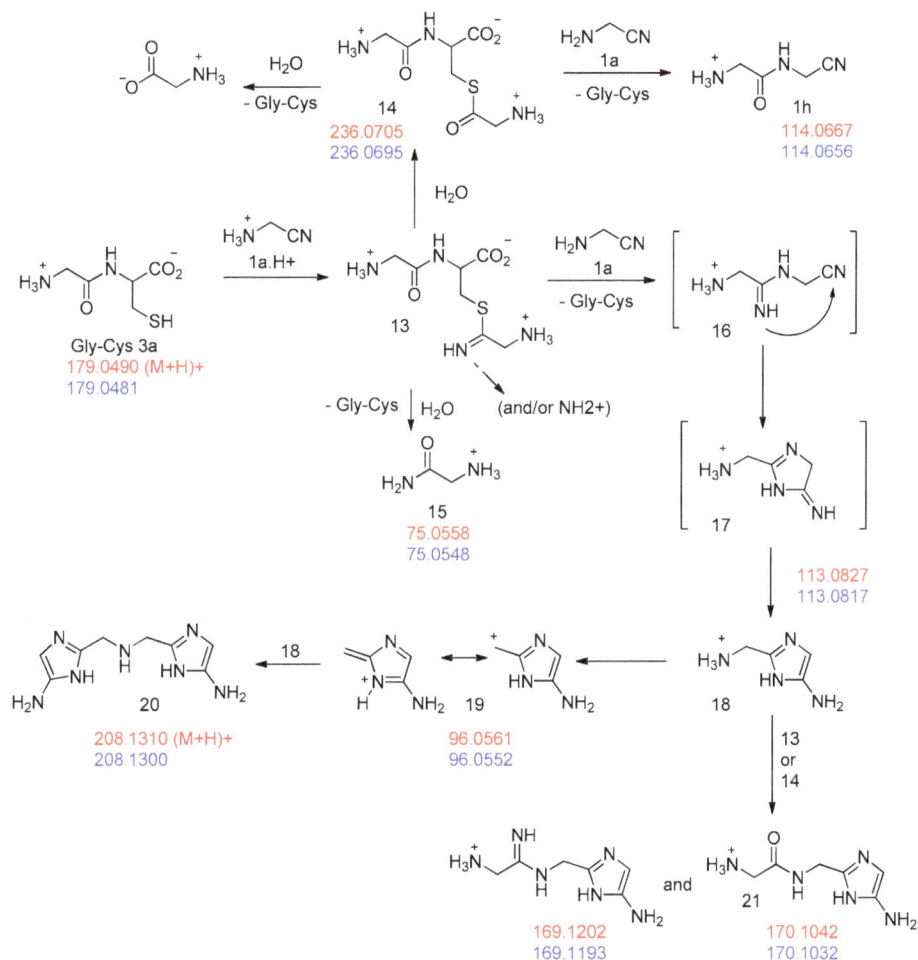

Figure 9. Proposed pathways in the reaction of excess GlyCN **1a** with GlyCys **3a**. Mass attribution (red, calculated; blue, found).

However, the formation of other products was also evidenced in the reaction of cysteine with excess GlyCN. Thus, in the ^{13}C NMR spectra, peaks at 130–140 ppm were observed (Figure 10). The mass spectrum of a reaction mixture in water also showed the formation of various products (Figure 11, see Supplementary Materials for complete spectrum and further mass attributions). This mass spectrum first confirmed the presence of Gly-GlyCN **1h** (protonated, found 114.0656, calcd 114.0667). Another mass was detected at 113.0817. It could be attributed to the amidine **16** (calcd 113.0827). However, in accordance with the observed ^{13}C NMR of the mixture, we propose that this amidine cyclized and that this mass peak should be attributed to the imidazole **18**. Indeed, the cyclization of an amidinonitrile similar to **16** into an aminoimidazole (5-amino-2-methyl-1H-imidazole) has already been reported [37]. At least some of the other products observed in the mass spectrum would be evolution products of **18**. For instance, this imidazole could lose ammonia to give the stabilized cation **19** that would in turn react with **18** (in its free amine form) to yield the bis-imidazole **20** (M+H, found 208.1300, calcd 208.1310). **18** itself could react with thiolester **14** to give the amide **21**. Other structures are possible (see Supplementary Materials).

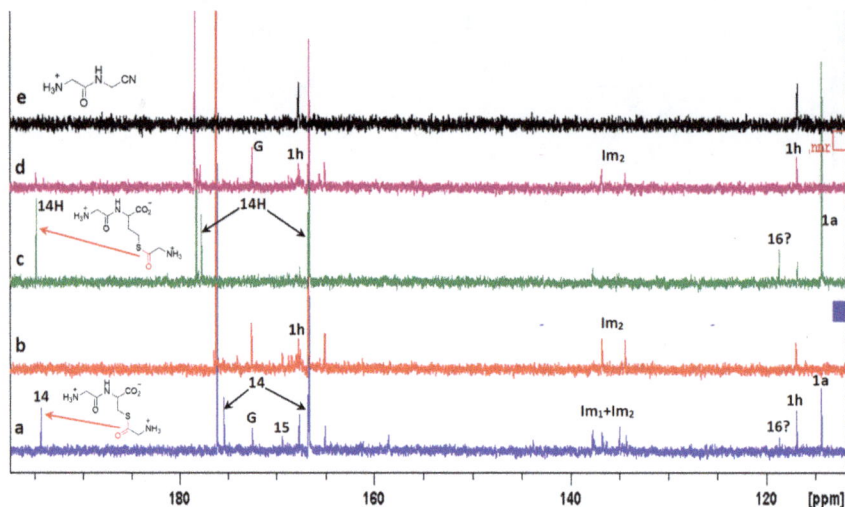

Figure 10. ^{13}C NMR spectra recorded during the reaction of an excess GlyCN **1a** with GlyCys **3a** (from **1a**. HCl + cysteine) or GlyHcy **5a** (from **1a**. HCl + homocysteine) at 45 °C, pH 6.5. (**a**) with **3a** 20 h after mixing **1a**. HCl and cysteine; (**b**) after 70h; (**c**) with **5a** 20 h after mixing **1a**. HCl and homocysteine; (**d**) after 70h; (**e**) reference spectrum of **1h**. **G**: glycine. Peaks corresponding to at least two products were detected near 135–140 ppm. They might correspond to two different imidazoles (named **Im₁** and **Im₂**). **14H**: the homocysteine thioester analogue of **14**. Big peaks at 166.65 (166.66) and 176.07 (178.33) belong to **3a** (and **5a**). One peak of both **14** and **14H** sticks to the foot of the 166.6 ppm peak of **3a** and **5a**.

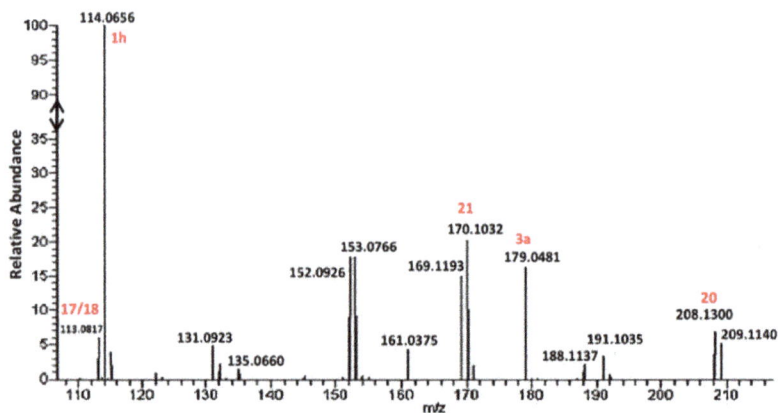

Figure 11. Mass spectrum of a reaction of an excess GlyCN **1a** with GlyCys.

In order to further ascertain the cyclization of the imino-compound **16**, we calculated its stability in comparison to cyclized forms. We used the strategy described previously for Table 1 results, with the 6-31+G(d,p) basis set (Figure 12). Not surprisingly, **18** was calculated to have free enthalpy 12.7 kcal/mol lower than **16** and 6.9 kcal/mol more stable than **17**.

Interestingly, it was found that the dissociation of **18** into **19** + NH₃ only costs around 7 kcal/mol. The ΔH of dissociation is nearly 19 kcal/mol (18.67 kcal/mol), but due to the dissociative character of the process, the ΔG value drops to 7.01 kcal/mol. This process does not show a well-defined TS. Thus, we think that the formation of cation **19** proposed in Figure 9 is a plausible event.

We studied more precisely the cyclization step from **16** to **17** (Figure 13). H_3O^+ was used to promote the reaction and to give a proton to the nitrogen atom of the nitrile group, which becomes part of the exocyclic imine of **17**. Two explicit molecules of water were introduced, in addition to the water continuum. It appeared that the cyclization step should be exocyclic and quick, with a low level TS (activation energy of 6.6 kcal/mol). This is another confirmation that the compound of mass 113.0817 we observed was not **16**, but indeed the imidazole **18** (resulting from a simple proton migration from **17**). It is noticeable that similar calculations for the potential cyclization of the amide **1h** into an oxazole indicated that this reaction should not happen. Indeed, we never observed it experimentally. In sharp contrast with **16**, **1h** (GlyGlyCN) is stable.

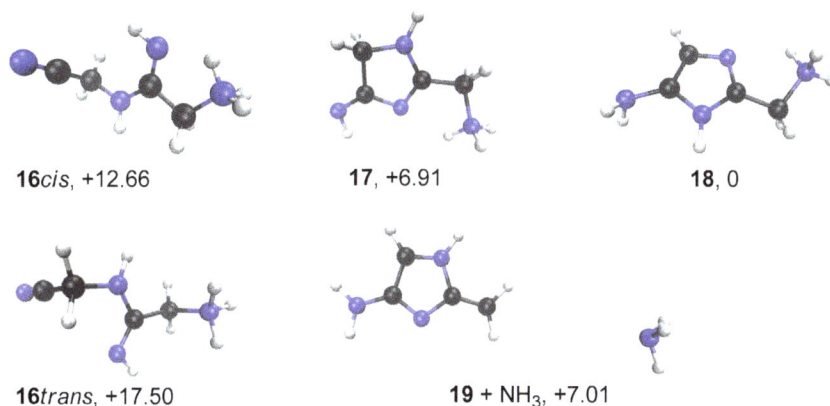

16cis, +12.66   **17**, +6.91   **18**, 0

16trans, +17.50   **19** + NH_3, +7.01

Figure 12. Minimized conformation for compounds **16** to **19**. B3LYP/6-31+G** with a continuum to mimic the solvent effect of water. ΔG's relative to **18** in kcal/mol.

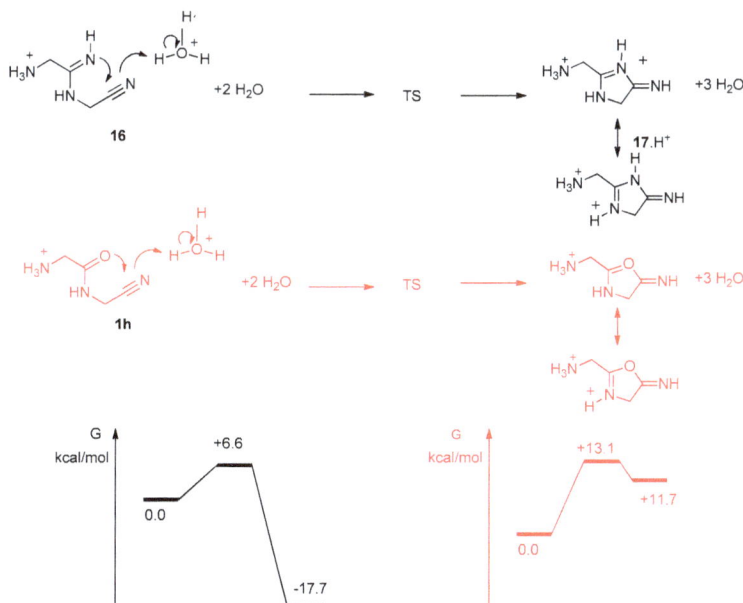

Figure 13. Reaction pathways for the cyclisation step from **16** (black), and **1h** (red). Level of calculation B3LYP/6-31+G**/SCRF(water).

4. Conclusions

A world containing small peptides and heterocycles, based on the chemistry of thiols and nitriles, can be delineated. It could have persisted as long as a significant amount of HCN was present in the ocean and permitted the synthesis of aminonitriles from aldehydes. In this "cyano-sulfidic" world [14], thiol-containing peptides would have been the most important molecules. We propose to name it the "Thiol Rich Peptide (TRP) world" [8]. In such a world, not only dipeptides, but also tripeptides, would have been formed. For instance, any dipeptide nitrile AA1-AA2CN (the simplest example being Gly-GlyCN) produced from the reaction of $H_2NAA2CN$ with a thioester of AA1, would react with cysteine to give the tripeptide AA1-AA2-Cys, and with homocysteine to give AA1-AA2-Hcy. Could some of these tripeptides have been the very first catalytic triads [38]? Indeed, we have demonstrated that even dipeptides like GlyCys (but not monomeric cysteine) are able to promote the formation of peptide bonds from nitriles. They are also able to promote the formation of imidazoles. Such heterocycles play an important role in today's biochemistry. Of special interest is the simplest aminoimidazole, which, as its ribonucleotide derivative (AIR) [39], is an intermediate in the de-novo synthesis of inosine monophosphate (IMP), hence of purine nucleotides. Thus, imidazoles could have established a bridge from peptides to nucleic acids.

Supplementary Materials: The following are available online at http://www.mdpi.com/2075-1729/8/4/47/s1. HRMS of **1a** + cysteine and + homocysteine reaction mixtures. NMR spectra of **11** and **12**. ESI mass spectrum and ^{13}C NMR of a cysteine + excess GlyCN reaction mixture. Additional data for theoretical calculations.

Author Contributions: Conceptualization, Y.V.; methodology, I.S., A.M. and Y.V.; validation, Y.V. and A.M.; investigation, I.S., S.C. and K.D.L.; writing—original draft preparation, Y.V.; writing—review and editing, Y.V.; visualization, Y.V. and A.M.; supervision, Y.V.; project administration, Y.V.; funding acquisition, Y.V. and A.M.

Funding: This research was funded by the Labex ARCANE and CBH-EUR-GS (ANR-17-EURE-0003), and the French National Research Agency (*Investissements d'Avenir* ANR-15-IDEX-02).

Acknowledgments: The authors thank *le Ministère des Affaires Etrangères* (*Consulat Général de France à Jérusalem* and *Ambassade de France au Mali*) for I.S. and S.C. grants, and *l'Université de Bamako, Mali*.

Conflicts of Interest: The authors declare no conflict of interest.

References

1. Hashimoto, S.I. Ribosome-independent peptide synthesis and their application to dipeptide production. *J. Biol. Macromol.* **2008**, *8*, 28–37.
2. Chalmet, S.; Harb, W.; Ruiz-Lòpez, M.F. Computer simulation of amide bond formation in aqueous solution. *J. Phys. Chem. A* **2001**, *105*, 11574–11581. [CrossRef]
3. Gajewski, E.; Steckler, D.K.; Goldberg, R.N. Thermodynamics of the hydrolysis of adenosine 5'-triphosphate to adenosine 5'-diphosphate. *J. Biol. Chem.* **1988**, *261*, 12733–12737.
4. Ruiz-Mirazo, K.; Briones, C.; de la Escosura, A. Prebiotic systems chemistry: New perspectives for the origins of life. *Chem. Rev.* **2014**, *114*, 285–366. [CrossRef] [PubMed]
5. Strecker, A. Ueber die künstliche bildung der milchsäure und einen neuen, dem glycocoll homologen körper. *Liebigs Ann. Chem.* **1850**, *75*, 27–45. [CrossRef]
6. Guthrie, J.P.; Yim, J.C.H.; Wang, Q. Hydration of nitriles: An examination in terms of no barrier theory. *J. Phys. Org. Chem.* **2014**, *27*, 27–37. [CrossRef]
7. Krimmer, H.P.; Drauz, K.; Kleemann, A. Umsetzung von β-mercapto-α-aminosäuren mit nitrilen. *Chemiker-Zeitung* **1987**, *111*, 357–361.
8. Vallée, Y.; Shalayel, I.; Dung, L.K.; Raghavendra Rao, K.V.; de Paëpe, G.; Märker, K.; Milet, A. At the very beginning of life on Earth: The thiol rich peptide (TRP) world hypothesis. *Int. J. Dev. Biol.* **2017**, *61*, 471–478. [CrossRef] [PubMed]
9. Ferris, J.P.; Hagan, W.J., Jr. HCN and chemical evolution: The possible role of cyano compounds in prebiotic synthesis. *Tetrahedron* **1984**, 1093–1120. [CrossRef]
10. Schilke, P.; Menten, K.M. Detection of a second, strong submillimeter HCN laser line toward carbon stars. *Astrophys. J.* **2003**, *583*, 446–450. [CrossRef]

11. Thorwirth, S.; Wyrowski, F.; Schilke, P.; Menten, K.M.; Brünken, S.; Müller, H.S.P.; Winnewisser, G. Detection of HCN direct L-type transitions probing hot molecular gas in the proto-planetary nebula CRL 618. *Astrophys. J.* **2003**, *586*, 338–343. [CrossRef]

12. Irvine, W.M.; Dickens, J.E.; Lovell, A.J.; Schoerb, F.P.; Senay, M.; Bergin, E.A.; Jewitt, D.; Matthews, H.E. The HNC/HCN ratio in comets. *Earth Moon Planets* **1997**, *78*, 29–35. [CrossRef] [PubMed]

13. Schloerb, F.P.; Kinzel, W.M.; Swade, D.A.; Irvine, W.M. Observations of HCN in comet P/Halley. *Astron. Astrophys.* **1987**, *187*, 475–480. [PubMed]

14. Patel, B.H.; Percivalle, C.; Ritson, D.J.; Duffy, C.D.; Sutherland, J.D. Common origins of RNA, protein and lipid precursors in a cyanosulfidic protometabolism. *Nature Chem.* **2015**, *7*, 301–307. [CrossRef] [PubMed]

15. Zahnle, K.J. Photochemistry of methane and the formation of hydrocyanic acid (HCN) in the Earth's early atmosphere. *J. Geophys. Res.* **1986**, *91*, 2819–2834. [CrossRef]

16. Martin, R.S.; Mather, T.A.; Pyle, D.M. Volcanic emissions and the early Earth atmosphere. *Geochim. Cosmochim. Acta* **2007**, *71*, 3673–3685. [CrossRef]

17. Mukhin, L.M. Volcanic processes and synthesis of simple organic compounds on primitive Earth. *Orig. Life Evol. Biosph.* **1976**, *7*, 355–368. [CrossRef]

18. Martin, W.; Baross, J.; Kelley, D.; Russell, M.J. Hydrothermal vents and the origin of life. *Nat. Rev. Microbiol.* **2008**, *6*, 805–814. [CrossRef] [PubMed]

19. Thaddeus, P.; Kutner, M.L.; Penzias, A.A.; Wilson, R.W.; Jefferts, K.B. Interstellar hydrogen sulfide. *Astrophys. J.* **1972**, *176*, L73–L76. [CrossRef]

20. Minh, Y.C.; Ziurys, L.M.; Irvine, W.M.; McGonagle, D. Abundances of hydrogen sulfide in star-forming regions. *Astrophys. J.* **1991**, *366*, 192–197. [CrossRef] [PubMed]

21. Minh, Y.C.; Irvine, W.M.; Ziurys, L.M. Detection of interstellar hydrogen sulfide in cold, dark clouds. *Astrophys. J.* **1989**, *345*, L63–L66. [CrossRef] [PubMed]

22. Eberhardt, P.; Meier, R.; Krankowsky, D.; Hodges, R.R. Methanol and hydrogen sulfide in comet P/Halley. *Astron. Astrophys.* **1994**, *288*, 315–329.

23. Carapezza, M.L.; Badalamenti, B.; Cavarra, L.; Scalzo, A. Gas hazard assessment in a densely inhabited area of Colli Albany Volcano (Cava dei Selci, Roma). *J. Volcanol. Geotherm. Res.* **2003**, *123*, 81–94. [CrossRef]

24. Oppenheimer, C.; Scaillet, B.; Martin, R.S. Sulfur degassing from volcanoes: Source conditions, surveillance, plume chemistry and Earth system impacts. *Rev. Min. Geochem.* **2011**, *73*, 363–421. [CrossRef]

25. Parker, E.T.; Cleaves, H.J.; Dworkin, J.P.; Glavin, D.P.; Callahan, M.; Aubrey, A.; Lazcano, A.; Bada, J.L. Primordial synthesis of amines and amino acids in a 1958 Miller H_2S-rich spark discharge experiment. *Proc. Natl. Acad. Sci. USA* **2011**, *108*, 5526–5531. [CrossRef] [PubMed]

26. Mennucci, B.; Cancès, E.; Tomasi, J. Evaluation of the solvent effects in isotropic and anisotropic dielectrics and in ionic solutions with a unified integral equation method: Theoretical bases, computational implementation, and numerical applications. *J. Phys. Chem. B* **1997**, *101*, 10506–10517. [CrossRef]

27. Bounama, C.; Franck, S.; von Bloh, W. The fate of Earth's ocean. *Hydrol. Earth Syst. Sci.* **2001**, *5*, 569–575. [CrossRef]

28. Pinti, D.L. The origin and evolution of the oceans. In *Lectures in Astrobiology, Vol. 1*; Gargaud, M., Barbier, B., Martin, H., Reisse, J., Eds.; Springer-Verlag: Berlin/Heidelberg, Germany, 2005; pp. 83–112.

29. Boger, D.L.; Keim, H.; Oberhauser, B.; Schreiner, E.P.; Foster, C.A. Total synthesis of HUN-7293. *J. Am. Chem. Soc.* **1999**, *121*, 6197–6205. [CrossRef]

30. Xiang, Y.B.; Drenkard, S.; Baumann, K.; Hickey, D.; Eschenmoser, A. Chemie von α-aminonitrilen. Sondierungen über thermische umwandlungen von α-aminonitrilen. *Helv. Chem. Acta* **1994**, *77*, 2209–2250. [CrossRef]

31. Léger, S.; Bayly, C.I.; Black, W.C.; Desmarais, S.; Falgueyret, J.P.; Massé, F.; Percival, M.D.; Truchon, J.F. Primary amides as selective inhibitors of cathepsin K. *Bioorg. Med. Chem. Lett.* **2007**, *17*, 4328–4332. [CrossRef] [PubMed]

32. Guillemin, J.C.; Denis, J.M. Synthèse d'imines linéaires non-stabilisées par réactions gaz-solide sous vide. *Tetrahedron* **1988**, *44*, 4431–4446. [CrossRef]

33. Martin, T.A.; Causey, D.H.; Sheffner, A.L.; Wheeler, A.G.; Corrigan, J.R. Amides of N-acylcysteines as mucolytic agents. *J. Med. Chem.* **1967**, *10*, 1172–1176. [CrossRef] [PubMed]

34. De Duve, C. A research proposal on the origin of life. *Orig. Life Evol. Biosph.* **2003**, *33*, 559–574. [CrossRef] [PubMed]

35. Song, B.D.; Jencks, W.P. Aminolysis of benzoyl fluorides in water. *J. Am. Chem. Soc.* **1989**, *111*, 8479–8484. [CrossRef]

36. Lam, A.K.Y.; Ryzhov, V.; O'Hair, R.A.J. Mobile protons versus mobile radicals: Gas-phase unimolecular chemistry of radical cations of cysteine-containing peptides. *J. Am. Soc. Mass Spectrom.* **2010**, *21*, 1296–1312. [CrossRef] [PubMed]

37. Elkholy, Y.M.; Erian, A.W. An aminoimidazole and its utility in heterocyclic synthesis. *Heteroatom Chem.* **2003**, *14*, 503–508. [CrossRef]

38. Buller, A.R.; Townsend, C.A. Intrinsic evolutionary constraints on protease structure, enzyme acylation, and the identity of the catalytic triad. *Proc. Natl. Acad. Sci. USA* **2013**, *110*, E653–E661. [CrossRef] [PubMed]

39. Bhat, B.; Groziak, M.P.; Leonard, N.J. Nonenzymatic synthesis and properties of 5-aminoimidazole ribonucleotide (AIR). Synthesis of specifically ^{15}N-labeled 5-aminoimidazole ribonucleoside (AIRs) derivatives. *J. Am. Chem. Soc.* **1990**, *112*, 4891–4897. [CrossRef]

Article

Microfluidic Reactors for Carbon Fixation under Ambient-Pressure Alkaline-Hydrothermal-Vent Conditions

Victor Sojo [1,2,3,*], Aya Ohno [1], Shawn E. McGlynn [1,4,5], Yoichi M.A. Yamada [1] and Ryuhei Nakamura [1,4]

[1] RIKEN Center for Sustainable Resource Science, 2-1 Hirosawa, Wako, Saitama 351-0198, Japan; aohno@riken.jp (A.O.); mcglynn@elsi.jp (S.E.M.); ymayamada@riken.jp (Y.M.A.Y.); ryuhei.nakamura@riken.jp (R.N.)

[2] Systems Biophysics, Ludwig-Maximilian University of Munich, Munich 80799, Germany

[3] Institute for Advanced Study, Berlin. Wallotstr. 19, Berlin 14193, Germany

[4] Earth-Life Science Institute, Tokyo Institute of Technology, 2-12-1 Ookayama, Meguro-ku, Tokyo 152-8550, Japan

[5] Blue Marble Space Institute of Science, Seattle, WA 98154, USA

* Correspondence: victor.sojo@wiko-berlin.de and v.sojo.11@ucl.ac.uk; Tel.: +49-30-89001-0

Received: 20 December 2018; Accepted: 25 January 2019; Published: 1 February 2019

Abstract: The alkaline-hydrothermal-vent theory for the origin of life predicts the spontaneous reduction of CO_2, dissolved in acidic ocean waters, with H_2 from the alkaline vent effluent. This reaction would be catalyzed by Fe(Ni)S clusters precipitated at the interface, which effectively separate the two fluids into an electrochemical cell. Using microfluidic reactors, we set out to test this concept. We produced thin, long Fe(Ni)S precipitates of less than 10 μm thickness. Mixing simplified analogs of the acidic-ocean and alkaline-vent fluids, we then tested for the reduction of CO_2. We were unable to detect reduced carbon products under a number of conditions. As all of our reactions were performed at atmospheric pressure, the lack of reduced carbon products may simply be attributable to the low concentration of hydrogen in our system, suggesting that high-pressure reactors may be a necessity.

Keywords: origin of life; abiogenesis; carbon fixation; hydrothermal vents; electrochemistry; reduction

---

1. Introduction

From its very start, life required reduced organic molecules. In a minimalistic scenario for abiogenesis (i.e. the emergence of life), one source of such molecules was the reduction of CO_2, in a process overall similar (and potentially homologous) to the modern enzyme-facilitated pathways of extant autotrophic cells [1–8]. A number of reducing agents (i.e. sources of electrons) for reducing CO_2 were possible on the early Earth, but multiple reasons make hydrogen (H_2) a good candidate. This is discussed at length elsewhere (see [5,8–10] and references therein), but two points are worth mentioning here, namely: (1) hydrogen is formed spontaneously in the Earth's crust via the "serpentinization" of ultramafic-rock minerals such as olivine [9,11–13]; and (2) it is used to reduce CO_2 by members of both archaea and bacteria, in their respective versions of the Wood–Ljungdahl (WL) or acetyl Co-A pathway [5,14,15].

This process results in CO_2 fixation and the production of ATP within a single, linear, metabolic pathway [5,16,17], so it has been suggested as a potential candidate for the metabolism of the last universal common ancestor (LUCA). However, one problem with extrapolating this scenario towards the origin of life is that, while the overall pathway of carbon fixation via the WL pathway is exergonic, the initial reaction between H_2 and CO_2 is not spontaneous under standard abiotic conditions [18].

Confirming these thermodynamic predictions, numerous experimental electrochemical results show that CO_2 reduction is indeed disfavored under most observational conditions, requiring overpotentials of at least 180 mV in order to overcome the initial endergonic steps [19].

However, under putative ancient alkaline-vent conditions, CO_2 would have been dissolved in slightly acidic ocean waters (pH 5~7), whereas H_2 would have been a product of serpentinization, emanating as part of the efflux of the alkaline vent, itself rich in OH^- (pH 9~12).

The geologically sustained pH difference across the vent minerals provided an additional electrochemical driving force, potentially circumventing the lack of reducing power of H_2 for CO_2 reduction [19]. The reduction of CO_2 to formic acid (HCOOH, its first 2-electron reduction product) involves protonation, so it would have been favored in acidic ocean waters. In turn, the oxidation of H_2 releases protons (H^+), which would have been favored in the alkaline waters that contained the dissolved H_2 [20]. The two fluids would have been separated by the mineral precipitates of the vent, which included iron (and nickel) sulfides (Fe(Ni)S) as well as silicates (as reviewed in [21]). Reduced on the inside and oxidized on the outside, a situation analogous to an electrochemical cell would have existed between the two sides. The electrons released from H_2 would then hypothetically travel through the electrically conductive Fe(Ni)S network [19], and drive the reduction of CO_2 on the other side (Figure 1). This contrast between the pH of the two solutions matches the polarity of modern cells, and it has been suggested as a potential driver of the origins of both membrane bioenergetics and carbon fixation [6–8,20,22–24].

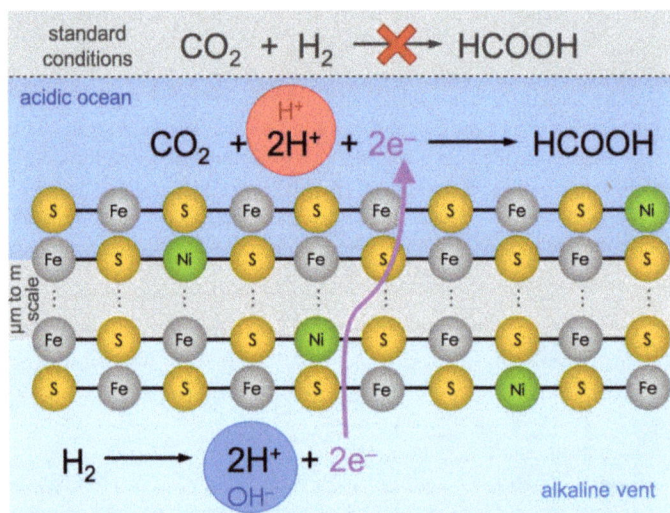

Figure 1. Under standard conditions (top, grey background), CO_2 reduction with H_2 is not viable thermodynamically. Conversely, in alkaline vents, the reaction is hypothetically split into two halves, effectively producing an electrochemical cell. On the alkaline-vent side (bottom, light-blue background), the oxidation of H_2 to H^+ is favored because of the alkaline pH in the vent fluid (symbolized by the blue circle with "OH^-"). The electrons would travel through the micrometer-to-meter-scale catalytic Fe(Ni)S precipitate network and meet CO_2 at the ocean side (dark-blue background), where the relatively acidic pH (red circle with "H^+") would favor the reduction and protonation towards formic acid (HCOOH).

The pH gradients have also recently been shown to hold in the microscale, at up to six pH-unit differences [25], suggesting the potential of microfluidic devices to study the reduction of CO_2 with H_2 under these conditions.

The immobilization of a catalytic boundary by the meeting of two fluids has been demonstrated using a microfluidic reactor [26], so we envisioned that this methodology could be applied to the formation of catalytic Fe(Ni)S clusters at the interface, elaborating on previous results [25]. This effectively mimics the ancient alkaline-vent conditions by the in-situ creation of an electrochemical cell between the oxidized CO_2 in the acidic-ocean side and the reduced H_2 in the alkaline-vent side (Figure 1).

Here, we present preliminary results in our study of the potential reduction of CO_2 with H_2. We used microfluidics to simulate the mixing of oceanic and serpentinizing fluids under the putative conditions of ancient alkaline hydrothermal vents—although notably at atmospheric pressure.

We simulated the two sides of the vent system by mixing fluids containing combinations of Fe^{2+}/Ni^{2+} and CO_2 for the acidic-ocean side; whereas the alkaline-vent simulant contained HS^- and bubbled H_2 (Figure 2).

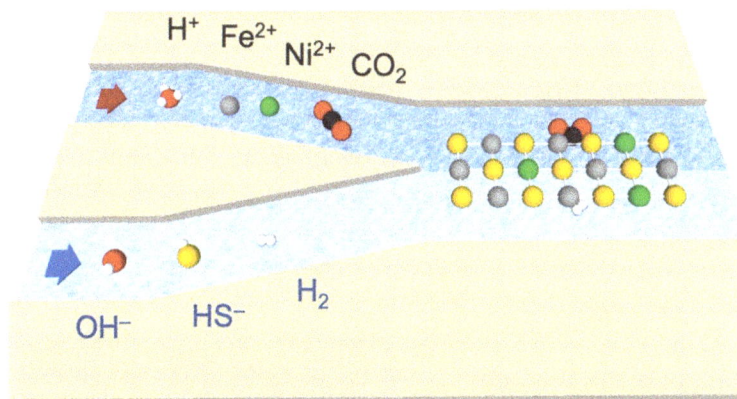

Figure 2. Diagram of the reaction system, depicting the input of acidic fluid (top, left half) containing H^+, Fe^{2+}, Ni^{2+}, and dissolved CO_2. The alkaline fluid (bottom, left half) contained OH^-, HS^-, and dissolved H_2. Upon meeting, the fluids form Fe(Ni)S precipitates (represented by the reticulation in the right half), which may serve as catalysts for the indirect redox reaction between H_2 and CO_2.

2. Methods

The microfluidic reactor systems were assembled using custom chips with a Y-shape design, etched from glass by the Institute of Microchemical Technology Co., Ltd. The channels in the system were half-pipes with a width of 100 ± 2.5 µm and a maximum depth of 40 ± 1 µm.

At one tip of the "Y", we input the analog of the acidic-ocean fluid, with the alkaline-vent analog being input at the other tip (as summarized in the diagram of Figure 2). The compositions of the fluids, presented in Table 1 and detailed in the Results, are similar to those reported elsewhere [10].

Because of the sensitivity of Fe^{2+} to oxygen in air, and in order to mimic the anoxic Hadean conditions at the origin of life more closely, the water in all of the experiments was de-gassed by boiling for 5 min and then cooling under constant argon bubbling for 30 min. The salts were weighed and then kept as solids under positive pressure of argon. The necessary amounts of de-aerated water were added, and the solutions bubbled with argon for another 15 minutes, with the exception of the H_2-containing solutions, which were bubbled with H_2. The final pH of the H_2-containing solution was ~11, whereas that of the carbonic solution was ~6. Gas-tight syringes were then filled (all with a maximum volume of 1 mL, from Hamilton USA).

Table 1. Conditions tested. The concentrations of $NaHCO_3$ were inevitably lower than shown after acidification to pH 6 with 1 M HCl. In alternative experiments we used CO_2 bubbled at atmospheric pressure instead of $NaHCO_3$.

ACIDIC-SIDE CONCENTRATIONS		ALKALINE-SIDE CONCENTRATIONS	
[FeCl$_2$]	50 mM	[Na$_2$S]	10 and 100 mM
[NiCl$_2$]	0 and 10 mM	[K$_2$HPO$_4$]	10 mM
[NaHCO$_3$]	10, 50, and 100 mM (acidified to pH ~6)	[Na$_2$Si$_3$O$_7$]	0 and 10 mM
		[Na$_2$MoO$_4$]	0 and 1 mM
CO$_2$	Bubbled at atmospheric pressure (final pH ~6)	[H$_2$]	Bubbled at atmospheric pressure (final pH ~11)
OTHER CONDITIONS			
Reaction durations	$^1/_2$, 1, 2, 5, 12, and 24 h	Temperature	~25 (room), 40, 50, 60, and 70 °C

The full system setup is presented in Figure 3. The simulants of the acidic-ocean and alkaline-vent fluids were driven into the system using syringe pumps at adjustable flow rates, generally between 0.2 and 20 μL/min. The pumps were modular BabyBee Syringe Drive units from Bioanalytical Systems Inc. (BASi, West Lafayette, IN, USA), regulated by a BeeHive Syringe Drive Controller, also from BASi. The temperature was measured using an infrared thermometer, and regulated using a standard heating plate. A stainless-steel block was laid directly onto the heating plate, with the glass reactor laid on top of the block (Figure 3c, middle). The reactor chip was held in a custom-made stainless-steel casing (Figure 3a–c). The formation of the precipitates was followed using an inline USB microscope (Figure 3b, middle) connected to a standard laptop computer.

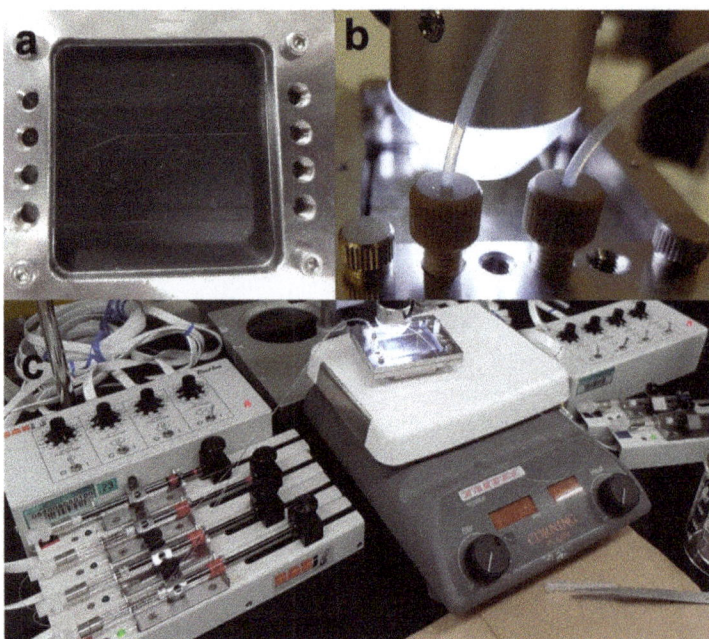

Figure 3. Reactor setup. (**a**) Microfluidic Y-shaped reactor chip with ~100 μm-width channels, fitted into a stainless-steal holder with screw-in inlets. (**b**) The two inlets adjusted into position (bottom). A microscope (middle and top) was used to follow the precipitation reaction. (**c**) Left and far right: BASi drive controller and syringe pumps with Hamilton syringes. Center: reaction chip on the heating plate, with a USB microscope adjusted on top to follow the precipitation.

Reaction blanks were taken by running water through the chip on both inlets, after the precipitation reaction had taken place, but before adding any $CO_2/NaHCO_3$.

NMR spectra (^1H and ^{13}C) were determined using a JEOL spectrometer with a 600 MHz magnet.

3. Results

To facilitate microfluidic mixing and simulate the mixing of alkaline-vent and oceanic fluids, we replicated previous concentrations [10] and separated each of the fluids into independently controlled gas-tight syringes. For the acidic side, three inflows were used (see Table 1 for further details), as follows:

- De-aerated water (to achieve parallel flow and take sample blanks).
- $FeCl_2$ (50 mM) and $NiCl_2$ (5 mM).
- De-aerated water bubbled with CO_2 (at atmospheric pressure), or alternatively dissolved $NaHCO_3$ (100 mM), acidified with HCl (1 M) to pH 6.

Conversely, the two alkaline-side syringes contained the following fluids:

- De-aerated water.
- De-aerated water with Na_2S (10 mM), K_2HPO_4 (10 mM), and $Na_2Si_3O_7$ (10 mM), bubbled with H_2 (at atmospheric pressure), and at a final pH of ~11.

After attaining parallel flow by letting water run from both inlets for 20 min, the mixing of the metal and sulfide fluids produced a thin dark precipitate at the interface. By changing either or both of the two fluids back to the water syringes, the thickness of the precipitate could be closely controlled (Figure 4).

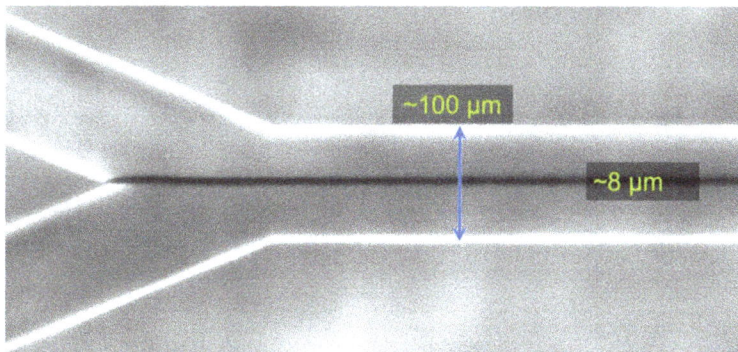

Figure 4. Precipitation of Fe(Ni)S at the interface between the acidic and alkaline fluids. By replacing the flow of either (or both) the sulfide or the metals with water, the precipitate could be kept arbitrarily thin.

Once the precipitates were formed at the interface, the inflows were swapped to the respective syringes containing CO_2 and H_2 (which, in the latter case, was the same as that for the sulfide in most of our reactions). The reaction conditions were controlled to last between 30 min and 24 h, and temperatures between laboratory conditions (~25 °C) and 70 °C (Table 1 and Methods).

Analysis using ^1H-NMR showed a peak in the formic acid region (~8.3 ppm, Figure A1). To assess whether the peak in this result was indeed due to formic acid, ^{13}C-labelled sodium bicarbonate ($NaH^{13}CO_3$), acidified to pH 6, was used. A peak in the formic-acid region (~164 ppm) of the ^{13}C-NMR spectrum (Figure A2) was however shown to correspond to unreacted $NaH^{13}CO_3$ itself. This was confirmed by spiking with ^{13}C-labelled formic acid ($H^{13}COOH$, Figure A3). The ^1H spectrum for ^{13}C-labelled formic acid should show a splitting of the original singlet into a doublet. However, the doublet is not observed in the reaction efflux with $NaH^{13}CO_3$, appearing only when ^{13}C-labelled formic acid was added externally (Figure A4). The original peak at ~8.3 ppm in the ^{1}H spectrum remains unidentified.

Overall, we did not detect reduced carbon products under the conditions that we tested. Varyin conditions (Table 1), including concentrations, thickness of the precipitates, reaction temperature, reaction times, or doping the Fe(Ni)S precipitates with heteroatoms such as Mo(VI), produced no detectable difference.

4. Discussion

We aimed to probe the reduction of CO_2 with H_2 under putative ancient alkaline-hydrothermal-vent conditions.

In contrast to previous work [10], in which formate and formaldehyde were reported, we do not detect any soluble reduced carbon products under our experimental conditions. A separate set of experiments conducted with a larger-scale system modeled after previous work [27] also failed to yield detectable CO_2 reduction (Chang and McGlynn, unpublished).

In view of this, it is important to stress that our experimental conditions were not exhaustive, and they failed to replicate alkaline vents in at least one crucial aspect—pressure.

The solubility of hydrogen is extremely low at ambient pressures, and decreases sharply as temperature rises towards the boiling point of water [28]. We bubbled H_2 at atmospheric pressure, prior to the reaction, and did not continue bubbling during the reaction once the desired volume was stored in a gas-tight syringe. Since the H_2-containing fluid in our experiments is at pH 11 and the CO_2-containing fluid at pH 6, 1 bar of H_2 at room temperature is predicted to be sufficient to reduce CO_2, a reaction whose favorability would be enhanced with greater pressure as a result of the gain in concentrations [29,30]. The magnitude of the overpotential needed to overcome any kinetic barriers however remains unknown, and it is possible that high-pressure reactors are a necessity to overcome these barriers and evaluate the possibility of H_2-powered reduction of CO_2 under alkaline-vent conditions. Similarly, continuous bubbling of H_2 (as in previous work [10]) may be necessary, given the high volatility of H_2 gas.

Notably, recent results show that reduced carbon products are undetectable under similar reaction conditions using pure metals as catalysts [31,32], instead staying bound to the catalysts until concentrated KOH is used to remove them. It is therefore plausible (although it would need to be shown) that any reduced products that we may have formed remained bound to the Fe(Ni)S precipitates. We also note that we did not investigate potential gas-phase products in our study, leaving these as possibilities to be explored.

The electrical potential from the pH gradient under alkaline-vent conditions has been measured at the microscale [25], but it remains to be shown that it can indeed drive otherwise unfavorable redox reactions. Thus, to test the validity of the electrochemical-cell concept—irrespective of its relevance to the origin of life—further (potentially less geologically relevant) experiments could include using stronger reducing and oxidizing agents, at varying concentrations. The utilization of high-pressure reactors to achieve higher concentrations of dissolved gas may be especially important to test for H_2-driven reductions.

Broadly, our scheme for overcoming the exergonic steps of carbon reduction relies on the separation of solutions at different conditions coupled to the ability to transfer electrons, which is not unique to pH gradients at alkaline vents—thermal gradients and reducing agents other than H_2 are also possible drivers of reduction [29,30].

The alkaline-hydrothermal-vent theory has come under criticism in recent years [33–35]. These issues have been addressed elsewhere [36], but it is important to note here that we do not see our results as either disproving the alkaline-vent theory, or providing support for alternative theories for the origin of life. Most simply, they suggest that the catalysts that appropriately lower the kinetic barriers have not been implemented as of yet, or that more realistic conditions—crucially higher pressures, particularly for H_2—need to be evaluated to more closely in order ascertain the viability of the hypotheses tested here. These are lines of enquiry that we are pursuing, as are other researchers in the field.

Author Contributions: V.S. planned and performed the research with experimental support from A.O. All authors designed the experiments and analyzed the results. V.S., S.E.M., Y.M.A.Y., and R.N. wrote the manuscript.

Funding: V.S. acknowledges financial support from the Japan Society for the Promotion of Science (JSPS) (FY2016-PE-16721), the European Commission Marie Skłodowska—Curie Actions/European Molecular Biology Organization (EMBO) (ALTF-1455-2015), and the Institute for Advanced Study in Berlin. S.E.M. is supported by NSF Award 1724300.

Acknowledgments: The authors thank Reuben Hudson, Nick Lane, Yamei Li, Hideshi Ooka, and Alexandra Whicher for their suggestions.

Conflicts of Interest: The authors declare no conflict of interest.

Appendix A

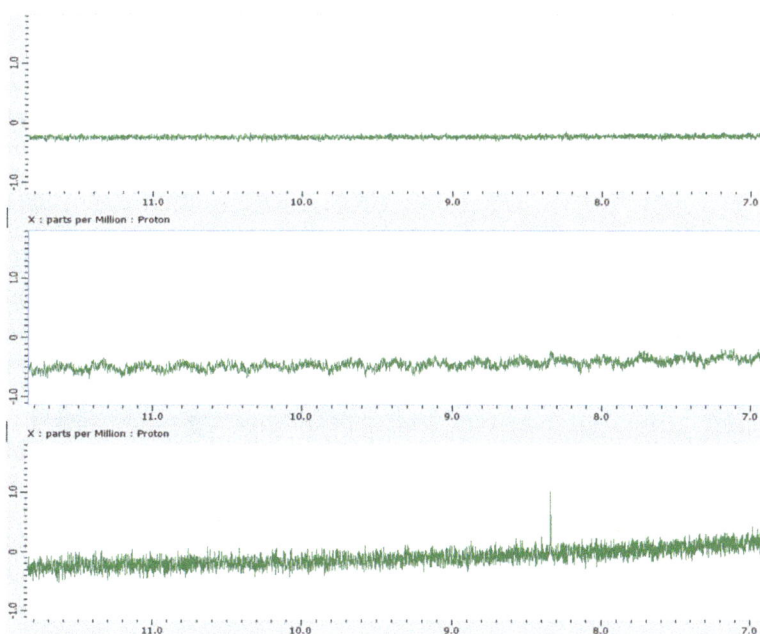

Figure A1. ^{1}H spectrum of dissolved $NaHCO_3$ (top), reaction blank (middle), and reaction efflux (bottom). The reaction efflux shows a peak in the formic acid region, slightly over 8.3 ppm.

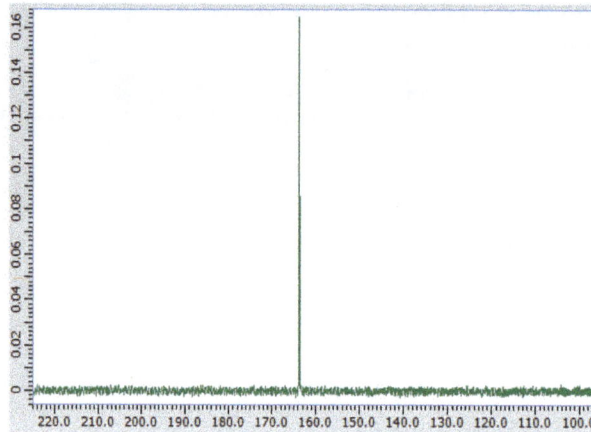

Figure A2. ^{13}C spectrum of reaction efflux, showing a peak in the formic acid region (~164 ppm).

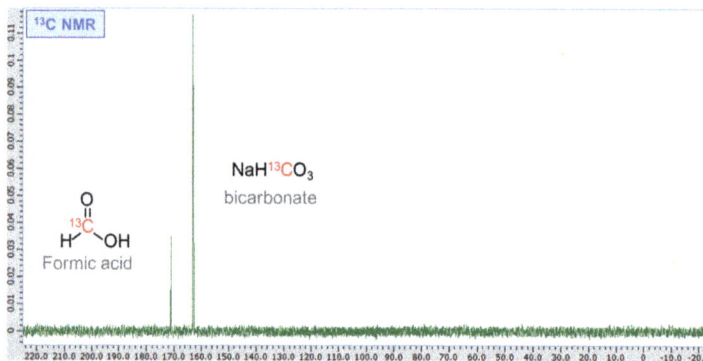

Figure A3. ^{13}C spectrum of reaction efflux with added ^{13}C-labelled formic acid. Instead of a single increased peak, two peaks are visible, thus revealing that formic acid was not produced and the peak to the right corresponds to unreacted bicarbonate in the efflux.

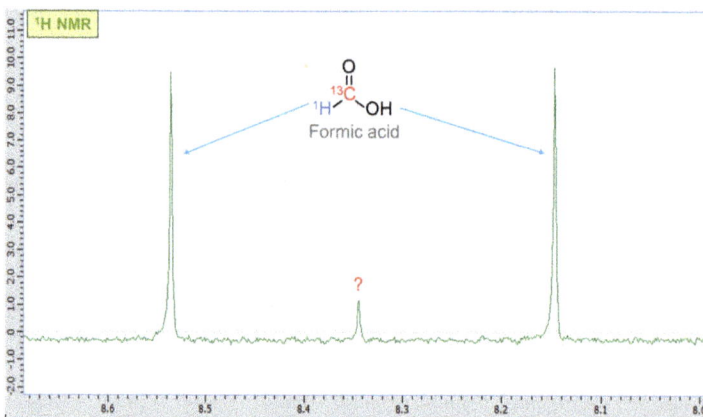

Figure A4. Adding ^{13}C-labelled formic acid gives a doublet in the ^{1}H spectrum, indicated by the arrows. The peak in the middle (~8.34 ppm) remains unidentified.

References

1. Wächtershäuser, G. Pyrite formation, the first energy source for life: A hypothesis. *Syst. Appl. Microbiol.* **1988**, *10*, 207–210. [CrossRef]
2. Russell, M.J.; Hall, A.J.; Turner, D. *In vitro* growth of iron sulphide chimneys: Possible culture chambers for origin-of-life experiments. *Terra Nova* **1989**, *1*, 238–241. [CrossRef]
3. Wächtershäuser, G. Groundworks for an evolutionary biochemistry: The iron-sulphur world. *Prog. Biophys. Mol. Biol.* **1992**, *58*, 85–201. [CrossRef]
4. Maden, B. No soup for starters? Autotrophy and the origins of metabolism. *Trends Biochem. Sci.* **1995**, *20*, 337–341. [CrossRef]
5. Russell, M.J.; Martin, W. The rocky roots of the acetyl-CoA pathway. *Trends Biochem. Sci.* **2004**, *29*, 358–63. [CrossRef] [PubMed]
6. Martin, W.; Russell, M.J. On the origin of biochemistry at an alkaline hydrothermal vent. *Phil. Trans. R. Soc. London B* **2007**, *362*, 1887–1925. [CrossRef] [PubMed]
7. Lane, N.; Martin, W.F. The origin of membrane bioenergetics. *Cell* **2012**, *151*, 1406–1416. [CrossRef]
8. Sojo, V.; Herschy, B.; Whicher, A.; Camprubí, E.; Lane, N. The origin of life in alkaline hydrothermal vents. *Astrobiology* **2016**, *16*, 181–197. [CrossRef]
9. Sleep, N.H.; Bird, D.K.; Pope, E.C. Serpentinite and the dawn of life. *Phil. Trans. R. Soc. London B. Biol. Sci.* **2011**, *366*, 2857–2869. [CrossRef]
10. Herschy, B.; Whicher, A.; Camprubi, E.; Watson, C.; Dartnell, L.; Ward, J.; Evans, J.R.G.; Lane, N. An origin-of-life reactor to simulate alkaline hydrothermal vents. *J. Mol. Evol.* **2014**, *79*, 213–227. [CrossRef]
11. Sleep, N.H.; Meibom, A.; Fridriksson, T.; Coleman, R.G.; Bird, D.K. H_2-rich fluids from serpentinization: Geochemical and biotic implications. *Proc. Natl. Acad. Sci.* **2004**, *101*, 12818–12823. [CrossRef] [PubMed]
12. Kelley, D.S.; Baross, J.A.; Delaney, J.R. Volcanoes, fluids, and life at mid-ocean ridge spreading centers. *Annu. Rev. Earth Planet. Sci.* **2002**, *30*, 385–491. [CrossRef]
13. Kelley, D.S.; Karson, J.A.; Früh-Green, G.L.; Yoerger, D.R.; Shank, T.M.; Butterfield, D.A.; Hayes, J.M.; Schrenk, M.O.; Olson, E.J.; Proskurowski, G.; et al. A serpentinite-hosted ecosystem: The Lost City hydrothermal field. *Science* **2005**, *307*, 1428–1434. [CrossRef] [PubMed]
14. Wood, H.G. Life with CO or CO_2 and H_2 as a source of carbon and energy. *FASEB J.* **1991**, *5*, 156–163. [CrossRef] [PubMed]
15. Nitschke, W.; Russell, M. Beating the acetyl coenzyme A-pathway to the origin of life. *Phil. Trans. R. Soc. B.* **2013**, *368*, 20120258. [CrossRef] [PubMed]
16. Lane, N.; Allen, J.F.; Martin, W. How did LUCA make a living? Chemiosmosis in the origin of life. *BioEssays* **2010**, *32*, 271–280. [CrossRef] [PubMed]
17. Poehlein, A.; Schmidt, S.; Kaster, A.-K.; Goenrich, M.; Vollmers, J.; Thürmer, A.; Bertsch, J.; Schuchmann, K.; Voigt, B.; Hecker, M.; et al. An ancient pathway combining carbon dioxide fixation with the generation and utilization of a sodium ion gradient for ATP synthesis. *PLoS ONE* **2012**, *7*, e33439. [CrossRef] [PubMed]
18. Maden, B.E. Tetrahydrofolate and tetrahydromethanopterin compared: Functionally distinct carriers in C1 metabolism. *Biochem. J.* **2000**, *350*, 609–629. [CrossRef]
19. Yamaguchi, A.; Yamamoto, M.; Takai, K.; Ishii, T.; Hashimoto, K.; Nakamura, R. Electrochemical CO_2 reduction by Ni-containing iron sulfides: How is CO_2 electrochemically reduced at bisulfide-bearing deep-sea hydrothermal precipitates? *Electrochim. Acta* **2014**, *141*, 311–318. [CrossRef]
20. Lane, N. *The Vital Question: Energy, Evolution, and the Origins of Complex Life*; WW Norton & Company: New York, NY, USA, 2015.
21. Li, Y.; Kitadai, N.; Nakamura, R. Chemical diversity of metal sulfide minerals and its implications for the origin of life. *Life* **2018**, *8*, 1–26. [CrossRef]
22. Martin, W.; Baross, J.; Kelley, D.; Russell, M.J. Hydrothermal vents and the origin of life. *Nat. Rev. Microbiol.* **2008**, *6*, 805–814. [CrossRef] [PubMed]
23. Sojo, V.; Pomiankowski, A.; Lane, N. A bioenergetic basis for membrane divergence in archaea and bacteria. *PLoS Biol.* **2014**, *12*, e1001926. [CrossRef] [PubMed]
24. Lane, N. Bioenergetic constraints on the evolution of complex life. *Cold Spring Harb. Perspect. Biol.* **2014**, *6*, a015982. [CrossRef]

25. Möller, F.M.F.M.; Kriegel, F.; Kieß, M.; Sojo, V.; Braun, D. Steep pH Gradients and Directed Colloid Transport in a Microfluidic Alkaline Hydrothermal Pore. *Angew. Chemie Int. Ed.* **2017**, *56*, 1–6. [CrossRef] [PubMed]
26. Yamada, Y.M.A.; Ohno, A.; Sato, T.; Uozumi, Y. Instantaneous click chemistry by a copper-containing polymeric-membrane-installed microflow catalytic reactor. *Chem. A Eur. J.* **2015**, *21*, 17269–17273. [CrossRef]
27. Batista, B.C.; Steinbock, O. Growing inorganic membranes in microfluidic devices: Chemical gardens reduced to linear walls. *J. Phys. Chem. C* **2015**, *119*, 27045–27052. [CrossRef]
28. Pray, H.A.; Schweickert, C.E.; Minnich, B.H. Solubility of hydrogen, oxygen, nitrogen, and helium in water at elevated temperatures. *Ind. Eng. Chem.* **1952**, *44*, 1146–1151. [CrossRef]
29. Kitadai, N.; Nakamura, R.; Yamamoto, M.; Takai, K.; Li, Y.; Yamaguchi, A.; Gilbert, A.; Ueno, Y.; Yoshida, N.; Oono, Y. Geoelectrochemical CO production: Implications for the autotrophic origin of life. *Sci. Adv.* **2018**, *4*, eaao7265. [CrossRef]
30. Ooka, H.; Mcglynn, S.E.; Nakamura, R. Electrochemistry at deep-sea hydrothermal vents: Utilization of the thermodynamic driving force towards the autotrophic origin of life. *ChemElectroChem* **2018**. [CrossRef]
31. Muchowska, K.B.; Varma, S.J.; Chevallot-Beroux, E.; Lethuillier-Karl, L.; Li, G.; Moran, J. Metals promote sequences of the reverse Krebs cycle. *Nat. Ecol. Evol.* **2017**, *1*, 1716–1721. [CrossRef]
32. Varma, S.J.; Muchowska, K.B.; Chatelain, P.; Moran, J. Native iron reduces CO_2 to intermediates and end-products of the acetyl-CoA pathway. *Nat. Ecol. Evol.* **2018**, *2*, 1019–1024. [CrossRef] [PubMed]
33. Jackson, J.B. Natural pH gradients in hydrothermal alkali vents were unlikely to have played a role in the origin of life. *J. Mol. Evol.* **2016**, *83*, 1–11. [CrossRef] [PubMed]
34. Wächtershäuser, G. In praise of error. *J. Mol. Evol.* **2016**, *82*, 75–80. [CrossRef] [PubMed]
35. Sutherland, J.D. Studies on the origin of life—the end of the beginning. *Nat. Rev. Chem.* **2017**, *1*, 0012. [CrossRef]
36. Lane, N. Proton gradients at the origin of life. *BioEssays* **2017**, *39*, 1600217. [CrossRef] [PubMed]

Review

How Prebiotic Chemistry and Early Life Chose Phosphate

Ziwei Liu [1,2], Jean-Christophe Rossi [1] and Robert Pascal [1,*]

[1] UMR5247, CNRS—University of Montpellier—ENSCM, Place E. Bataillon, 34095 Montpellier CEDEX 5, France; zliu@mrc-lmb.cam.ac.uk (Z.L.); jean-christophe.rossi@umontpellier.fr (J.-C.R.)
[2] MRC Laboratory of Molecular Biology, Cambridge Biomedical Campus, Cambridge CB2 0QH, UK
* Correspondence: robert.pascal@umontpellier.fr; Tel.: +33-467-14-4229

Received: 5 February 2019; Accepted: 25 February 2019; Published: 3 March 2019

Abstract: The very specific thermodynamic instability and kinetic stability of phosphate esters and anhydrides impart them invaluable properties in living organisms in which highly efficient enzyme catalysts compensate for their low intrinsic reactivity. Considering their role in protein biosynthesis, these properties raise a paradox about early stages: How could these species be selected in the absence of enzymes? This review is aimed at demonstrating that considering mixed anhydrides or other species more reactive than esters and anhydrides can help in solving the paradox. The consequences of this approach for chemical evolution and early stages of life are analysed.

Keywords: phosphoryl transfer; metabolism; energy currency; mixed anhydride

1. Introduction

Preceding the discovery of the double-helical structure of DNA by more than a decade, the understanding of the metabolic roles of ATP and phosphoryl transfers was an essential step in the disclosure of the foundations of biochemistry by clarifying how energy is distributed and serves as a fuel for the achievement of the different functions of the cell [1]. Biochemists observed that the phosphoryl group is essentially kept within the boundaries of the cell and constantly recycled through the energy-rich intermediates of the metabolism [1]. The determining factor of the properties of phosphate derivatives lies in their negatively charged character, responsible for both their sequestration within compartments delimited by phospholipid membranes as a result of electrostatic forces and for their stability towards hydrolysis and other nucleophilic attacks. Such properties make them suited for their biochemical duties [2]. Westheimer also emphasized "this remarkable combination of thermodynamic instability and kinetic stability" [3]. He expressed how the specific physicochemical properties of these anions are so important for ensuring the different functions played by phosphoryl groups in living organisms: (1) as leaving groups for nucleophilic substitution universal in biology; and (2) as ionized groups useful to conserve metabolites within a compartment having a negatively charged boundary. The values of the pKa for the first ionization (ca. 1–2) [4] are such that a negligible amount of the corresponding biochemicals remains uncharged at physiological pH values. As a result, nucleic acids are conserved within the vesicle as well as many other nucleotide derivatives in such a way that most components of their metabolism remain confined together in a limited volume. From a kinetic perspective, the rates of base-catalysed hydrolysis of phosphoryl derivatives are lowered by electrostatic destabilization of a negatively charged transition state. A noteworthy consequence of this lack of reactivity can be found in the fact that certain phosphate diesters undergo hydrolysis through C-O bond cleavage rather than by that of the P-O bonds [5].

Though their physicochemical properties are adapted to modern biology in which enzymes are able to compensate for the low intrinsic reactivity of negatively charged phosphates mentioned above [3,6], they usually result in sluggish spontaneous phosphoryl transfer reactions, constituting a drawback that probably hampered their selection as reactive intermediates at the chemical stage preceding the evolution of catalytic polymers. In the literature, several authors indeed considered the modern role of ATP in energy exchange as an acquired function and concluded that earlier chemical processes for energy transfer were needed for life to start [7,8].

Building a reasonable scenario for the role of phosphate in the origin and early evolution of life needs therefore to answer two questions, namely: Which kind of reactions could have been prevalent at those stages? What kind of phosphate reactivity could take place spontaneously in the absence of enzymes? These questions become even more crucial taking into account the fact that early membranes, probably more permeable than those based on phospholipids, were less efficient in sequestrating anionic metabolites. No obvious selective advantage of derivatives involving phosphoryl groups can therefore be foreseen at the prebiotic stage when no efficient coded catalyst could compensate for the low intrinsic rates of most phosphoryl transfers. Low-valence phosphorus derivatives have been proposed as an alternative to provide a higher reactivity [9–12]. Without prejudice to the actual relevance of this attractive possibility, our work is aimed at determining which chemical pathways could have been critical for the introduction of phosphate anhydrides and esters as intermediates in early biochemistry. Our main tenet is related to the importance of mixed anhydrides that can be formed from phosphates and high-energy carboxylic acid derivatives and that may have contributed to the distribution of energy in early metabolisms. The fact that phosphates could play a role in chemical and early biochemical evolution could therefore be related to a very peculiar chemistry having a limited relationship to the usual biochemical role of phosphate derivatives. Though the importance of phosphate chemistry in the structure and stability of biomolecules and biopolymers will be mentioned, this review is mainly focused on reactivity issues related to phosphoryl transfers and their potential contribution to the distribution of energy in protometabolisms or early biological metabolisms.

2. Phosphoryl Transfer Pathways

2.1. Phosphate Esters

At moderate pH, chemical transformations at the phosphorus centre of phosphoryl groups usually take place from the monoanion. However, there is a profound difference depending on the degree of substitution at phosphorus. The situation is clearly illustrated by the difference in reactivity between diesters and monoesters. Phosphodiesters are highly stable to hydrolysis largely because the presence of a negative charge at moderate and alkaline pH values constitutes a barrier towards nucleophilic attack, which can be appreciated by considering the alkaline hydrolysis of the simplest model of phosphodiester, dimethyl phosphate (Figure 1). In an unexpected way, this reaction takes place through a nucleophilic substitution at carbon rather than at phosphorus [5]. The P-O bond of phosphodiesters is therefore almost unreactive towards hydrolysis at physiological pH values explaining why a phosphodiester backbone could have been selected by evolution for the long-term storage of information in DNA [13]. We can conclude that the relevance of phosphate diesters to the origin or the early developments of life rather lies in their chemical resistance than in their reactivity. Making this reaction compatible with a protometabolism time scale would require the lifetime of phosphodiesters to be reduced from tens of million years into days [14]. Such values of rate enhancement exceeding 10^8 are only accessible through catalysis by enzymes. Simple chemical catalysts could hardly reach that efficiency except intramolecular reactions in which the proximity of reacting groups can compensate for the kinetic barriers [15]. The well-known instability of RNA compared to that of DNA precisely lies in the presence of a hydroxyl group at the ribose 2'-position capable of provoking a cleavage of the internucleotidic linkage intramolecularly. This observation supports the importance of intramolecular processes before the advent of enzymes [16–18].

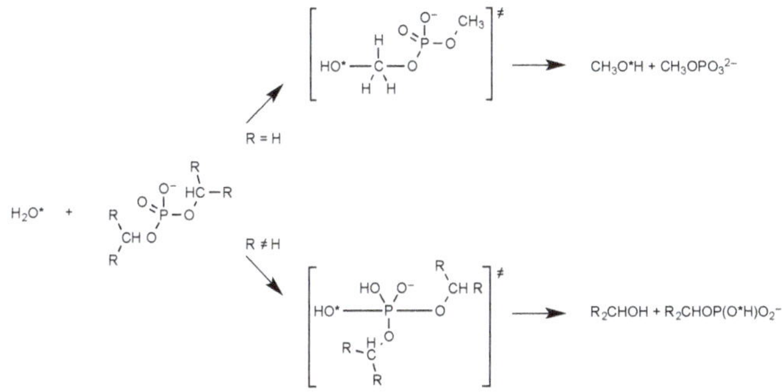

Figure 1. Hydrolysis of phosphate diesters. Nucleophilic attack can take place at carbon or phosphorus depending on the degree of substitution at carbon.

Monoesters such as methyl phosphate [5] behave in a completely different way and are much less stable than diesters at neutral and mildly acidic pH values. In contrast with the reaction of phosphodiesters corresponding to an associative mechanism, monoesters are cleaved through dissociative transition states resembling the resonance-stabilized metaphosphate ion (PO_3^-, Figure 2) [19,20]. There has been a long-lasting debate on the actual lifetime of the metaphosphate ion, which may not be sufficient for it to be considered as a true intermediate and the dissociative nature of the reaction pathway has been disputed on theoretical grounds [21]. Anyway, it can be acknowledged that resonance stabilization plays a role at the transition state so that the hydrolysis of phosphate monoesters is ca. 6 orders of magnitude faster than the corresponding reaction of diesters assessed using substrates unable to undergo substitution at carbon [13].

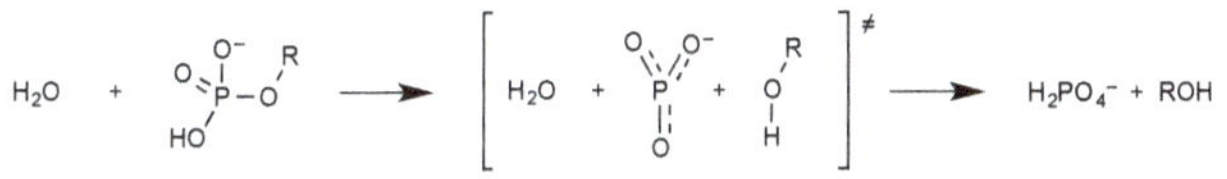

Figure 2. The dissociative pathway of phosphoryl transfer in the hydrolysis of phosphate monoesters. A metaphosphate ion (PO_3^-) intermediate or at least resonance stabilization at the transition state is involved in the reaction.

In spite of a faster reaction, phosphate mono-alkyl esters still present lifetimes (measured in tenth of years at moderate values of pH and temperature [13]) incompatible with a role of reactive intermediates of a metabolism. However, this stability is the basis of their role of constituents of basic structures of the cell such as phosphatidic acids as components of the membranes and other metabolites bearing an anionic charge, allowing them to remain sequestrated within the boundaries of the cell [3].

2.2. Phosphate Anhydrides

The presence of a much better leaving group in phosphate anhydrides tends to increases their reactivity. However, it must be taken into account that this effect is offset in anhydrides such pyrophosphate and ATP (Figure 3) by the presence of 3 or 4 negative charges at moderate pH values, rendering these activated species spontaneously almost non-reactive towards nucleophilic attack since it hinders the development of more negatively charged transition states. Therefore, most reactions of these high-energy intermediates require catalysis to take place at rates compatible with the time scale of a metabolism. In the living world, this limited reactivity results in a kinetic stabilization and the reactivity of ATP can be orientated towards specific paths by enzymes like kinases, which are responsible for many cellular functions and for the role of ATP as an energy currency. With regards to the origin of life, an advantage can hardly be expected from this exceedingly limited reactivity due to the lack of selective catalysts for many phosphorylation processes. Pyrophosphate and more generally polyphosphates would suffer from a similar lack of reactivity making their involvement in prebiotic chemistry and early biological evolution questionable even though that contribution to the chemistry of the origins of life has been proposed in many instances [22–30]. Independently of the limitations to their possible role of energy currency, a similar difficulty related to the lack of enzyme catalysts has been raised when considering the biochemical use of nucleoside triphosphates as activated monomers for RNA oligomerization [31].

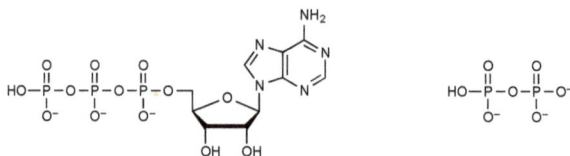

Figure 3. ATP and pyrophosphate have been proposed as prebiotic energy currencies, in spite of the kinetic barrier hindering their reactions with nucleophiles at moderate pH values at which they are negatively charged.

2.3. Phosphate Mixed Anhydrides

Acyl phosphates are among the more potent activated biochemicals [32]. Values of their free energy of hydrolysis at pH 7 ($\Delta G^{\circ\prime}$) reach -43 kJ mol^{-1} for acetyl phosphate [33]. By contrast with ATP and pyrophosphate, acyl phosphates and acyl adenylates bear a single negative charge at mildly acidic pH values and benefit to a much lesser degree from the kinetic stabilization that inhibits the increase of negative charge at the transition state of the reactions with nucleophiles including water and hydroxide ion. This limitation is likely to be even less stringent for mixed anhydrides of inorganic phosphate that can be cleaved through a dissociative mechanism, in which resonance stabilization occurring within a transition state resembling metaphosphate ion replaces the interaction with the nucleophile as the main driving force [34].

Indeed, aminoacyl phosphates (Figure 4) were shown to undergo a cleavage of the P-O bond and constitute efficient phosphorylating agents [35]. Mixed anhydrides with phosphate monoesters like acyl adenylates remain hydrolytically unstable and susceptible to spontaneously undergoing reactions with other nucleophiles at the carboxyl moiety. These mixed anhydrides can be formed as intermediates in the reactions of various acyl donors including activated esters [36–38], thioesters [39,40], or anhydrides [37,41]. The biochemically essential acetyl phosphate can for instance be formed photochemically by oxidation of the thioacid [42]. A similar reaction of thioacetate has recently been reported to occur in limited yields in a hydrothermal context [43] without mention of the possibility of photo-oxidation [42]. Amino acyl adenylates (Figure 4) deserve a particular mention because of their role in protein biosynthesis. Their degree of activation has been assessed in the case of Tyr-AMP to a value of $\Delta G^{\circ\prime} = -70$ kJ mol^{-1} [44] much higher than that observed for simple acyl phosphates and that exceeds the values observed for the main intermediates of energy metabolism including phosphoenol pyruvate. Amino acyl adenylates are formed biochemically by the reaction of amino acids with ATP. However, because of the endergonic character of the reaction, amino acyl adenylates usually remain sequestrated in the active site of aminoacyl-tRNA synthetases, the enzymes that are responsible for their formation from ATP and for the further aminoacylation of tRNA [45]. Abiotically, the formation of mixed anhydrides through a similar reaction of phosphate anhydrides with unprotected amino acids can therefore be considered as unlikely for thermodynamic reasons in addition to the sluggish kinetic availability of ATP due to its multiple negative charges. Therefore, science has to solve the paradox of the initial formation of aminoacyl-adenylates required for the evolution of translation but impossible from ATP without enzymes. That paradox requires the occurrence of alternative pathways [46,47]. This possibility has experimentally been supported by the observation that two categories of prebiotically plausible activated derivatives of α-amino acids undergo spontaneous conversion into aminoacyl adenylates or related mixed anhydrides. 5(4*H*)-Oxazolones, formed as a result of the strong activation of acylated amino acids or peptides, have demonstrated an ability to be converted spontaneously into mixed anhydrides in the presence of inorganic phosphate or phosphate esters [48,49]. These reactions yield peptidyl- or acyl-substituted products but derivatives with a free amino group can be obtained directly by the analogous reaction of amino acids *N*-carboxyanhydrides (NCAs) [35,50,51]. NCAs have been proposed as plausible activated forms of amino acids under prebiotic conditions and several potential pathways are available for their formation [46,52–55]. It is worth emphasizing that any reaction involving phosphate as well as its monoesters as nucleophiles and activated carboxylic acids would be facilitated rather than kinetically inhibited by phosphate negative charges, which avoids the need for catalysis for an abiotic process. A potential role for these intermediates in the chemical processes associated with the development of life is therefore highly likely, provided that carboxylic acid activation into high-energy intermediates is possible in that environment. Since NCAs are formed rapidly from most other forms of activated α-amino acids having a free amino group in aqueous media containing carbon dioxide [52], the prebiotic relevance of phosphate mixed anhydrides of amino acids should be recognized provided that phosphate is available. However, the fast reverse reaction of carbon dioxide also prevails from phosphate mixed anhydrides of α-amino acids, which are converted back into NCAs rapidly [48,56]. Though present to a lesser degree than in polyphosphates like ATP or pyrophosphate, the negative charge of phosphate esters mixed anhydrides reduces their reactivity with nucleophiles so that the reaction pathway may involve a prior conversion into neutral (and therefore highly reactive) NCAs rather than a direct reaction of amino acid phosphate anhydrides (Figure 5). Accordingly, the polymerization of aminoacyl adenylates into peptides takes place through the NCA pathway [48] rather than from a direct polymerization as proposed earlier [57].

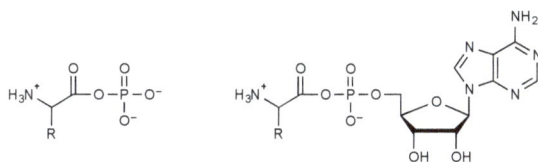

Figure 4. Amino acyl phosphates and aminoacyl adenylates are highly activated biochemicals.

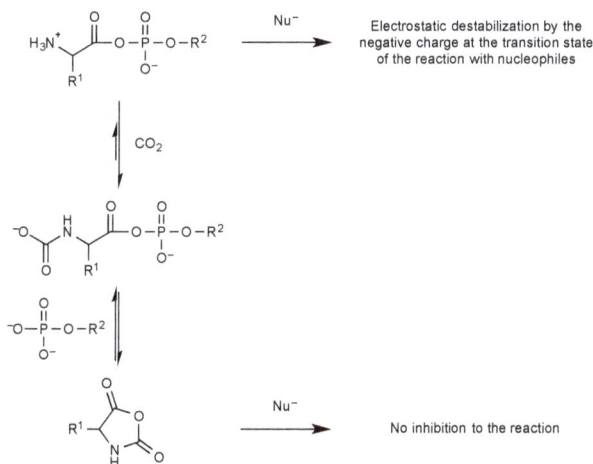

Figure 5. Phosphate esters mixed anhydrides suffer from stabilization against reaction with nucleophiles as other phosphate derivatives. The pathway involving *N*-carboxyanhydrides (NCAs) as intermediates must be taken into consideration as soon as CO_2 is present in the atmosphere, even at low levels.

The role of phosphate mixed anhydrides in the development of life should therefore be analysed by taking into account two counteracting factors: (1) a less important kinetic stabilization by negative charges as compared to polyphosphates; and (2) but a relative stabilization compared to neutral highly reactive activated acyl precursors. Generally no kinetic advantage is therefore to be expected from a reaction of mixed anhydrides compared to activated acyl precursors as in the case of the formation of peptides in which the fast polymerization of NCAs competes favourably with the polymerization of mixed anhydrides [48]. However, in some cases their reactivity could be advantageous as probably in the case of the aminoacylation of the 3'(2')-end of RNA for which NCA proved to be inefficient [58]. Examples of an advantageous role of mixed anhydrides have been observed from their involvement as intermediates undergoing a fast intramolecular acyl transfer as in the formation of esters with ribonucleotides [50,56,59,60]. Phosphate moieties could indeed act as handles capable of reacting with activated acyl moieties and then to intramolecularly transfer the acyl group to a poor nucleophile thanks to the entropic advantage of intramolecular processes [16,61]. This property provides a rationale for the selection of mixed anhydrides in the evolutionary process. On the other hand, the easy conversion of activated acyl derivatives including those of α-amino acids into phosphate mixed anhydrides might be considered as an early example of how free energy could be exchanged between the chemistries of α-amino acids and that of nucleotides predating the role of ATP as an energy currency [46,47]. Thioesters constitute other activated acyl derivatives that yield phosphate mixed anhydrides by interaction with phosphate. Pathways leading to their formation from carbon chemistry have been proposed [62]. This contribution could be important for many thioesters with the exception of α-amino acid thioesters that are rapidly converted into NCAs in the presence of CO_2 or bicarbonate [52] so that their reactivity cannot be considered as different from that of other activated α-amino acid derivatives.

2.4. Phosphoramidates

The chemical interaction of free α-amino acids with activated phosphates can also yield phosphoramidate derivatives by nucleophilic reaction of the amino group (Figure 6). By contrast with the behaviour of carboxylic acid derivatives, phosphoramidates are more reactive than phosphate esters and correspond to an activated state of phosphate, which has been illustrated by the ability of some of them to behave as polymerase substrates for the synthesis of DNA [63–68]. In addition to that ability in enzyme reactions, *N*-phosphoryl amino acids proved to be capable of yielding both phosphate esters and polypeptides through spontaneous reactions in aqueous solution [69–71]. Lastly an intermediate role has been proposed for phosphoramidates in the polymerization of amino acid promoted by EDC (1-ethyl-3-(3-dimethylaminopropyl)-carbodiimide) in the presence of nucleotides [72,73]. A contribution of phosphoramidates to prebiotic chemistry and early biochemistry can therefore be considered as highly likely as soon as powerful activating agents were present. Interestingly, activating agents based on phosphoramidate moieties have been proposed in an origin of life context [74–77]. In addition to being involved as intermediates in the formation of biopolymers, it is worthy to note that chemical ligations as well as template-directed polymerization proved to proceed more easily using modified nucleotides bearing an amine nucleophile instead of the the 3'-hydroxyl group yielding phosphoramidate linkages owing to the increased nucleophilic power of amines compared to alcohols [78–82]. The facilitated nucleotide polymerization has allowed major studies of the replication process proceeding in the absence of enzymes [31,83–85]. It could also be considered as a basis for the formation of mixed structures [60,86] involving both amino acids and nucleotides bound through ester and phosphoramidate linkage with an unexpected lifetime for aminoacyl esters (Figure 6) [60].

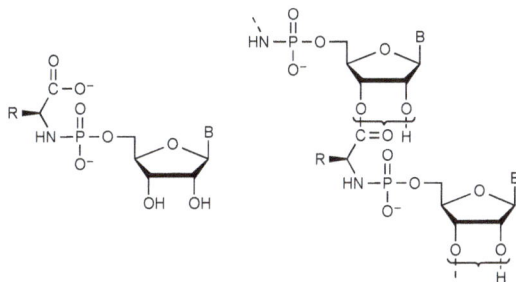

Figure 6. Phosphoramidate derivatives of amino acids and the ester- and phosphoramidate-based structures constituting the basis of prebiotically plausible copolymers.

3. Which Phosphate Derivatives Could Play A Role as Early Energy Currencies?

The availability of free energy is crucial for self-organization to maintain a system in a far from equilibrium state [87]. However, this energy must not be dissipated directly through a linear spontaneous process in order that work can be carried out. In other words, as Eschenmoser [88,89] emphasized using a different terminology, the chemical environment must be held far from equilibrium by kinetic barriers. From this point of view, the kinetic stability of ATP makes it a unique component of metabolism. ATP is well known for its ability to act as an energy currency that it is constantly synthesized and used up by hydrolysis into ADP and inorganic phosphate [1]. An open question with respect to early metabolism is related to the probable inability of ATP to play this role and, consequently, to the possible existence of others chemicals acting as substitutes. In earlier reports, a body of evidence was gathered to support the idea that ATP could not be involved as an energy source for the development of translation [46,47]. This conclusion was mainly based on the observation that there is no chemical (non-enzymatic) path available for the conversion of ATP into amino acid adenylates for both thermodynamic and kinetic reasons. Namely, the free energy potential of ATP is unable to afford significant concentrations of adenylates at equilibrium with amino acids in the presence of ATP and the thermodynamically favourable reverse reactions yielding ATP from adenylates and pyrophosphate do not spontaneously take place and require the presence of enzymes.

Considering the properties required for a chemical species to act as an energy currency (Figure 7) should be helpful in identifying alternatives. A first requirement corresponds to a far from equilibrium state meaning that the thermodynamic potential of the currency makes it able to dissipate energy in the environment. Potential energy currencies can therefore be considered on the basis of their thermodynamic potential (see Table 1). However, dissipation must be hindered by kinetic barriers so that the energy currency can act as an activating agent able to produce work by delivering energy to other components of the system (Figure 7). This second condition, which could seem somewhat contradictory with the preceding one, corresponds to the need for kinetic stability of the potential candidate that must be able to transfer its energy to a recipient chemical system with rates faster than, or at least competing with, those at which its potential is dissipated in the environment through breakdown processes (e.g., by direct hydrolysis). As far as non-living systems are concerned, limited possibilities of *selective* catalytic pathways are available to make reactions rates consistent with the time scale of the system *without increasing those of dissipation pathways*. In spite of the fact that they have been proposed as early analogues of ATP, polyphosphates including pyrophosphate fail to fulfil that latter kinetic requirement. Therefore, both the above-mentioned inability to activate amino acids into adenylates and a poor spontaneous reactivity can be considered as indications that pyrophosphate and other polyphosphates could hardly play a role in energy transduction in early metabolisms unless efficient catalytic pathways for the transfer of their energy are found in the future. Anhydrides bearing less negative charges would react faster, which supports a potential role of carboxylic-phosphoric mixed anhydrides. From an energy perspective, a thermodynamic potential sufficient to allow for the formation of aminoacyl adenylates was required for the emergence of translation and more precisely for the evolution of aminoacyl-tRNA synthetases (aaRS) that use amino acids activated as adenylates. Amino acid *N*-carboxy anhydrides (NCAs) have been proposed as essential intermediates in this context [46,47,90]. NCAs were identified as reagents capable of providing adenylates without requiring catalysis by enzymes [50,51]. The value of their free energy of hydrolysis at pH 7 ($\Delta G^{\circ\prime}$ = ca. −60 kJ mol^{-1} [46]) associated with a spontaneous reaction with phosphate and phosphate mono-esters makes them likely precursors of mixed anhydrides, including adenylates. The inability of ATP to provide adenylates in a similar way shows that another reagent played its role or that no reagent played the role of universal energy currency. However, some of the species of Table 1 having a high potential, could be formed abiotically or at least without requiring catalysis, some of them, including acetyl phosphate (as other acyl and aminoacyl phosphates) and carbamyl phosphate, indeed still play a role in biochemistry. They could be considered as possible alternative energy shuttles between different systems, notably able to yield mixed anhydrides required for different metabolic functions in early living organisms, without reaching the status of universal energy currency as ATP in evolved living system.

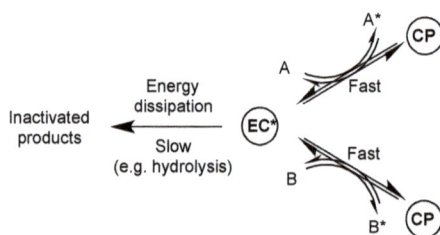

Figure 7. An energy currency (activated form EC*) formed from a currency precursor (CP) requires a high free energy potential and pathways available to transfer energy between different processes faster than the dissipation of energy. Energy currencies must therefore comply with kinetic and thermodynamic requirements.

Table 1. Values of the free energy of hydrolysis at pH 7 for different phosphate-based energy-rich biochemical metabolites.

Reagent	Product(s)	$\Delta G^{\circ\prime}$ kJ mol^{-1}	Reference
PPi	2 Pi	−19	[33]
ATP	AMP + PPi	−32.2	[33]
ATP	ADP + Pi	−30.5	[33]
Acetyl phosphate	AcOH + Pi	−43.1	[33]
Carbamyl phosphate	CO_2 + NH_3 + Pi	ca. −51 [1]	[33]
Aminoacyl phosphate	Amino acid + Pi	ca. −50	[91]
Aminoacyl adenylate	Amino acid + AMP	−70	[44]
Phosphoenol pyruvate	Pyruvate + Pi	−62	[33]

[1] Value determined at pH 9.5.

4. The Question of Prebiotic Phosphorylation

The abiotic formation of phosphorylated metabolites is a central issue in prebiotic chemistry and comprehensive reviews dealing with this question and providing a list of reagents relevant to the origin of life context have been published [92,93]. The possibility of a contribution of phosphates to prebiotic chemistry and the origin of life should have been limited by the availability of phosphate or other phosphorus containing intermediates (including low valence derivatives). Solution phosphorylation would for instance be limited by the solubility of phosphate, which is strongly reduced in the presence of di- or tri-valent cations [30]. As these ions were likely present in the environment on the prebiotic Earth, the low content of phosphate in solution should be considered as unfavourable to phosphorylation. However, the low availability of phosphate in an ocean could be compensated in some cases by the favourable effect of cations on the phosphorylation reaction. A phosphorylation process involving cyanate as an activating agent and precipitated apatite was reported as a realistic pathway in prebiotic chemistry, which means that the reaction can take place on the surface of the solid [23]. The activation of inorganic phosphate can take place by reaction with energy-rich chemicals (Table 2).

Table 2. Values of the free energy of hydrolysis at pH 7 for different potential phosphate activating agents available in the literature.

Reagent	Product(s)	$\Delta G^{\circ\prime} \text{ kJ mol}^{-1}$	Reference
HNCO	$CO_2 + NH_3$	−54	[91]
Urea	$CO_2 + NH_3$	−28	[91]
Cyanamide	Isourea	−83	[94]
Carbodiimide	Isourea	−97	[94]
Acetic anhydride	Acetic acid	−91	[33]
NCA	Amino acid + CO_2	−60	[46]

Cyanamide dimer [95] or cyanate [96,97] are able to promote the formation of reactive adducts with inorganic phosphate that subsequently act as phosphoryl donors, very probably through a dissociative pathway involving a resonance-stabilized transition state (Figure 8).

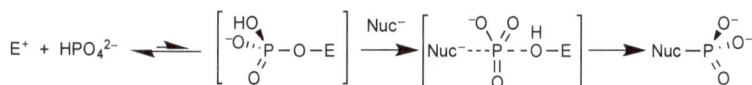

Figure 8. Phosphorylation can by promoted by electrophilic activating agent capable of generation an intermediate capable of transferring the phosphoryl group to an acceptor nucleophile through a metaphosphate or, at least, resonance-stabilized intermediate.

The reaction of cyanate is well-documented, yielding carbamyl phosphate as a transient species upon reaction with inorganic phosphate [96,97]. Then the intermediate decays either through hydrolysis yielding eventually CO_2 and NH_3 or through an elimination pathway specific of the mono-anion [98,99] and reverting cyanate (Figure 9).

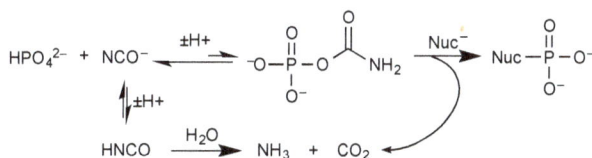

Figure 9. Cyanate-promoted phosphorylation through a carbamyl phosphate intermediate.

The overall process constitutes a catalytic pathway of hydrolysis of cyanate quite similar to that observed for carbonate and dicarboxylic acids, initially reported to involve general acid catalysis [100], but later proven to actually correspond to nucleophilic catalysis [101]. Carbamyl phosphate can also be prepared photochemically from $Fe(CN)_6^{3-}$ and is able to promote the formation of ATP or acetyl phosphate [102–104].

The most important limitation for the formation of phosphate monoesters in diluted aqueous solution lies in the usual low selectivity of the reaction of alcohols compared to that of water in large excess that outcompetes that of diluted substrates. A very attractive possibility to solve this issue lies in the use of chemical catalysis. The condensation of aldehydes with diamidophosphate provides a pathway to regioselectively phosphorylate glycoaldehyde and other aldoses very efficiently through an induced intramolecular pathway [74–77]. Phosphoryl transfer in a supramolecular environment has also been useful to promote selective phosphorylation [105,106].

Performing the reaction under dehydrating conditions under the effect of heat has been considered as another possibility to avoid dilution in aqueous solution. Heating mixtures of reagents to temperatures above 80 °C in the presence of ammonium formate [27], or formamide [107] proved to be efficient though the regioselectivity was limited. Since its first mention [108], urea has been used in many instances to perform phosphorylation of nucleotides [109] or long-chain alcohols [110] under the effect of heat. It is worth noting that reactions in urea-inorganic phosphate mixtures proceed faster than with the other additives. No definitive answer has been given to the actual pathway through which urea promotes phosphorylation. Phosphoramidate or carbamyl phosphate intermediates [108] have been mentioned as possible actual phosphorylating species. An activated intermediate of unknown nature has also been proposed [109]. Other explanations involve nucleophilic [93,111] or acid–base [112] catalyses. A more likely explanation has been proposed [113] that takes into account the easy breakdown of urea into cyanate at high temperature [114]. This ability of urea independently accounts for the formation of amino acids *N*-carboxyanhydrides in hot aqueous solutions from urea [55]. It would however mean that the activity of urea for promoting phosphorylation is the result of a stoichiometric rather than catalytic reaction involving cyanate as an activating intermediate and carbamyl phosphate as the actual phosphorylating agent (Figure 9).

5. Conclusions

This review focuses on the specific features of phosphoryl group reactivity that raise constraints on the prebiotic and early biochemical pathways involved in the origin of life and its early developments. The charge of phosphate moieties constituted a determining advantage for sequestrating substrates as soon as phospholipids, fatty acids or other negatively charged amphiphiles were present and able to form membrane-delimited compartments. Another essential biochemical consequence of this charge is the resistance of phosphate moieties to nucleophilic reactions and most notably to base-catalysed hydrolysis that is hindered by repulsive electrostatic interactions at the transition state. The later advantage is fully operational in modern biology because of the evolution of highly effective and selective enzymes. However, it constituted very probably a strong limitation in chemical systems having limited possibilities of selective catalysis. Therefore, the early role of phosphate-derived species is more likely to be the result of their lack of reactivity than that of possibilities of transferring energy between metabolic subsystems. We therefore conclude that the role of ATP as a universal energy currency is unlikely to be an early invention of life. In spite of these limitations, it is possible to depict the possibilities opened by phosphate chemistry at an early stage just by considering its specific reactivity. Though their role could be limited to specific processes, mixed anhydrides could have played a role in transferring energy from the chemistry of amino acids to that of nucleotides being essential in the emergence of translation. More generally, pathways for the phosphorylation of nucleosides and hydrophobic alcohols are available in an origin of life context. As mentioned above, a very important property of phosphate derivatives such as phosphate esters is their reduced kinetic reactivity. This property has certainly been selected for information storage and is a major reason for the selection of the phosphodiester-based nucleic acid backbone, which is expressed at the highest degree in DNA. It could additionally be considered that the lack of reactivity of phosphate esters is also revealed by the difficulty in building the phosphodiester bond. Imidazolides and their derivatives have been considered in many RNA world experiments as convenient activated monomers for RNA polymerization [115] rather than the biochemical triphosphate substrates. This possibility is supported by new reports on the relevance of the abiotic synthesis of imidazole derivatives under early Earth conditions as well as to their specific reactivity [31,116].

Author Contributions: R.P. conceived the research project; All authors wrote the manuscript.

Funding: The work was supported by grants from the Agence Nationale de la Recherche (PeptiSystems project ANR-14-CE33-0020) and the Simons Foundation (Grant Number 293,065 to Z.L.).

Conflicts of Interest: The authors declare no conflict of interest.

References

1. Lipmann, F. Metabolic generation and utilization of phosphate bond energy. *Adv. Enzymol. Relat. Areas Mol. Biol.* **1941**, *1*, 99–162.

2. Lipmann, F. *Phosphorus Metabolism*; McElroy, W.D., Glass, H.B., Eds.; Johns Hopkins Press: Baltimore, MD, USA, 1951; Volume 1, p. 521.

3. Westheimer, F.H. Why nature chose phosphate. *Science* **1987**, *235*, 1173–1178. [CrossRef] [PubMed]

4. Kumler, W.D.; Eiler, J.J. The Acid Strength of Mono and Diesters of Phosphoric Acid. The n-Alkyl Esters from Methyl to Butyl, the Esters of Biological Importance, and the Natural Guanidine Phosphoric Acids. *J. Am. Chem. Soc.* **1943**, *65*, 2355–2361. [CrossRef]

5. Wolfenden, R.; Ridgway, C.; Young, G. Spontaneous hydrolysis of ionized phosphate monoesters and diesters and the proficiencies of phosphatases and phosphodiesterases as catalysts. *J. Am. Chem. Soc.* **1998**, *120*, 833–834. [CrossRef]

6. Bowler, M.W.; Cliff, M.J.; Waltho, J.P.; Blackburn, G.M. Why did Nature select phosphate for its dominant roles in biology? *New J. Chem.* **2010**, *34*, 784–794. [CrossRef]

7. Goldford, J.E.; Hartman, H.; Smith, T.F.; Segrè, D. Remnants of an Ancient Metabolism without Phosphate. *Cell* **2017**, *168*, 1126–1134. [CrossRef] [PubMed]

8. Martin, W.F.; Thauer, R.K. Energy in Ancient Metabolism. *Cell* **2017**, *168*, 953–955. [CrossRef] [PubMed]

9. Pasek, M.A.; Kee, T.P.; Bryant, D.E.; Pavlov, A.A.; Lunine, J.I. Production of Potentially Prebiotic Condensed Phosphates by Phosphorus Redox Chemistry. *Angew. Chem. Int. Ed.* **2008**, *47*, 7918–7920. [CrossRef] [PubMed]

10. Bryant, D.E.; Greenfield, D.; Walshaw, R.D.; Evans, S.M.; Nimmo, A.E.; Smith, C.L.; Wang, L.; Pasek, M.A.; Kee, T.P. Electrochemical studies of iron meteorites: Phosphorus redox chemistry on the early Earth. *Int. J. Astrobiol.* **2009**, *8*, 27–36. [CrossRef]

11. Bryant, D.E.; Marriott, K.E.R.; Macgregor, S.A.; Kilner, C.; Pasek, M.A.; Kee, T.P. On the prebiotic potential of reduced oxidation state phosphorus: The H-phosphinate–pyruvate system. *Chem. Commun.* **2010**, *46*, 3726–3728. [CrossRef] [PubMed]

12. Kaye, K.; Bryant, D.; Marriott, K.; Ohara, S.; Fishwick, C.; Kee, T. Selective Phosphonylation of 5′-Adenosine Monophosphate (5′-AMP) via Pyrophosphite [PPi(III)]. *Orig. Life Evol. Biosph.* **2016**, *46*, 425–434. [CrossRef] [PubMed]

13. Schroeder, G.K.; Lad, C.; Wyman, P.; Williams, N.H.; Wolfenden, R. The time required for water attack at the phosphorus atom of simple phosphodiesters and of DNA. *Proc. Natl. Acad. Sci. USA* **2006**, *103*, 4052–4055. [CrossRef] [PubMed]

14. Wolfenden, R. Benchmark reaction rates, the stability of biological molecules in water, and the evolution of catalytic power in enzymes. *Annu. Rev. Biochem.* **2011**, *80*, 645–667. [CrossRef] [PubMed]

15. Kirby, A.J. Effective molarities for intramolecular reactions. *Adv. Phys. Org. Chem.* **1980**, *17*, 183–278.

16. Pascal, R. Catalysis through Induced Intramolecularity: What Can Be Learned by Mimicking Enzymes with Carbonyl Compounds that Covalently Bind Substrates? *Eur. J. Org. Chem.* **2003**, *2003*, 1813–1824. [CrossRef]

17. Pascal, R. Kinetic Barriers and the Self-organization of Life. *Isr. J. Chem.* **2015**, *55*, 865–874. [CrossRef]

18. Li, B.-J.; El-Nachef, C.; Beauchemin, A. Organocatalysis Using Aldehydes: The Development and Improvement of Catalytic Hydroaminations, Hydrations and Hydrolyses. *Chem. Commun.* **2017**, *53*, 13192–13204. [CrossRef] [PubMed]

19. Kirby, A.J.; Varvoglis, A.G. The Reactivity of Phosphate Esters. Monoester Hydrolysis. *J. Am. Chem. Soc.* **1967**, *89*, 415–423. [CrossRef]

20. Westheimer, F.H. Monomeric metaphosphate ion. *Chem. Rev.* **1981**, *81*, 313–326. [CrossRef]

21. Kamerlin, S.C.; Sharma, P.K.; Prasad, R.B.; Warshel, A. Why nature really chose phosphate. *Q. Rev. Biophys.* **2013**, *46*, 1–132. [CrossRef] [PubMed]

22. Lipmann, F. Projecting backward from the present stage of evolution of biosynthesis. In *The Origins of Prebiological Systems and of Their Molecular Matrices*; Fox, S.W., Ed.; Academic Press: New York, NY, USA, 1965; pp. 259–280.

23. Miller, S.L.; Parris, M. Synthesis of pyrophosphate under primitive Earth conditions. *Nature* **1964**, *204*, 1248–1250. [CrossRef]

24. Vieyra, A.; Meyer-Fernandes, J.R.; Gama, O.B. Phosphorolysis of acetyl phosphate by orthophosphate with energy conservation in the phosphoanhydride linkage of pyrophosphate. *Arch. Biochem. Biophys.* **1985**, *238*, 574–583. [CrossRef]

25. Vieyra, A.; Gueiros-Filho, F.; Meyer-Fernandes, J.R.; Costa-Sarmento, G.; De Souza-Barros, F. Reactions involving carbamyl phosphate in the presence of precipitated calcium phosphate with formation of pyrophosphate: A model for primitive energy-conservation pathways. *Orig. Life Evol. Biosph.* **1995**, *25*, 335–350. [CrossRef] [PubMed]

26. Keefe, A.D.; Miller, S.L. Are polyphosphate or phosphate esters prebiotic reagents? *J. Mol. Evol.* **1995**, *41*, 693–702. [CrossRef] [PubMed]

27. Keefe, A.D.; Miller, S.L. Potentially prebiotic synthesis of condensed phosphates. *Orig. Life Evol. Biosph.* **1996**, *26*, 15–25. [CrossRef] [PubMed]

28. Deamer, D.W. The First Living Systems: A Bioenergetic Perspective. *Microbiol. Mol. Biol. Rev.* **1997**, *61*, 239–261. [PubMed]

29. Russell, M.J. The Alkaline Solution to the Emergence of Life: Energy, Entropy and Early Evolution. *Acta Biotheor.* **2007**, *55*, 133–179. [CrossRef] [PubMed]

30. Hagan, W.J., Jr.; Parker, A.; Steuerwald, A.; Hathaway, M. Phosphate Solubility and the Cyanate-Mediated Synthesis of Pyrophosphate. *Orig. Life Evol. Biosph.* **2007**, *37*, 113–122. [CrossRef] [PubMed]

31. Szostak, J. The Origin of Life on Earth and the Design of Alternative Life Forms. *Mol. Front. J.* **2017**, *1*, 121–131. [CrossRef]

32. Lipmann, F. Acetyl Phosphate. *Adv. Enzymol. Relat. Areas Mol. Biol.* **1946**, *6*, 231–267.

33. Jencks, W.P. Free energies of hydrolysis and decarboxylation. In *Handbook of Biochemistry and Molecular Biology*, 3rd ed.; Fasman, G.D., Ed.; CRC Press: Cleveland, OH, USA, 1976; Volume 1, pp. 296–304.

34. Di Sabato, G.; Jencks, W.P. Mechanism and catalysis of reactions of acyl phosphates II. Hydrolysis. *J. Am. Chem. Soc.* **1961**, *83*, 4400–4405. [CrossRef]

35. Biron, J.-P.; Pascal, R. Amino acid N-carboxyanhydrides: Activated peptide monomers behaving as phosphate-activating agents in aqueous solution. *J. Am. Chem. Soc.* **2004**, *126*, 9198–9199. [CrossRef] [PubMed]

36. O'Connor, C.J.; Wallace, R.G. A phosphate-catalysed acyl transfer reaction. Hydrolysis of 4-nitrophenyl acetate in phosphate buffers. *Aust. J. Chem.* **1984**, *37*, 2559–2569. [CrossRef]

37. Andrés, G.O.; Granados, A.M.; de Rossi, R.H. Kinetic Study of the Hydrolysis of Phthalic Anhydride and Aryl Hydrogen Phthalates. *J. Org. Chem.* **2001**, *66*, 7653–7657. [CrossRef] [PubMed]

38. El Seoud, O.A.; Ruasse, M.-F.; Rodrigues, W.A. Kinetics and mechanism of phosphate-catalyzed hydrolysis of benzoate esters: Comparison with nucleophilic catalysis by imidazole and o-iodosobenzoate. *J. Chem. Soc. Perkin Trans. 2* **2002**, 1053–1058. [CrossRef]

39. Gill, M.S.; Neverov, A.A.; Brown, R.S. Dissection of nucleophilic and general base roles for the reaction of phosphate with p-nitrophenylthiolacetate, p-nitrophenylthiolformate, phenylthiolacetate. *J. Org. Chem.* **1997**, *62*, 7351–7357. [CrossRef] [PubMed]

40. Thomas, G.L.; Payne, R.J. Phosphate-assisted peptide ligation. *Chem. Commun.* **2009**, 4260–4262. [CrossRef] [PubMed]

41. Higuchi, T.; Flynn, G.L.; Shah, A.C. Reversible Formation and Hydrolysis of Phthaloyl and Succinyl Monophosphates in Aqueous Solution. *J. Am. Chem. Soc.* **1967**, *89*, 616–622. [CrossRef]

42. Hagan, W.J., Jr. Uracil-Catalyzed Synthesis of Acetyl Phosphate: A Photochemical Driver for Protometabolism. *ChemBioChem* **2010**, *11*, 383–387. [CrossRef] [PubMed]

43. Whicher, A.; Camprubi, E.; Pinna, S.; Herschy, B.; Lane, N. Acetyl Phosphate as a Primordial Energy Currency at the Origin of Life. *Orig. Life Evol. Biosph.* **2018**, *48*, 159–179. [CrossRef] [PubMed]

44. Wells, T.N.C.; Ho, C.K.; Fersht, A.R. Free Energy of Hydrolysis of Tyrosyl Adenylate and Its Binding to Wild-Type and Engineered Mutant Tyrosyl-tRNA Synthetases. *Biochemistry* **1986**, *25*, 6603–6608. [CrossRef] [PubMed]

45. Ribas de Pouplana, L.; Schimmel, P. Aminoacyl-tRNA synthetases: Potential markers of genetic code development. *Trends Biochem. Sci.* **2001**, *26*, 591–596. [CrossRef]

46. Pascal, R.; Boiteau, L.; Commeyras, A. From the Prebiotic Synthesis of α-Amino Acids Towards a Primitive Translation Apparatus for the Synthesis of Peptides. *Top. Curr. Chem.* **2005**, *259*, 69–122.

47. Pascal, R.; Boiteau, L. Energetic constraints on prebiotic pathways: Application to the emergence of translation. In *Origin and Evolution of Life: An Astrobiology Perspective*; Gargaud, M., Lopez-Garcia, P., Martin, H., Eds.; Cambridge University Press: Cambridge, UK, 2011; pp. 247–258.

48. Liu, Z.; Beaufils, D.; Rossi, J.-C.; Pascal, R. Evolutionary importance of the intramolecular pathways of hydrolysis of phosphate ester mixed anhydrides with amino acids and peptides. *Sci. Rep.* **2014**, *4*, 7440. [CrossRef] [PubMed]

49. Liu, Z.; Rigger, L.; Rossi, J.-C.; Sutherland, J.D.; Pascal, R. Mixed Anhydride Intermediates in the Reaction of 5(4*H*)-Oxazolones with Phosphate Esters and Nucleotides. *Chem. Eur. J.* **2016**, *22*, 14940–14949. [CrossRef] [PubMed]

50. Biron, J.-P.; Parkes, A.L.; Pascal, R.; Sutherland, J.D. Expeditious, potentially primordial, aminoacylation of nucleotides. *Angew. Chem. Int. Ed. Engl.* **2005**, *44*, 6731–6734. [CrossRef] [PubMed]

51. Leman, L.; Orgel, L.; Ghadiri, M.R. Amino Acid Dependent Formation of Phosphate Anhydrides in Water Mediated by Carbonyl Sulfide. *J. Am. Chem. Soc.* **2006**, *128*, 20–21. [CrossRef] [PubMed]

52. Brack, A. Selective emergence and survival of early polypeptides in water. *Orig. Life* **1987**, *17*, 367–379. [CrossRef]

53. Commeyras, A.; Taillades, J.; Collet, H.; Boiteau, L.; Vandenabeele-Trambouze, O.; Pascal, R.; Rousset, A.; Garrel, L.; Rossi, J.-C.; Biron, J.-P.; et al. Dynamic co-evolution of peptides and chemical energetics, a gateway to the emergence of homochirality and the catalytic activity of peptides. *Orig. Life Evol. Biosph.* **2004**, *34*, 35–55. [CrossRef] [PubMed]

54. Leman, L.; Orgel, L.; Ghadiri, M.R. Carbonyl Sulfide–Mediated Prebiotic Formation of Peptides. *Science* **2004**, *306*, 283–286. [CrossRef] [PubMed]

55. Danger, G.; Boiteau, L.; Cottet, H.; Pascal, R. The peptide formation mediated by cyanate revisited. N-carboxyanhydrides as accessible intermediates in the decomposition of N-carbamoylamino acids. *J. Am. Chem. Soc.* **2006**, *128*, 7412–7413. [PubMed]

56. Wickramasinghe, N.S.M.D.; Staves, M.P.; Lacey, J.C. Stereoselective, nonenzymatic, intramolecular transfer of amino acids. *Biochemistry* **1991**, *30*, 2768–2772. [CrossRef] [PubMed]

57. Paecht-Horowitz, M.; Berger, J.; Katchalsky, A. Prebiotic synthesis of polypeptides by heterogeneous polycondensation of aminoacid adenylates. *Nature* **1970**, *228*, 636–639. [CrossRef]

58. Liu, Z.; Hanson, C.; Ajram, G.; Boiteau, L.; Rossi, J.-C.; Danger, G.; Pascal, R. 5(4*H*)-Oxazolones as Effective Aminoacylation Reagents for the 3′-Terminus of RNA. *Synlett* **2017**, *28*, 73–77.

59. Bowler, F.R.; Chan, C.K.; Duffy, C.D.; Gerland, B.; Islam, S.; Powner, M.W.; Sutherland, J.D.; Xu, J. Prebiotically plausible oligoribonucleotide ligation facilitated by chemoselective acetylation. *Nat. Chem.* **2013**, *5*, 383–389. [CrossRef] [PubMed]

60. Liu, Z.; Ajram, G.; Rossi, J.-C.; Pascal, R. The chemical likelihood of ribonucleotide-α-amino acid copolymers as players for early stages of evolution. *J. Mol. Evol.* **2019**, in press. [CrossRef]

61. Page, M.I.; Jencks, W.P. Entropic contribution to rate accelerations in enzymic and intramolecular reactions and the chelate effect. *Proc. Natl. Acad. Sci. USA* **1971**, *68*, 1678–1683. [CrossRef] [PubMed]

62. De Duve, C. A research proposal on the origin of life. *Orig. Life Evol. Biosph.* **2003**, *33*, 559–574. [CrossRef] [PubMed]

63. Adelfinskaya, O.; Herdewijn, P. Amino Acid Phosphoramidate Nucleotides as Alternative Substrates for HIV-1 Reverse Transcriptase. *Angew. Chem. Int. Ed.* **2007**, *46*, 4356–4358. [CrossRef] [PubMed]

64. Adelfinskaya, O.; Terrazas, M.; Froeyen, M.; Marlière, P.; Nauwelaerts, K.; Herdewijn, P. Polymerase-catalyzed synthesis of DNA from phosphoramidate conjugates of deoxynucleotides and amino acids. *Nucleic Acids Res.* **2007**, *35*, 5060–5072. [CrossRef] [PubMed]

65. Giraut, A.; Herdewijn, P. Influence of the Linkage between Leaving Group and Nucleoside on Substrate Efficiency for Incorporation in DNA Catalyzed by Reverse Transcriptase. *ChemBioChem* **2010**, *11*, 1399–1403. [CrossRef] [PubMed]

66. Song, X.-P.; Bouillon, C.; Lescrinier, E.; Herdewijn, P. Iminodipropionic acid as the leaving group for DNA polymerization by HIV-1 reverse transcriptase. *ChemBioChem* **2011**, *12*, 1868–1880. [CrossRef] [PubMed]

67. Song, X.-P.; Bouillon, C.; Lescrinier, E.; Herdewijn, P. Dipeptides as Leaving Group in the Enzyme-Catalyzed DNA Synthesis. *Chem. Biodivers.* **2012**, *9*, 2685–2700. [CrossRef] [PubMed]

68. Giraut, A.; Abu El-Asrar, R.; Marlière, P.; Delarue, M.; Herdewijn, P. 2′-Deoxyribonucleoside phosphoramidate triphosphate analogues as alternative substrates for E. coli polymerase III. *ChemBioChem* **2012**, *13*, 2439–2444. [CrossRef] [PubMed]

69. Cheng, C.M.; Liu, X.H.; Li, Y.M.; Ma, Y.; Tan, B.; Wan, R.; Zhao, Y.F. N-Phosphoryl amino acids and biomolecular origins. *Orig. Life Evol. Biosph.* **2004**, *34*, 455–464. [CrossRef] [PubMed]

70. Gao, X.; Deng, H.; Tang, G.; Liu, Y.; Xu, P.; Zhao, Y. Intermolecular Phosphoryl Transfer of N-Phosphoryl Amino Acids. *Eur. J. Org. Chem.* **2011**, *2011*, 3220–3228. [CrossRef]

71. Ying, J.; Fu, S.; Li, X.; Fen, L.; Pengxiang, X.; Liu, Y.; Gao, X.; Zhao, Y. A plausible model correlates prebiotic peptide synthesis with the primordial genetic code. *Chem. Commun.* **2018**, *54*, 8598–8601. [CrossRef] [PubMed]

72. Griesser, H.; Tremmel, P.; Kervio, E.; Pfeffer, C.; Steiner, U.E.; Richert, C. Ribonucleotides and RNA Promote Peptide Chain Growth. *Angew. Chem. Int. Ed.* **2017**, *56*, 1219–1223. [CrossRef] [PubMed]

73. Jauker, M.; Griesser, H.; Richert, C. Spontaneous Formation of RNA Strands, Peptidyl RNA, and Cofactors. *Angew. Chem. Int. Ed.* **2015**, *54*, 14564–14569. [CrossRef] [PubMed]

74. Krishnamurthy, R.; Arrhenius, G.; Eschenmoser, A. Formation of glycolaldehyde phosphate from glycolaldehyde in aqueous solution. *Orig. Life Evol. Biosph.* **1999**, *29*, 333–354. [CrossRef] [PubMed]

75. Krishnamurthy, R.; Guntha, S.; Eschenmoser, A. Regioselective α-phosphorylation of aldoses in aqueous solution. *Angew. Chem. Int. Ed.* **2000**, *39*, 2281–2285. [CrossRef]

76. Karki, M.; Gibard, C.; Bhowmik, S.; Krishnamurthy, R. Nitrogenous Derivatives of Phosphorus and the Origins of Life: Plausible Prebiotic Phosphorylating Agents in Water. *Life* **2017**, *7*, 32. [CrossRef] [PubMed]

77. Gibard, C.; Bhowmik, S.; Karki, M.; Kim, E.-K.; Krishnamurthy, R. Phosphorylation, oligomerization and self-assembly in water under potential prebiotic conditions. *Nat. Chem.* **2017**, *10*, 212–217. [CrossRef] [PubMed]

78. Lohrmann, R.; Orgel, L.E. Template-directed synthesis of high molecular weight polynucleotide analogues. *Nature* **1976**, *261*, 342–344. [CrossRef] [PubMed]

79. Zielinski, W.S.; Orgel, L.E. Oligomerization of activated derivatives of 3′-amino-3′-deoxyguanosine on poly(C) and poly(dC) templates. *Nucleic Acids Res.* **1985**, *13*, 2469–2484. [CrossRef] [PubMed]

80. Zielinski, W.S.; Orgel, L.E. Autocatalytic synthesis of a tetranucleotide analogue. *Nature* **1987**, *327*, 346–347. [CrossRef] [PubMed]

81. Zielinski, W.S.; Orgel, L.E. Oligoaminonucleoside phosphoramidates. Oligomerization of dimers of 3′-amino-3′-deoxynudeotides (GC and CG) in aqueous solution. *Nucleic Acids Res.* **1987**, *15*, 1699–1715. [CrossRef] [PubMed]

82. Zielinski, W.S.; Orgel, L.E. Polymerization of a Monomeric Guanosine Derivative in a Hydrogen-Bonded Aggregate. *J. Mol. Evol.* **1989**, *29*, 367–369. [CrossRef] [PubMed]

83. Sievers, D.; von Kiedrowski, G. Self-replication of complementary nucleotide-based oligomers. *Nature* **1994**, *369*, 221–224. [CrossRef] [PubMed]

84. Mansy, S.S.; Schrum, J.P.; Krishnamurthy, M.; Tobé, S.; Treco, D.A.; Szostak, J.W. Template-directed synthesis of a genetic polymer in a model protocell. *Nature* **2008**, *454*, 122–125. [CrossRef] [PubMed]

85. Zhang, S.; Blain, C.; Zielinska, D.; Gryaznov, S.; Szostak, J. Fast and accurate nonenzymatic copying of an RNA-like synthetic genetic polymer. *Proc. Natl. Acad. Sci. USA* **2013**, *110*, 17732–17737. [CrossRef] [PubMed]

86. Shim, J.L.; Lohrmann, R.; Orgel, L.E. Poly(U)-Directed Transamidation between Adenosine 5′-Phosphorimidazolide and 5′ Phosphoadenosine 2′(3′)-Glycine Ester. *J. Am. Chem. Soc.* **1974**, *96*, 5283–5284. [CrossRef] [PubMed]

87. Nicolis, G.; Prigogine, I. *Self-Organization in Nonequilibrium Systems: From Dissipative Structures to Order through Fluctuations*; Wiley: New York, NY, USA, 1977.

88. Eschenmoser, A. Chemistry of potentially prebiological natural products. *Orig. Life Evol. Biosph.* **1994**, *24*, 389–423. [CrossRef]

89. Eschenmoser, A. Question 1: Commentary Referring to the Statement "The Origin of Life can be Traced Back to the Origin of Kinetic Control" and the Question "Do You Agree with this Statement; and How Would You Envisage the Prebiotic Evolutionary Bridge Between Thermodynamic and Kinetic Control?" Stated in Section 1.1. *Orig. Life Evol. Biosph.* **2007**, *37*, 309–314. [PubMed]

90. Danger, G.; Plasson, R.; Pascal, R. Pathways for the formation and evolution of peptides in prebiotic environments. *Chem. Soc. Rev.* **2012**, *41*, 5416–5429. [CrossRef] [PubMed]

91. Boiteau, L.; Pascal, R. Energy sources, self-organization, and the origin of life. *Orig. Life Evol. Biosph.* **2011**, *41*, 23–33. [CrossRef] [PubMed]

92. Schwartz, A.W. Phosphorus in prebiotic chemistry. *Philos. Trans. R. Soc. Lond. Ser. B* **2006**, *361*, 1743–1749. [CrossRef] [PubMed]

93. Fernández-García, C.; Coggins, A.J.; Powner, M.W. A Chemist's Perspective on the Role of Phosphorus at the Origins of Life. *Life* **2017**, *7*, 31. [CrossRef] [PubMed]

94. Tordini, F.; Bencini, A.; Bruschi, M.; De Gioia, L.; Zampella, G.; Fantucci, P. Theoretical Study of Hydration of Cyanamide and Carbodiimide. *J. Phys. Chem. A* **2003**, *107*, 1188–1196. [CrossRef]

95. Steinman, G.; Lemmon, R.M.; Calvin, M. Cyanamide: A possible key compound in chemical evolution. *Proc. Natl. Acad. Sci. USA* **1964**, *52*, 27–30. [CrossRef] [PubMed]

96. Jones, M.E.; Spector, L.; Lipmann, F. Carbamyl phosphate, the carbamyl donor in enzymatic citrulline synthesis. *J. Am. Chem. Soc.* **1955**, *77*, 819–820. [CrossRef]

97. Jones, M.E.; Lipmann, F. Chemical and enzymatic synthesis of carbamyl phosphate. *Proc. Natl. Acad. Sci. USA* **1960**, *46*, 1194–1205. [CrossRef] [PubMed]

98. Halmann, M.; Lapidot, A.; Samuel, D. Kinetic and tracer studies of the reaction of carbamoyl phosphate in aqueous solution. *J. Chem. Soc.* **1962**, 1944–1957. [CrossRef]

99. Allen, C.M., Jr.; Jones, M.E. Decomposition of Carbamylphosphate in Aqueous Solutions. *Biochemistry* **1964**, *3*, 1238–1247. [CrossRef] [PubMed]

100. Vogels, G.D.; Uffink, L.; Van der Drift, C. Cyanate decomposition catalyzed by certain divalent anions. *Recl. Trav. Chim. Pays-Bas* **1970**, *89*, 500–508. [CrossRef]

101. Danger, G.; Charlot, S.; Boiteau, L.; Pascal, R. Activation of carboxyl group with cyanate: Peptide bond formation from dicarboxylic acids. *Amino Acids* **2012**, *42*, 2331–2341. [CrossRef] [PubMed]

102. Saygin, Ö. Non-enzymatic photophosphorylation with visible-light—A possible mode of prebiotic ATP. *Naturwissenschaften* **1981**, *68*, 617–619. [CrossRef]

103. Saygin, Ö. Non-enzymatic phosphorylation of acetate by carbamyl-phosphate—A model reaction for prebiotic activation of carboxyl groups. *Orig. Life* **1983**, *13*, 43–48. [CrossRef]

104. Saygin, Ö. Photochemical carbamylphosphate formation and metal ion-catalysed transphosphorylations between carbamylphosphate and adenine nucleotides or carboxyl groups. *Orig. Life* **1984**, *14*, 131–137. [CrossRef]

105. Hosseini, M.W.; Lehn, J.M. Supramolecular catalysis of phosphoryl transfer: Pyrophosphate synthesis from acetyl phosphate mediated by macrocyclic polyamines. *J. Am. Chem. Soc.* **1987**, *109*, 7047–7058. [CrossRef]

106. Hosseini, M.W.; Lehn, J.M. Supramolecular catalysis of adenosine triphosphate synthesis in aqueous solution mediated by a macrocyclic polyamine and divalent metal cations. *J. Chem. Soc. Chem. Commun.* **1991**, 451–453. [CrossRef]

107. Costanzo, G.; Saladino, R.; Crestini, C.; Ciciriello, F.; Di Mauro, E. Nucleoside Phosphorylation by Phosphate Minerals. *J. Biol. Chem.* **2007**, *282*, 16729–16735. [CrossRef] [PubMed]

108. Lohrmann, R.; Orgel, L.E. Urea-Inorganic Phosphate Mixtures as Prebiotic Phosphorylating Agents. *Science* **1971**, *171*, 490–494. [CrossRef] [PubMed]

109. Reimann, R.; Zubay, G. Nucleoside phosphorylation: A feasible step in the prebiotic pathway to RNA. *Orig. Life Evol. Biosph.* **1999**, *29*, 229–247. [CrossRef] [PubMed]

110. Albertsen, A.; Duffy, C.; Sutherland, J.; Monnard, P. Self-Assembly of Phosphate Amphiphiles in Mixtures of Prebiotically Plausible Surfactants. *Astrobiology* **2014**, *14*, 462–472. [CrossRef] [PubMed]

111. Orgel, L.E.; Lohrmann, R. Prebiotic Chemistry and Nucleic Acid replication. *Acc. Chem. Res.* **1974**, *7*, 368–377. [CrossRef]

112. Osterberg, R.; Orgel, L.E. Polyphosphate and Trimetaphosphate Formation under Potentially Prebiotic Conditions. *J. Mol. Evol.* **1972**, *1*, 241–248. [CrossRef] [PubMed]

113. Fiore, M.; Madanamoothoo, W.; Berlioz-Barbier, A.; Maniti, O.; Girard-Egrot, A.; Buchet, R.; Strazewski, P. Giant vesicles from rehydrated crude mixtures containing unexpected mixtures of amphiphiles formed under plausibly prebiotic conditions. *Org. Biomol. Chem.* **2017**, *15*, 4231–4240. [CrossRef] [PubMed]

114. Shaw, W.H.R.; Bordeaux, J.J. The Decomposition of Urea in Aqueous Media. *J. Am. Chem. Soc.* **1955**, *77*, 4729–4733. [CrossRef]

115. Orgel, L.E. Prebiotic chemistry and the origin of the RNA world. *Crit. Rev. Biochem. Mol. Biol.* **2004**, *39*, 99–123. [PubMed]
116. Fahrenbach, A.C.; Giurgiu, C.; Tam, C.P.; Li, L.; Hongo, Y.; Aono, M.; Szostak, J.W. Common and Potentially Prebiotic Origin for Precursors of Nucleotide Synthesis and Activation. *J. Am. Chem. Soc.* **2017**, *139*, 8780–8783. [CrossRef] [PubMed]

Article

Phosphates as Energy Sources to Expand Metabolic Networks

Tian Tian [1], Xin-Yi Chu [1], Yi Yang [1], Xuan Zhang [2], Ye-Mao Liu [1], Jun Gao [1], Bin-Guang Ma [1] and Hong-Yu Zhang [1,*]

[1] Hubei Key Laboratory of Agricultural Bioinformatics, College of Informatics, Huazhong Agricultural University, Wuhan 430070, China; ttlym1989@163.com (T.T.); chuxy@webmail.hzau.edu.cn (X.-Y.C.); yyphoenix@163.com (Y.Y.); lym25517235@163.com (Y.-M.L.); gaojun@mail.hzau.edu.cn (J.G.); mbg@mail.hzau.edu.cn (B.-G.M.)

[2] Beijing National Center for Molecular Sciences, Institute of Theoretical and Computational Chemistry, College of Chemistry and Molecular Engineering, Peking University, Beijing 100871, China.; xuanz@pku.edu.cn

* Correspondence: zhy630@mail.hzau.edu.cn; Tel.: +86-27-87285085

Received: 18 April 2019; Accepted: 21 May 2019; Published: 22 May 2019

Abstract: Phosphates are essential for modern metabolisms. A recent study reported a phosphate-free metabolic network and suggested that thioesters, rather than phosphates, could alleviate thermodynamic bottlenecks of network expansion. As a result, it was considered that a phosphorus-independent metabolism could exist before the phosphate-based genetic coding system. To explore the origin of phosphorus-dependent metabolism, the present study constructs a protometabolic network that contains phosphates prebiotically available using computational systems biology approaches. It is found that some primitive phosphorylated intermediates could greatly alleviate thermodynamic bottlenecks of network expansion. Moreover, the phosphorus-dependent metabolic network exhibits several ancient features. Taken together, it is concluded that phosphates played a role as important as that of thioesters during the origin and evolution of metabolism. Both phosphorus and sulfur are speculated to be critical to the origin of life.

Keywords: origin of life; metabolism; phosphates; network expansion simulation; thermodynamic bottleneck; molecular clocks

---

1. Introduction

Phosphates are basic components of many biomolecules and essential for modern biochemical reactions, but it is still not clear how phosphates play the critical role in metabolism in the origin of life. Phosphate minerals existing on the early Earth or in meteorites are thought to be the main sources of prebiotic phosphorus [1–3]. However, most of these phosphates are either insoluble in water or have low reactivity and thus they are considered to be problematic for primordial biological use [1].

To solve this 'phosphorus problem', Goldford and coworkers proposed a phosphorus-independent scenario for the emergence of protometabolism [4]. They constructed a phosphorus-independent protometabolism network starting from a set of prebiotically abundant compounds excluding phosphates. The obtained metabolic network contained various important metabolites and metabolic reactions, and exhibited the features of an ancient origin. Then, the researchers found that sulfur compounds (i.e., pantetheine) could alleviate the thermodynamic bottlenecks of the network expansion while phosphates (i.e., pyrophosphate or acetyl-phosphate) could not. Based on these findings, Goldford et al. proposed that a phosphorus-independent metabolism could exist before the emergence of the phosphate-based genetic coding system. However, this phosphorus-independent network could not produce nucleobases or ribose, which means that this network is unlikely a possible source of RNA.

What is more, phosphorus was thought to play a crucial role both in the prebiotic synthesis of important precursors of RNA and proteins [5–7] and in primordial energy metabolism [8–10]. These viewpoints support the importance of phosphorus in the origin of life. Moreover, the 'phosphorus problem' itself might have been solved by recent findings. Phosphite, a kind of water-soluble, reactive reduced-state phosphorus, was recently proposed to be an available prebiotic phosphorus source [10]. It has been proven that phosphite can be produced from the extraterrestrial phosphide mineral-schreibersite and is present in early Archean marine carbonates at significant levels [11,12]. This reduced-state phosphorus could generate orthophosphate, pyrophosphate and trimetaphosphate in plausible prebiotic environments [13]. With the presence of trimetaphosphate, diamidophosphate (DAP) could have formed and enable the phosphorylation and oligomerization of biologically meaningful molecules [14].

Goldford et al.'s new opinion stimulated our interest in the systematical exploration of the role that phosphorus played in the origin of metabolism. We noticed that Goldford et al. considered mere pyrophosphate and acetyl-phosphate but ignored other forms of primitive phosphates when investigating the thermodynamic bottlenecks of network expansion. This work started with alternative phosphates and adopted the network expansion algorithm [4,15–17] to simulate the expansion of metabolic networks. Then, the feasibility of this metabolic network on the early Earth was explored. Finally, biological features of this network were fully analyzed. This study reveals that: (i) some phosphorylated intermediates could efficiently alleviate thermodynamic bottlenecks; (ii) phosphorous-dependent metabolic network exhibits ancient biological features.

2. Materials and Methods

2.1. Data Sources

All KEGG reactions (i.e. chemical reactions in the KEGG database), compounds, the enzymes of reactions were downloaded from the KEGG database [18] (Release: 84.0, October 1, 2017). The putative LUCA genes (i.e. genes of the last universal common ancestor) were downloaded from the LUCApedia webpage [19]. Cofactor and PDB structures of the enzymes were downloaded from the Uniport database (Release: 2017_09). The folds and fold families of enzymes were downloaded from the SCOP database [20]. The architectures of enzymes were downloaded from the CATH database [21] (version 4.2).

2.2. Reconstruction of the Background Metabolism Pool

KEGG Reactions that met the following conditions were removed: (i) reactions in which the molecular formulas of compounds were undefined; (ii) reactions that contained n-subunit polymers; (iii) reactions that contained the metabolites with "R" groups; (iv) reactions that were elementally imbalanced with exception of hydrogen. A stoichiometric matrix was then constructed using reaction equations. The final background metabolism pool consisted of 7376 reactions.

2.3. Network Expansion Simulation

The phosphorus-independent and -dependent metabolic networks were constructed by a network expansion algorithm which was first proposed by Ebenhöh et al. [15,16] and was described in detail in previous studies [4,15–17].

A seed set, namely, initial metabolite set, M_0 was first defined for the network expansion algorithm. Then, initial reaction set R_0 was constructed by identifying those reactions whose reactants were all included in M_0. Reaction products M_p of these reactions were further added into M_0 and subsequently a new metabolite set M ($M = M_0 \cup M_p$) was constructed. The new reactions R_n whose reactants were present in M were identified and added into the reaction set R ($R = R_0 \cup R_n$). At each iteration k, the reaction set R and metabolite set M were updated with new added reactions and the products of these reactions ($R = R \cup R_n$, $M = M \cup M_n$). This process was terminated when no new reactions and products could be added into M and R.

The network expansion algorithm with different seed sets were executed. Every time, the initial seed set contained: (i) the possible abundant gases on the early Earth (dinitrogen, vapor water, hydrogen sulfide, and carbon dioxide); (ii) a possible prebiotic nitrogen source, i.e., ammonia; (iii) possible prebiotic carbon sources (acetate and formate); (iv) a possible prebiotic phosphorus source (i.e., orthophosphate, pyrophosphate and trimetaphosphate) (only for phosphorus-dependent metabolic network expansion simulation). The identification of the seed set was based on previous studies [4,9,13,22–25].

All reactions in this study came from the background metabolism pool. In our study, the reactions containing molecular oxygen were removed during the network expansion due to the anaerobic environment of the early Earth [17,26].

2.4. Thermodynamically Constrained Network Expansion Simulation

The thermodynamically constrained network expansion simulation was based on the network expansion algorithm. In the simulation, the cutoff value τ was set. The endergonic reactions in which the required free energies were above τ were removed. In other words, reactions with $\Delta G_r^0 > \tau$ were removed. The ΔG_r^0 of the reactions was estimated using eQuilibrator [27,28]. It should be noted that there was a lack of free energy estimation in more than one third of KEGG reactions. Reactions with unknown free energies were assumed to be either all available or all unavailable. The results were very similar between these two treatments. In this paper, all reactions with unknown free energies were assumed as available. The rest steps in simulation were executed as they were in the network expansion algorithm.

To determine the potential thermodynamic bottlenecks of the network, the thermodynamically constrained network expansion algorithm was executed by adopting the same seed set used to construct the protometabolic network. To explore the influence of the phosphorylated intermediates from glycolysis, one of the prebiotically available intermediates (i.e., glucose 6-phosphate, glyceraldehyde 3-phosphate, glycerate 2-phosphate, glycerate 3-phosphate, and phosphoenolpyruvate) [29,30] was added into the seed set every time, and then the thermodynamically constrained network expansion algorithm was executed.

2.5. Scale-Limiting Reaction Detection

Reactions meeting the following conditions were defined as scale-limiting reactions: (i) the reactions triggering the dramatic expansion of networks; (ii) reactions whose removal will severely limit the expansion of networks.

In order to identify the network-scale-limiting reactions, the following steps were executed:

(1). Obtaining the potential reactions: Reaction set R_1 and the corresponding metabolite set M_1 at the thermodynamic threshold τ_1 limiting the network expansion and reaction set R_2 and corresponding metabolite set M_2 at τ_2 ($\tau_2 = \tau_1 + 1$) were obtained

(2). Identifying the reactions triggering the dramatic expansion of networks: At first, a metabolite m derived from difference set (M_3) of M_2 and M_1 ($M_3 = M_2 - M_1$) was added into the metabolite set M_1 to form a new metabolite set M_4 ($M_4 = M_1 + m$). Then, new networks at τ_1 were constructed with M_4. Once the network expands dramatically, the corresponding reactions of m were considered to cause the dramatic expansion of networks. These steps were repeated for every metabolite in M_3, and all of the expansion-triggering reactions were identified as the potential scale-limiting reactions.

(3). Identifying the scale-limiting reactions: Since there may be more than one reaction limiting the dramatic expansion of the network, all combinations of potential scale-limiting reactions should be tested. Every combination of potential scale-limiting reactions was blocked in turn. The networks at the thermodynamic constraint τ_2 were constructed based on these abridged reaction sets. When dramatic network expansion was no longer observed, the reaction combinations were considered as scale-limiting ones. Thus, the final scale-limiting reactions were identified by analyzing all the combinations.

2.6. Protein Domain Age Estimation

In this study, the protein domains are classified according to SCOP and CATH protein structure classification schemes. The node distance (*nd*) values based on SCOP fold family (FF) were the united set of the *nd* values derived from several previous studies [31–34], the *nd* values based on CATH architecture (A) were derived from Bukhari et al.'s work [35].

The previous studies reported that the *nd* values of structural domains were closely related to their geological ages. Based on this finding, several molecular clocks were constructed [31–34]. In this study, the geological ages of proteins were defined according to the united set of those molecular clocks.

3. Results

3.1. Construction of Phosphorus-Dependent Metabolic Network

First, we attempted to evaluate the reliability of the network expansion simulation. First, we reconstructed the background metabolism pool with reactions and compounds from the updated version of the KEGG database [18]. The final updated background metabolism pool contained 7376 reactions (full-balanced network, Table S1), and included 496 more reactions than the pool constructed by Goldford and coworkers [4]. Then, a phosphorus-independent metabolic network was reconstructed with updated background reaction data started with a pre-defined seed metabolite set. This seed set was the same as that defined by Goldford et al. and was composed of a set of prebiotically abundant compounds excluding phosphates. The final phosphorus-independent network included 329 reactions and 266 metabolites, containing almost all of the reactions of the network constructed by Goldford et al. (Table S2). This result validated the reliability of network expansion simulation.

The phosphorus-dependent metabolic network was constructed with the same method except that prebiotic phosphates were added into the seed set. Due to the lack of KEGG reaction data of phosphite, we did not directly introduce phosphite as a phosphorus source. Phosphates we introduced here are orthophosphate, pyrophosphate and trimetaphosphate. These phosphates were widely thought to be present on the early Earth and could be prebiotically synthesized by phosphite [13,22–25]. Each time we introduced one of these phosphates and performed the network expansion simulation. The obtained phosphorus-dependent metabolic networks were composed of the same reactions (596 reactions) and metabolites (471 metabolites) (Figure 1 and Table S3), implying that the network was robust to different phosphorus sources.

The reactions and metabolites increased obviously in the phosphorus-dependent network. Although nucleobases were not found in the new-produced metabolites, ribose, which is also an essential component of RNA, was indeed produced in the phosphorus-dependent metabolic network, indicating the importance of phosphorus in the evolution of RNA synthesis.

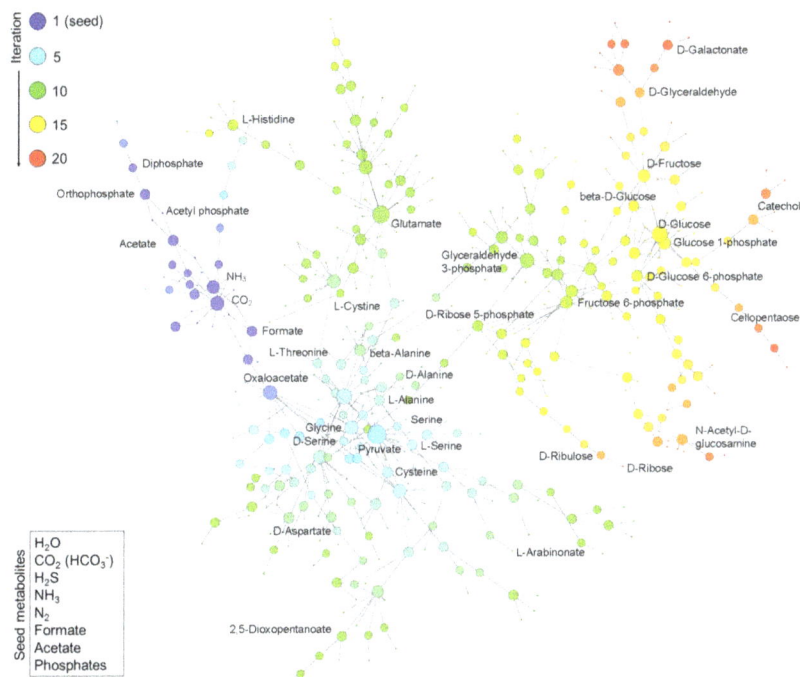

Figure 1. Construction of Phosphorus-Dependent Metabolic Network. Network expansion simulation was executed using a set of defined seed compounds (bottom left box) and all balanced reactions in the background metabolism pool derived from the updated KEGG reactions. The figure displays the obtained phosphorus-dependent network in which metabolites are linked if they have a reactant-product relationship during the expansion. The metabolites generated at different iteration steps during the network expansion process are represented by nodes in different colors. The size of node represents the degree of the node, i.e., the number of reactions added in the subsequent iteration.

3.2. Thermodynamic Bottleneck Alleviation by Primitive Phosphates

Thermodynamic constraints could limit the expansion of the metabolic network [36]. Phosphates play important roles in driving energetically uphill reactions. However, Goldford et al. claimed that phosphates such as pyrophosphate and acetyl-phosphate could not alleviate thermodynamic bottlenecks, while pantetheine could [4]. Their statement was supported by the updated simulation on the phosphorous-independent network (Figure 2A, Table S4). Nevertheless, we are still wondering whether there exist any other forms of primitive phosphates that could serve as alternative alleviators for the thermodynamic bottleneck?

Glycolysis-like reactions could spontaneously occur in a plausible ancient marine environment [37]. Many phosphorous intermediates of glycolysis, including glucose 6-phosphate, glyceraldehyde 3-phosphate, glycerate 2-phosphate, glycerate 3-phosphate, and phosphoenolpyruvate, were speculated to be prebiotically synthesized [29,30]. All of these phosphorous intermediates are present in the phosphorus-dependent metabolic network. Thus, we attempt to explore whether the glycolysis-generated metabolites can alleviate thermodynamic bottlenecks and can promote the expansion of the early metabolic network.

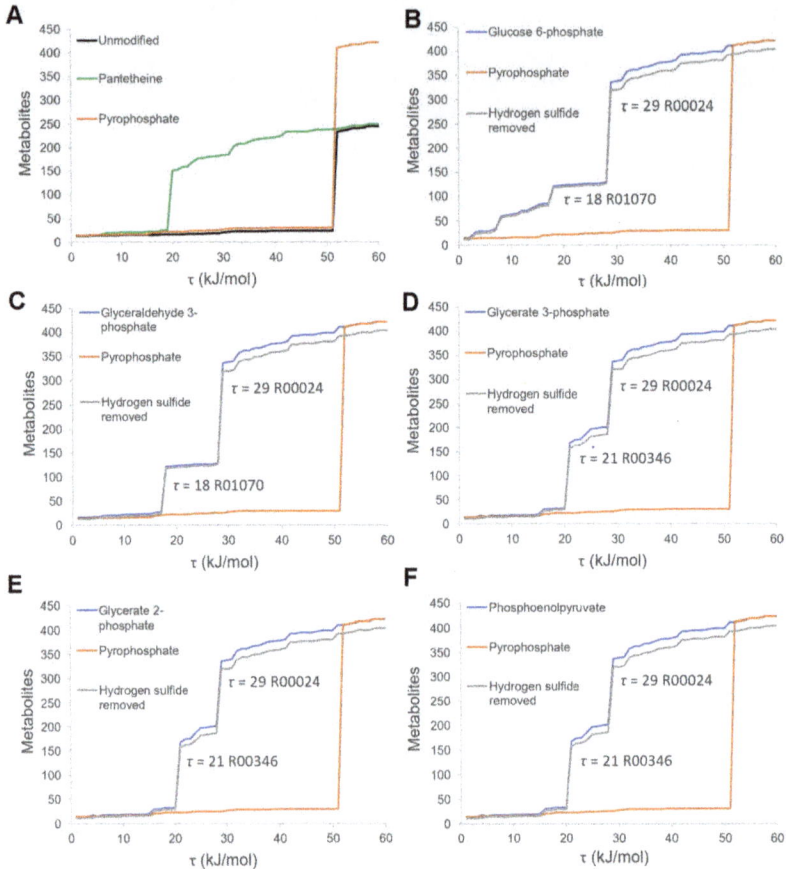

Figure 2. Thermodynamically Constrained Network Expansion. Thermodynamically constrained network expansion was simulated by using different seed sets. Endergonic reactions with ΔG_r^0 exceeding a thermodynamic threshold τ were defined as impossible. For each value of τ (x axis), we plotted the size of the final expanded network in terms of the number of metabolites (y axis). (**A**) displays the comparison of the network sizes of the unmodified phosphorus-dependent network (black line), the pyrophosphate-coupled network (with the addition of pyrophosphate in the seed set) (orange line), and the pantetheine-coupled network (with the addition of pantetheine in the seed set) (green line) at different thermodynamic thresholds, τ. (**B–F**) display the comparison of the network sizes of the pyrophosphate-coupled network (orange line), the phosphorylated intermediates-coupled network (blue line), and the phosphorylated intermediates-coupled network without hydrogen sulfide (gray line) at different thermodynamic thresholds, τ. The thermodynamic bottlenecks and the reactions limiting the scale of different phosphorylated intermediates-coupled networks are shown in corresponding figures. The used phosphorylated intermediates include: glucose 6-phosphate (**B**), glyceraldehyde 3-phosphate (**C**), glycerate 2-phosphate (**D**), glycerate 3-phosphate (**E**), phosphoenolpyruvate (**F**).

The expansion of phosphorus-dependent metabolic network was re-simulated under the thermodynamic constraints. During the simulation, endergonic reactions in which the required free energies were above a cutoff value, τ, were blocked during the expansion of the network. When τ was below 51 kJ/mol, the scale of the network was strictly limited with reactions and metabolites limited to <26 and <30, respectively (Table S4). When τ exceeded this threshold, the network expanded dramatically (Figure 2A, Table S4). It seemed impossible for early metabolism to overcome the energetic constraint of 51 kJ/mol because endergonic reactions with ΔG_r^0 (standard transformed Gibbs energies) above 30 kJ/mol needed to be activated by exergonic reaction like ATP hydrolysis [36]. However, this kind of exergonic reaction might be unavailable in the primitive world [36].

Then, the glycolysis-derived phosphorylated intermediates were introduced into the network during the thermodynamically constrained network expansion. With the addition of phosphorylated intermediates, the bottlenecks limiting the network expansion were reduced to below 30 kJ/mol (Figure 2B–F, Table S4), implying the expansion of these networks is thermodynamically feasible without other energy sources [36]. At the end of each simulation, the thermodynamically constrained networks contained at least 338 metabolites and 413 reactions.

To exclude the influence of sulfur, we removed hydrogen sulfide from the seed set and found that its removal had little impact on the thermodynamically constrained network expansion (Figure 2B–F, Table S4), suggesting that sulfur made no significant contribution to alleviating the thermodynamic bottlenecks of the phosphorus-dependent network expansion.

The reactions involved in the dramatic expansion of the metabolic networks were also investigated. The dramatic expansions of the networks disappeared when certain reactions (i.e., R00024, R01070 and R00346) were blocked, indicating that these reactions played a critical role in limiting the expansion of the networks (Figure 2 and Table S5). Then, the feasibility of these scale-limiting reactions at the early stage of evolutionary history of metabolism was explored. Reversible reaction R00024 was observed to be the most common scale-limiting reaction in five thermodynamically constrained networks with different phosphorous intermediates. In R00024, glycerate 3-phosphate and ribulose 1,5-bisphosphate were key metabolites. This reaction is catalyzed by RubisCO (D-ribulose 1,5-bisphosphate carboxylase/oxygenase, EC: 4.1.1.39), which was assumed to originate 3.5 Gy ago [38,39]. Besides, it was reported that RubisCO catalyzed this reaction by offering COO⁻, and H⁺ [40,41]. All these ions could exist in the primitive Earth environment, which suggested that reaction R00024 might occur before RubisCO appeared.

Reaction R01070 is catalyzed by beta-D-fructose-1,6-bisphosphate D-glyceraldehyde-3-phosphate-lyase (EC: 4.1.2.13) [42]. Inferred by protein-structure-based molecule clocks [31–34], this enzyme catalyzing appeared earlier than 3 Gy. However, the feasibility of R00346 in ancient time remains unknown due to the lack of knowledge of its structure of catalyzing enzyme (oxaloacetate carboxy-lyase, EC: 4.1.1.38). Taken together, it can be concluded that phosphorylated intermediates could have alleviated the thermodynamic bottlenecks of metabolic network expansion, at least the expansion constrained by R00024 and R01070.

3.3. Ancient Origin of Phosphorus-Dependent Metabolic Network

The ancient origin of the phosphorus-dependent metabolic network was evaluated by the biological characteristics of the network. We analyzed enzymes of the phosphorus-dependent network to explore the potential biological features associated with ancient metabolism. The phosphorus-dependent network was found to be enriched with the enzymes, orthologs and protein fold families of LUCA ($p < 10^{-2}$, Fisher's exact test, Figure 3A). This result implied that a great portion of the reactions in the phosphorus-dependent network existed in the early life. This network was also enriched with the enzymes which contained metal cofactors Mg^{2+}, Zn^{2+} and FeS ($p < 0.05$, Fisher's exact test, Figure 3B). The relative higher requirement for metal ions might be a remnant of prebiotic catalysts, because existent enzymes may still retain characters of prebiotic catalysts, such as usage of metal cofactors [43,44]. Moreover, pyridoxal phosphate was considered to play a critical role in prebiotic transamination [45]. The enzymes using pyridoxal phosphate as a cofactor were also enriched in the network ($p < 10^{-5}$, Fisher's exact test, Figure 3B).

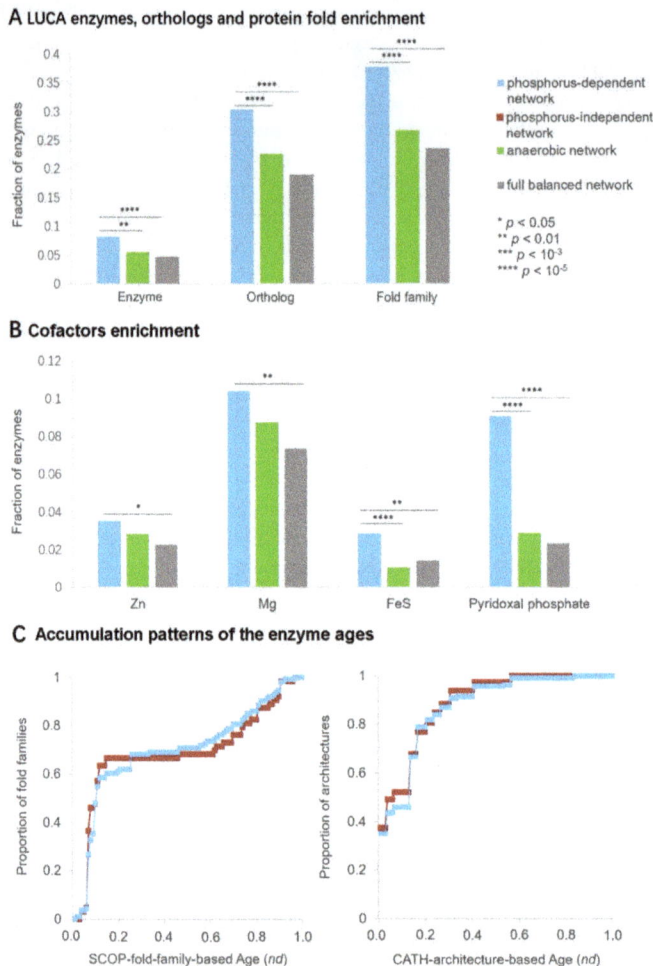

Figure 3. Biological Characteristics of Phosphorus-Dependent Metabolic Network. (**A**) The phosphorus-dependent network is enriched with enzymes, ortholog genes, and protein fold families that are thought to be present in LUCA, relative to all metabolic reactions in background metabolism pool (full-balanced network) or to the oxygen-independent (anaerobic) portion of the full network. (**B**) The phosphorus-dependent network is enriched with metal cofactors (Zn^{2+}, Mg^{2+} and FeS) and pyridoxal phosphate, relative to all metabolic reactions in background metabolism pool or to the oxygen-independent portion of the full network. (**C**) The accumulation patterns of the enzyme ages in two networks show no significant difference. All of these results show the ancient biological characteristics of phosphorus-dependent metabolic network, suggesting that both phosphorus-dependent network and phosphorus-independent network are at least as ancient as LUCA. The significance was analyzed by Fisher's exact test or Kolmogorov–Smirnov test: * $p < 0.05$; ** $p < 0.01$; *** $p < 10^{-3}$; **** $p < 10^{-5}$.

Besides, the structure of proteins was rather conserved during evolution and could serves as molecular fossils in the study of the early history of biochemistry evolution [31]. In previous research, molecular clocks based on different protein structure classification schemes (i.e., SCOP and CATH) were established and the relative ages of protein domains were characterized by node distances (*nd*) [31–35]. Node distance is the distance from the position of a taxon of protein domain structures on the phylogenetic tree to the root node, with the scale from 0 (most ancient) to 1 (most recent). It has been shown that *nd* values of protein domain structures correlate strongly with their geological times. In this study, the ages of enzymes in the phosphorus-dependent and -independent networks were inferred using these molecular clocks at both fold family level (SCOP classification) [31–34] and architecture level [35]. The accumulation patterns of the enzyme ages in two networks exhibited no significant difference (Kolmogorov–Smirnov test, $p > 0.05$, Figure 3C), suggesting that enzymes in phosphorus-independent network are not more ancient than phosphorus-dependent counterpart.

Taken together, the above results indicated that both phosphorus-dependent and phosphorus-independent networks are at least as ancient as LUCA. Besides, the phosphorus-dependent network retains a higher requirement for metal ions and pyridoxal phosphate, which might be remnants of prebiotic chemistry.

4. Discussion

It is undeniable that phosphorus is essential for modern metabolism from both material and energy perspectives. Phosphates are basic components of important biomolecules and play an important role in energy transduction, signal transmission and redox catalysis. Considering its critical role in metabolism, phosphorus is thought to make great contributions to the origin of life. To examine the role of phosphorus in the origin of metabolism, we constructed a metabolic network using a network expansion algorithm. The phosphorus-dependent metabolic network contains much more metabolites than the phosphorus-independent counterpart. Among the phosphorus-dependent network, ribose is produced, which is an essential component of RNA, indicating the significance of phosphorus for the primordial synthesis of RNA.

To explore the influence of phosphorus on the thermodynamic feasibility of ancient metabolic system, the thermodynamically constrained network expansion with various forms of phosphates was simulated. This study found that some phosphorous intermediates of the glycolytic pathway could dramatically alleviate the thermodynamic bottlenecks and promote the expansion of the network. Further study of scale-limiting reactions (i.e., R00024 and R01070) during the thermodynamically constrained network expansion showed that the expansion of ancient metabolic network might be feasible with the presence of phosphorous intermediates such as glucose 6-phosphate and glyceraldehyde 3-phosphate.

The biological characteristics of phosphorus-dependent network were comprehensively analyzed, and results showed that this network exhibits several ancient features. The enzymes in this network were enriched with LUCA elements and metal-based cofactors which were considered to be used in original biochemical reactions [4,44]. The ages of enzymes in phosphorus-dependent and -independent networks exhibited similar accumulation patterns. These results indicated that both phosphorus-dependent and phosphorus-independent networks are at least as ancient as LUCA. Moreover, phosphorus-dependent network exhibits some more 'primitive' traits, such as retaining a relative higher requirement for metal ions and pyridoxal phosphate.

In summary, our research demonstrates that (i) some high-energy phosphates can ensure the primitive metabolism under feasible energetic constraints; (ii) the phosphorous-dependent metabolism might originate in the very early stage of biochemical processes.

Therefore, it can be speculated that phosphates are as important as thioesters for the origin and evolution of metabolism. Both phosphorus and sulfur are critical to the origin of life on Earth. This has meaningful implications for extraterrestrial life detection. Recently, Enceladus, a satellite of Saturn, was reported to have a global liquid water ocean and the jets from this ocean contain simple organic chemicals, suggesting that Enceladus provides some basic conditions to fulfill the existence of life [46]. However, phosphorus and sulfur have not yet been detected in the ocean jets of Enceladus [46], that casts a shadow over the existence of Enceladus life.

It should bear in mind that this work is based on KEGG reactions. The premise of KEGG reactions is that there must be a cellular environment. Thus, the prebiotic reactions which might have been replaced during the evolution of life cannot be included in the current networks. As a consequence, the conclusion of this study may have some limitations. Besides, the phosphorus-dependent network does not produce nucleobases, which implied that there still is a gap to evolve RNA. Finally, why phosphorous intermediates of glycolysis mechanism could alleviate the thermodynamic bottlenecks remains to be elucidated, in particular considering the fact that addition of these triose phosphates may cause complex changes of metabolism.

Supplementary Materials: The following are available online at http://www.mdpi.com/2075-1729/9/2/43/s1, Table S1: All balanced reactions and metabolites, Table S2: Phosphorus-independent network, Table S3: Phosphorus-dependent network, Table S4: Number of reactions under different thresholds, Table S5: Key reactions in the network expansion.

Author Contributions: Conceptualization, H.-Y.Z.; resources, T.T.; methodology, T.T., X.-Y.C., Y.Y., X.Z., Y.-M.L. and J.G.; validation, T.T., X.-Y.C. and X.Z.; writing—original draft, T.T., X.-Y.C. and Y.-M.L.; Writing—review and editing, J.G., B.-G.M. and H.-Y.Z.; project administration, H.-Y.Z.; data curation, T.T.; funding acquisition, B.-G.M. and H.-Y.Z.

Funding: This work has been supported by the National Natural Science Foundation of China (31870837, 31670779 and 31570844), and the Fundamental Research Funds for the Central Universities (2662016PY094).

Acknowledgments: We are grateful for the help from the Daniel Segrè group, who generously provided the code for network construction.

Conflicts of Interest: The authors declare no conflict of interest.

References

1. Schwartz, A.W. Phosphorus in prebiotic chemistry. *Philos. Trans. R. Soc. B* **2006**, *361*, 1743–1749. [CrossRef]
2. Gedulin, B.; Arrhenius, G. Sources and geochemical evolution of RNA precursor molecules-the role of phosphate. *Early Life Earth* **1994**, *84*, 91–110.
3. Handschuh, G.J.; Orgel, L.E. Struvite and prebiotic phosphorylation. *Science* **1973**, *179*, 483–484. [CrossRef]
4. Goldford, J.E.; Hartman, H.; Smith, T.F.; Segrè, D. Remnants of an ancient metabolism without phosphate. *Cell* **2017**, *168*, 1126–1134. [CrossRef]
5. Islam, S.; Bučar, D.K.; Powner, M.W. Prebiotic selection and assembly of proteinogenic amino acids and natural nucleotides from complex mixtures. *Nat. Chem.* **2017**, *9*, 584–589. [CrossRef]
6. Zhang, W.; Tam, C.P.; Walton, T.; Fahrenbach, A.C.; Birrane, G.; Szostak, J.W. Insight into the mechanism of nonenzymatic RNA primer extension from the structure of an RNA-GpppG complex. *Proc. Natl. Acad. Sci. USA* **2017**, *114*, 7659–7664. [CrossRef] [PubMed]
7. Gao, X.; Liu, Y.; Xu, P.X.; Cai, M.Y.; Zhao, Y.F. Alpha-amino acid behaves differently from beta- or gamma-amino acids as treated by trimetaphosphate. *Amino Acids* **2008**, *34*, 47–53. [CrossRef]
8. Pascal, R.; Poitevin, F.; Boiteau, L. Energy sources for prebiotic chemistry and early life: Constraints and availability. *Orig. Life Evol. Biosph.* **2009**, *39*, 260–261.
9. Martin, W.; Russell, M.J. On the origin of biochemistry at an alkaline hydrothermal vent. *Philos. Trans. R. Soc. B* **2007**, *362*, 1887–1925. [CrossRef]
10. Piast, R.W.; Wieczorek, R.M. Origin of life and the phosphate transfer catalyst. *Astrobiology* **2017**, *17*, 277–285. [CrossRef] [PubMed]
11. Pasek, M.A. Schreibersite on the early Earth: Scenarios for prebiotic phosphorylation. *Geosci. Front.* **2017**, *8*, 329–335. [CrossRef]
12. Pasek, M.A.; Harnmeijer, J.P.; Buick, R.; Gull, R.; Atlas, Z. Evidence for reactive reduced phosphorus species in the early Archean ocean. *Proc. Natl. Acad. Sci. USA* **2013**, *110*, 10089–10094. [CrossRef]
13. Pasek, M.; Kee, T.; Bryant, D.; Pavlov, A.A.; Lunine, J.I. Production of potentially prebiotic condensed phosphates by phosphorus redox chemistry. *Angew. Chem.* **2008**, *120*, 8036–8038. [CrossRef]
14. Gibard, C.; Bhowmik, S.; Karki, M.; Kim, E.-K.; Krishnamurthy, R. Phosphorylation, oligomerization and self-assembly in water under potential prebiotic conditions. *Nat. Chem.* **2018**, *10*, 212–217. [CrossRef]
15. Ebenhöh, O.; Handorf, T.; Heinrich, R. Structural analysis of expanding metabolic networks. *Genome Inf.* **2004**, *15*, 35–45.

16. Handorf, T.; Ebenhöh, O.; Heinrich, R. Expanding metabolic networks: Scopes of compounds, robustness, and evolution. *J. Mol. Evol.* **2005**, *61*, 498–512. [CrossRef] [PubMed]

17. Raymond, J.; Segrè, D. The effect of oxygen on biochemical networks and the evolution of complex life. *Science* **2006**, *311*, 1764–1767. [CrossRef]

18. Kanehisa, M.; Goto, S. KEGG: Kyoto encyclopedia of genes and genomes. *Nucleic Acids Res.* **2000**, *28*, 27–30. [CrossRef] [PubMed]

19. Goldman, A.D.; Bernhard, T.M.; Dolzhenko, E.; Landweber, L.F. LUCApedia: A database for the study of ancient life. *Nucleic Acids Res.* **2013**, *41*, D1079–D1082. [CrossRef]

20. Hubbard, T.J.; Murzin, A.G.; Brenner, S.E.; Chothia, C. SCOP: A structural classification of proteins database. *Nucleic Acids Res.* **1997**, *25*, 236–239. [CrossRef]

21. Dawson, N.L.; Lewis, T.E.; Das, S.; Lees, J.G.; Lee, D.; Ashford, P.; Orengo, C.A.; Sillitoe, I. CATH: An expanded resource to predict protein function through structure and sequence. *Nucleic Acids Res.* **2017**, *45*, D289–D295. [CrossRef]

22. Osterberg, R.; Orgel, L.E. Polyphosphate and trimetaphosphate formation under potentially prebiotic conditions. *J. Mol. Evol.* **1972**, *1*, 241–248. [CrossRef]

23. Yamagata, Y.; Watanabe, H.; Saitoh, M.; Namba, T. Volcanic production of polyphosphates and its relevance to prebiotic evolution. *Nature* **1991**, *352*, 516–519. [CrossRef]

24. Hermes-Lima, M.; Vieyra, A. Pyrophosphate synthesis from phosphor-(enol)pyruvate catalyzed by precipitated magnesium phosphate with "enzyme-like activity". *J. Mol. Evol.* **1992**, *35*, 277–285. [CrossRef]

25. Zwart, I.I.D.; Meade, S.J.; Pratt, A.J. Biomimetic phosphoryl transfer catalysed by iron(II)-mineral precipitates. *Geochem. Cosmochim. Acta* **2004**, *68*, 4093–4098. [CrossRef]

26. Dodd, M.S.; Papineau, D.; Grenne, T.; Slack, J.F.; Pirajno, F.; O'Neil, J.; Little, C.T.S. Evidence for early life in Earth's oldest hydrothermal vent precipitates. *Nature* **2017**, *543*, 60–64. [CrossRef] [PubMed]

27. Flamholz, A.; Noor, E.; Bar-Even, A.; Milo, R. eQuilibrator-the biochemical thermodynamics calculator. *Nucleic Acids Res.* **2012**, *40*, D770–D775. [CrossRef]

28. Maheen, G.; Wang, Y.; Wang, Y.; Shi, Z.; Tian, G.; Feng, S. Mimicking the prebiotic acidic hydrothermal environment: One-pot prebiotic hydrothermal synthesis of glucose phosphates. *Heteroat. Chem.* **2011**, *22*, 186–191. [CrossRef]

29. Coggins, A.J.; Powner, M.W. Prebiotic synthesis of phosphoenol pyruvate by α-phosphorylation-controlled triose glycolysis. *Nat. Chem.* **2017**, *9*, 310–317. [CrossRef] [PubMed]

30. Noor, E.; Haraldsdóttir, H.S.; Milo, R.; Fleming, R.M. Consistent estimation of Gibbs energy using component contributions. *PLoS Comput. Biol.* **2013**, *9*, e1003098. [CrossRef] [PubMed]

31. Wang, M.; Jiang, Y.Y.; Kim, K.M.; Qu, G.; Ji, H.-F.; Mittenthal, J.E.; Zhang, H.-Y.; Caetano-Anolles, G. A universal molecular clock of protein folds and its power in tracing the early history of aerobic metabolism and planet oxygenation. *Mol. Biol. Evol.* **2011**, *28*, 567–582. [CrossRef] [PubMed]

32. Kelsey, C.A.; Gustavo, C.A. Structural phylogenomics reveals gradual evolutionary replacement of abiotic chemistries by protein enzymes in purine metabolism. *PLoS ONE* **2013**, *8*, e59300.

33. Kim, K.M.; Qin, T.; Jiang, Y.Y.; Chen, L.-L.; Xiong, M.; Caetano-Anollés, D.; Zhang, H.-Y.; Caetano-Anollés, G. Protein domain structure uncovers the origin of aerobic metabolism and the rise of planetary oxygen. *Structure* **2012**, *20*, 67–76. [CrossRef] [PubMed]

34. Caetano-Anollés, G.; Kim, K.M.; Caetano-Anollés, D. The phylogenomic roots of modern biochemistry: Origins of proteins, cofactors and protein biosynthesis. *J. Mol. Evol.* **2012**, *74*, 1–34. [CrossRef]

35. Bukhari, S.A.; Caetano-Anollés, G. Origin and evolution of protein fold designs inferred from phylogenomic analysis of CATH domain structures in proteomes. *PLoS Comput. Biol.* **2013**, *9*, e1003009. [CrossRef] [PubMed]

36. Bar-Even, A.; Flamholz, A.; Noor, E.; Milo, R. Thermodynamic constraints shape the structure of carbon fixation pathways. *Biochim. Biophys. Acta* **2012**, *1817*, 1646–1659. [CrossRef] [PubMed]

37. Keller, M.A.; Turchyn, A.V.; Ralser, M. Non-enzymatic glycolysis and pentose phosphate pathway-like reactions in a plausible Archean ocean. *Mol. Syst. Biol.* **2014**, *10*, 725. [CrossRef]

38. Ślesak, I.; Ślesak, H.; Kruk, J. RubisCO early oxygenase activity: A kinetic and evolutionary perspective. *Bioessays* **2017**, *39*, 1700071. [CrossRef]

39. Caetano-Anollés, G. RubisCO and the search for biomolecular culprits of planetary change. *Bioessays* **2017**, *39*, 1700074. [CrossRef]

40. Nascimento, J.M.; Shi, L.Z.; Tam, J.; Chandsawangbhuwana, C.; Durrant, B.; Botvinick, E.L.; Berns, M.W. Comparison of glycolysis and oxidative phosphorylation as energy sources for mammalian sperm motility, using the combination of fluorescence imaging, laser tweezers, and real-time automated tracking and trapping. *J. Cell. Physiol.* **2008**, *217*, 745–751. [CrossRef]
41. Saito, Y.; Ashida, H.; Sakiyama, T.; de Marsac, N.T.; Danchin, A.; Sekowska, A.; Yokota, A. Structural and functional similarities between a ribulose-1,5-bisphosphate carboxylase/oxygenase (rubisco)-like protein from bacillus subtilis and photosynthetic rubisco. *J. Biol. Chem.* **2009**, *284*, 13256–13264. [CrossRef]
42. Kai, T. Sweet siblings with different faces: The mechanisms of FBP and F6P aldolase, transaldolase, transketolase and phosphoketolase revisited in light of recent structural data. *Bioorg. Chem.* **2014**, *57*, 263–280.
43. Nitschke, W.; McGlynn, S.E.; Milner-White, E.J.; Russell, M.J. On the antiquity of metalloenzymes and their substrates in bioenergetics. *Biochim. Biophys. Acta* **2013**, *1827*, 871–881. [CrossRef]
44. Sousa, F.L.; Martin, W.F. Biochemical fossils of the ancient transition from geoenergetics to bioenergetics in prokaryotic one carbon compound metabolism. *Biochim. Biophys. Acta* **2014**, *1837*, 964–981. [CrossRef] [PubMed]
45. Zabinski, R.F.; Toney, M.D. Metal ion inhibition of nonenzymatic pyridoxal phosphate catalyzed decarboxylation and transamination. *J. Am. Chem. Soc.* **2001**, *123*, 193–198. [CrossRef] [PubMed]
46. Waite, J.H.; Glein, C.R.; Perryman, R.S.; Teolis, B.D.; Magee, B.A.; Miller, G.; Grimes, J.; Perry, M.E.; Miller, K.E.; Bouquet, A.; et al. Cassini finds molecular hydrogen in the Enceladus plume: Evidence for hydrothermal processes. *Science* **2017**, *356*, 155–159. [CrossRef] [PubMed]

life

MDPI

Hypothesis

Making Molecules with Clay: Layered Double Hydroxides, Pentopyranose Nucleic Acids and the Origin of Life

Harold S. Bernhardt

Department of Chemistry, University of Otago, P.O. Box 56 Dunedin, New Zealand;
harold.bernhardt@otago.ac.nz

Received: 2 December 2018; Accepted: 9 February 2019; Published: 15 February 2019

Abstract: A mixture of sugar diphosphates is produced in reactions between small aldehyde phosphates catalysed by layered double hydroxide (LDH) clays under plausibly prebiotic conditions. A subset of these, pentose diphosphates, constitute the backbone subunits of nucleic acids capable of base pairing, which is not the case for the other products of these LDH-catalysed reactions. Not only that, but to date no other polymer found capable of base pairing—and therefore information transfer—has a backbone for which its monomer subunits have a plausible prebiotic synthesis, including the ribose-5-phosphate backbone subunit of RNA. Pentose diphosphates comprise the backbone monomers of pentopyranose nucleic acids, some of the strongest base pairing systems so far discovered. We have previously proposed that the first base pairing interactions were between purine nucleobase precursors, and that these were weaker and less specific than standard purine-pyrimidine interactions. We now propose that the inherently stronger pairing of pentopyranose nucleic acids would have compensated for these weaker interactions, and produced an informational polymer capable of undergoing nonenzymatic replication. LDH clays might also have catalysed the synthesis of the purine nucleobase precursors, and the polymerization of pentopyranose nucleotide monomers into oligonucleotides, as well as the formation of the first lipid bilayers.

Keywords: origin of life; layered double hydroxide (LDH) clay; pentose diphosphate; pentopyranose nucleic acid; arabinopyranose nucleic acid; RNA; base pairing; purine precursor; inosine; AICAR

---

1. Introduction

Demonstrating the potential utility of a unified 'systems chemistry' approach to the problem of the origin of life, Powner and colleagues [1–3] recently proposed a link between a plausible early metabolic pathway and amino acid synthesis, through reactions between the small aldehyde phosphates glycolaldehyde phosphate (GAP) and glyceraldehyde-2-phosphate (G2P). Previously, Krishnamurthy and colleagues [4] showed that these same two molecules react—at millimolar concentrations, low-to-moderate temperatures and near-neutral pH—in the presence of layered double hydroxide (LDH) clays to produce pentose-2,4-diphosphates in 7% yield. One of these—ribose-2,4-diphosphate—constitutes the backbone subunit of pyranosyl-RNA [5], a pentopyranose nucleic acid that differs from RNA only in the position of its phosphate groups on the ribose ring, and in the ribose ring being six-membered (as opposed to five-membered as it is in RNA) (Figure 1). Pentopyranosyl nucleic acid systems have been found to constitute stronger base-pairing systems than RNA, with α-arabinopyranose nucleic acids possessing especially remarkable pairing strength [6]. In addition, the four pentopyranose phosphate systems with $4' \rightarrow 2'$-backbone connectivity (including pyranosyl-RNA and α-arabinopyranose nucleic acid) are able to base pair with each other; in contrast, none of the four form base pairs with RNA. LDH-catalyzed reactions between GAP and G2P produce a mixture of different phosphorylated sugars, as shown in

Figure 1, including tetrose-2,4-diphosphates (Figure 1A), pentose-2,4-diphosphates (Figure 1B, *boxed*) and hexose-2,4,6-triphosphates (Figure 1C) [4]; of these, the pentose-2,4-diphosphates are unique in constituting the backbone subunit monomers of oligonucleotides capable of base pairing [7].

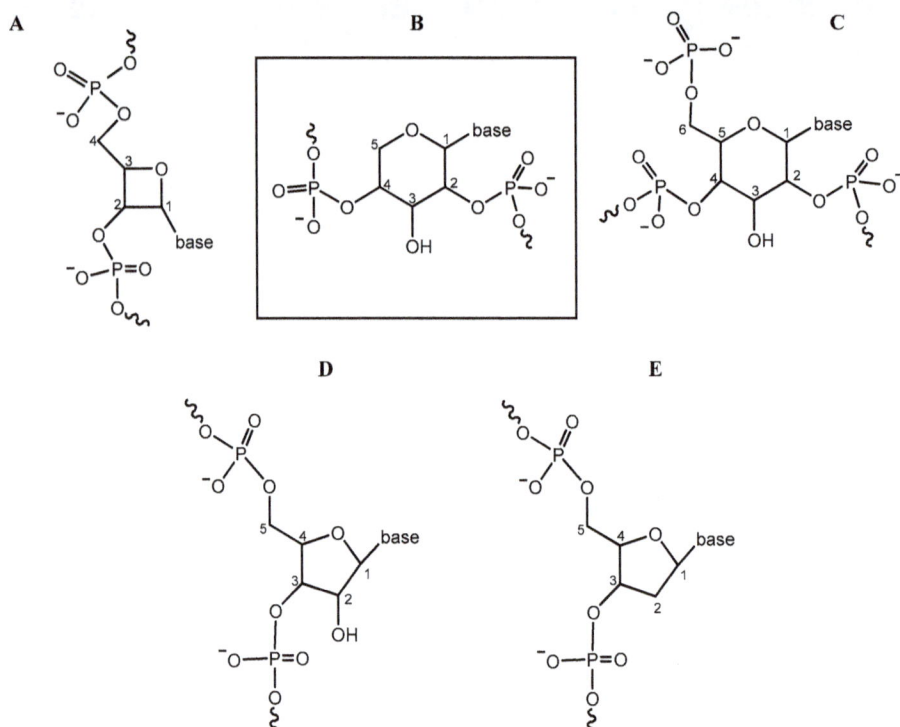

Figure 1. Nucleic acid backbone subunits, showing position of phosphate groups and attached (nucleo)base. (**A–C**) (minus base) are produced in reactions catalysed by layered double hydroxide (LDH) clay, as described in the text; of these, only (**B**) (boxed) (present in pentopyranose nucleic acids) is capable of base pairing. In contrast, the backbone subunits of RNA (**D**) and DNA (**E**) do not have a plausibly prebiotic synthesis. A = tetrooxetose nucleic acid, B = pentopyranose nucleic acid, C = hexopyranose nucleic acid, D = RNA, E = DNA. Numbering of carbohydrate carbon atoms is shown. Figure produced in ChemDraw® 18, PerkinElmer Informatics, Waltham, MA.

Layered double hydroxide (LDH) clays are mixed-valence metal hydroxides with positively charged layers that adsorb charge-balancing anions such as Cl^-, CO_3^{2-} and PO_4^{3-} within their aqueous interlayer [8,9]. These anions are exchangeable by diffusion with organic anions such as GAP and G2P, which bind through electrostatic interactions between the charged phosphate group and the charged layers, as well as through hydrogen bonding interactions with the LDH hydroxide groups. (A snapshot of a simulation of the interaction between a RNA 25-mer oligonucleotide and LDH [10] is shown in Figure 2). The distance between LDH sheets is remarkably flexible, with the LDH interlayer able to more than triple its width from ~7 Å to 24 Å in order to accommodate a polyanionic strand of DNA [11]. In catalysing the reaction between GAP and G2P, LDHs perform a number of key functions:

Concentrating GAP and G2P from millimolar—or even micromolar—concentrations in the bulk medium to ~10 molar within the interlayer [4,12].

Binding GAP and G2P molecules in close proximity, thereby promoting their reaction.

Stabilizing and sequestering the reaction products, with 95 % of the pentose-2,4-diphosphate products still being extractable from the LDH after three months [4].

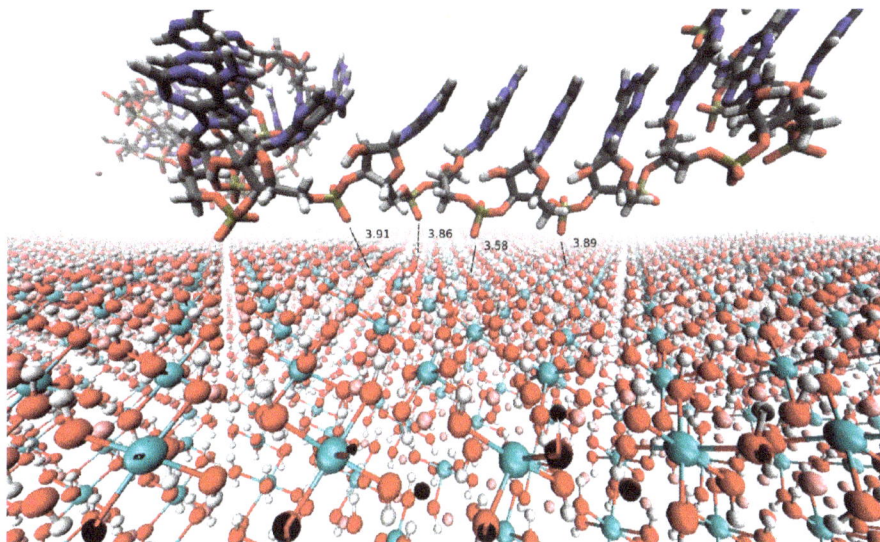

Figure 2. Computer simulation snapshot of the interaction between an RNA 25-mer and LDH interlayer surface, depicting the inner-sphere complexes between RNA phosphate oxygen atoms (red) and LDH metal hydroxide hydrogen atoms (white) in the absence of bridging water molecules. Numerical values shown in black indicate the distance between interaction donors and acceptors in angstroms. Color scheme: O (red), H (white), P (gold), C (grey), N (dark blue), Mg (cyan) and Al (pink). Reprinted with permission from [10]. Copyright 2013 American Chemical Society.

In these roles, the LDH clay is functioning as both a proto-enzyme and quasi-compartment, with Arrhenius [12] stating that, "Like cells, they retain phosphate-charged reactants against high concentration gradients and exchange matter with the surroundings by controlled diffusion through the 'pores' provided by the opening of the interlayers at the crystal edges" (p. 1580). LDHs also catalyse reactions between cyanide (CN^-) anions to form diaminomaleonitrile, a precursor to the purine nucleobase adenine [13], and appear able to stabilize base pairs between RNA nucleotides in the adsorption of guanosine monophosphate (GMP) at low to moderate temperatures (Figure 3) [14]. Clay minerals such as LDHs are thought to have been present early in Earth's history [15,16], with Arrhenius and colleagues suggesting that LDH clays such as hydrotalcite (a naturally-occurring Mg/Al-LDH clay) might have existed as "surface coatings on submerged weathering basalt, or in salt brine deposits in arid lakes" [17] p. 506, as occur on the Earth today. Another LDH mineral, green rust (an Fe^{2+}/Fe^{3+}-LDH), is thought to have been widespread on an early anoxic Earth, with the eventual rise in atmospheric oxygen causing its oxidation and precipitation from the ocean, giving rise to the ubiquitous banded iron formations. In relation to this, Russell has argued that, "the Hadean ocean crust ... was likely thick, relatively cool, and covered in carbonate green rust" [18] p. 6.

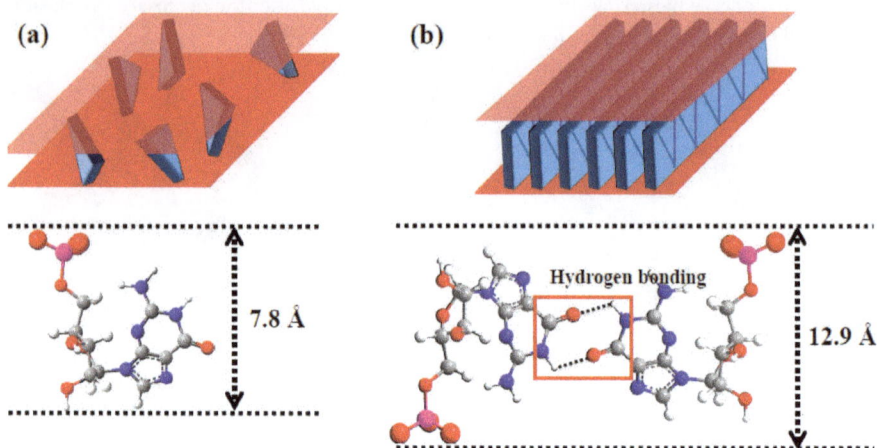

Figure 3. Schematic illustrations for interlayer structure of GMP/LDH hybrids according to molecular arrangement of GMPs: (**a**) single molecule arrangement (GL-S) and (**b**) ribbon II arrangement (GL-R). GL-S: 12.6 Å (d-spacing) − 4.8 Å (LDH layer thickness) = 7.8 Å; GL-R: 17.7 Å (d-spacing) − 4.8 Å (LDH layer thickness) = 12.9 Å. Adapted with permission from [14].

Almost a decade ago, Powner and colleagues demonstrated a possible prebiotic synthesis of the pyrimidine nucleotides [19]; however, the search for a plausibly prebiotic synthesis of the purine nucleotides remains ongoing [20]. Kim and Benner have reported the synthesis of adenosine and inosine from reactions between ribose-1,2-cyclophosphate and the purine nucleobases [21], while Carell and colleagues have demonstrated synthesis of both ribopyranose and standard and non-standard RNA nucleotides starting with a derivatized pyrimidine as a molecular scaffold [22,23]. Powner and Szostak have also found a potential link with pyrimidine synthesis, discovering a branch point in Powner's previously-discovered pyrimidine synthesis which leads to the non-standard 8-oxopurines, described as potential precursors to the standard purine nucleotides in the original report [24]. However, in subsequent work from the Szostak lab [25], the 8-oxopurines were found to be poor substrates in a nonenzymatic primer extension model system, which the authors now consider makes it unlikely that they played a role as purine precursors. They now believe that inosine—in contrast a good substrate in their model nonenzymatic system—is more likely to have played such a role [25]. We have previously proposed [26] that the prebiotic synthesis of RNA might have occurred through the progressive synthesis of the purine nucleobases on a pre-existing ribose-phosphate backbone, with the driving force for this synthesis being the increasing stabilization of the backbone through intermolecular interactions, including base pairing and—potentially—duplex formation. We also proposed that these nucleobase precursors were similar or identical to the intermediates of the modern *de novo* purine biosynthetic pathway, through which modern organisms synthesize purines [27], for example inosine and 5-aminoimidazole-4-carboxamide riboside (AICAR) (Figure 4). Furthermore, due to the plausibly prebiotic nature of many of the reactants in the modern purine biosynthetic pathway (such as glycine and CO_2) [28], as well as the fact that two of the reactions also occur by alternative nonenzymatic reactions [29,30], we proposed that the biosynthetic pathway might have originated from a series of uncatalysed prebiotic reactions [26]. Could the first informational molecules have contained purines (and/or purine precursors) without pyrimidines? As discussed above, Szostak and colleagues have shown that the purine (precursor) inosine participates efficiently in nonenzymatic primer extension using a model system, leading them to conclude, in the words of their article title, "Inosine, but none of the 8-oxo-purines, is a plausible component of a primordial version of RNA." [25] Crick [31] and Wachtershauser [32] have also previously argued in favour of an "all-purine precursor of nucleic acids", although Wachtershauser has proposed a different sugar-nucleobase connectivity than exists in modern RNA.

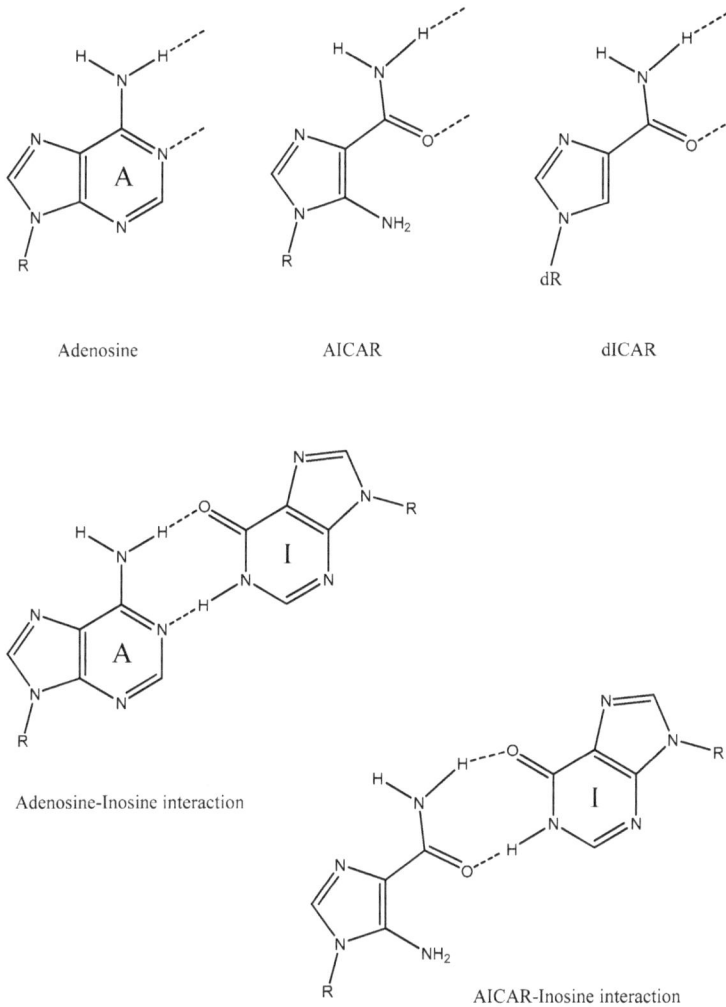

Figure 4. (**Top**) the potential of 5-aminoimidazole-4-carboxamide riboside (AICAR) (an intermediate in the modern *de novo* purine biosynthetic pathway) and its 2′-deoxy analogue 1-(2-deoxy-β-D-ribofuranosyl)-imidazole-4-carboxamide (dICAR) to form hydrogen-bonding interactions similar to those formed by adenosine. (**Bottom**) hydrogen-bonding (base pairing) interaction between adenosine and inosine (a later intermediate in the purine biosynthetic pathway), and proposed interaction between AICAR and inosine, examples of the weaker interactions between purine nucleobase precursors proposed to have preceded purine-pyrimidine base pairing. R = ribofuranose; dR = 2′-deoxyribofuranose, figure produced in ChemDraw® 18 (PerkinElmer Informatics, Waltham, MA).

2. Hypothesis

We propose that life has undergone (at least) two genetic transitions, the first from a mixed system of pentopyranose nucleic acids to RNA, and the second from RNA to a mixed RNA/DNA system. Because both transitions involved a *decrease* in the strength of base pairing, they must have been driven by selection for another property. RNA may have been selected for its catalytic ability. The relative flexibility of its backbone and consequent ability to adopt a variety of non-standard base-pairing interactions enables RNA to assume multiple complex 3D conformations, including binding motifs and catalytic sites [33,34]. In contrast, pentopyranose nucleic acids possess more rigid backbones and are relatively constrained in their base pairing interactions [7]; it therefore appears likely that these systems might possess a somewhat more modest catalytic repertoire than RNA. The expansion in the chemical landscape offered by RNA catalysis would see its ultimate achievement in the advent of coded peptide synthesis [35], which produced a virtual explosion in catalytic potential. The evolutionary driver from RNA to a mixed RNA/DNA system appears to have been DNA's greater chemical stability, which allowed for long-term storage of genetic information as well as a massive increase in genome size [36]. Therefore, the selection criteria would have been different for each genetic transition. In contrast, initial selection of pentopyranose nucleic acids would have been due to the availability of the pentose-2,4-diphosphate backbone monomers through LDH-catalysed synthesis. In addition, pentopyranose nucleic acids appear able to undergo facile nonenzymatic replication, required prior to the advent of—RNA, or possibly pentopyranose nucleic acid—replicase enzymes able to catalyse these reactions. The large angle of inclination ($\sim 45°$) between the base pairs and backbone in pentopyranose systems enables these oligonucleotides to undergo nonenzymatic replication using prebiotically-plausible $2',3'$-cyclic phosphate-activated monomers [7]. In contrast, in the case of RNA, the same $2',3'$-cyclic phosphate-activated monomers form unnatural $2',5'$-RNA phosphate linkages. It is unknown how a transition from a pentopyranose system to RNA would have occurred. It might have been directly, by—for example—the conversion of a six-membered ribopyranose ring to a five-membered RNA ring. However, if so, this would need to have occurred at the ribose phosphate level, as the transfer of the phosphate group from the $4'$- to the $5'$-OH first requires opening of the six-membered pyranose ring, which would have been prohibited at the nucleotide stage by the attachment of the nucleobase (precursor).

In addition to catalysing the synthesis of the monomer units of the pentopyranose–phosphate backbone, we propose that LDH clays played a role in the synthesis of the purine nucleobase precursors, either by catalysing reactions between cyanide anions [13] or through the promotion of reactions between prebiotically plausible molecules such as glycine, NH_3 and CO_2, similar to those which occur in the modern *de novo* purine biosynthetic pathway [26]. Greenwell and colleagues have demonstrated peptide bond synthesis catalysed by a ternary Mg/Cu/Al-LDH clay under wet–dry cycles [37], in an analogous reaction to the amide bond formation that occurs in the biosynthesis of GAR, the first stable intermediate in the purine biosynthetic pathway [26,27]. Krishnamurthy and colleagues have shown that LDH clays catalyse the formylation of GAP utilizing a sulfite–formaldehyde adduct [38], in a parallel to the two indirect formylation reactions that occur in the purine biosynthetic pathway. As described above, Gwak and colleagues [14] have produced evidence that GMP nucleotides in the presence of LDHs form non-standard base pairs at low to moderate temperatures (20–60 °C), whereas at higher temperatures (80–100 °C) only unpaired GMP is present (Figure 3). This suggests that the LDH-promoted polymerisation of pentopyranose nucleotide monomers to form oligomers might also be possible. In addition, montmorillonite, a phyllosilicate clay containing negatively charged sheets that bind Mg^{2+} and other cations, catalyses the polymerisation of RNA nucleotides to form up to 50-mer RNA oligos [39], suggesting such reactions might also be possible for LDH clays. Supporting this possibility, RNA nucleotides [14,40,41], oligonucleotides and duplexes [42] are strongly adsorbed by LDH clays through their charged phosphate backbones, as demonstrated both experimentally and in simulation experiments (Figure 2). Polymerisation of phosphorylated sugars lacking an attached nucleobase (precursor) may also be possible, as demonstrated by the putative polymerisation of fructose-1,6-bisphosphate (a sugar diphosphate with a similar structure to a pentose-2,4-diphosphate), in the presence of a Li^+-LDH [43]. The high affinity of LDH clays for phosphate anions is rather striking, and suggests that these clays might have played a critical role in extracting and concentrating phosphate from low background levels in the prebiotic environment; however, this raises the question of how the strongly bound pentopyranose oligonucleotides might have been released from the LDH interlayer. Russell has posited that LDH clays (including green rust) initially formed in the high pH environment of alkaline deep-sea hydrothermal vents [18], which suggests the possibility that release from the LDH interlayer might have occurred in the increasingly acidic conditions distal from the vent, as LDH clays largely disintegrate at acidic pH [8]. Exchange of the phosphate moieties with divalent carbonate anions [8], possibly through the interaction with high atmospheric levels of CO_2, would have been another possible mechanism for the eventual release of pentopyranose oligonucleotides from their LDH 'cells'.

Conceivably, the high pairing strength of pentopyranose systems might be considered a disadvantage for nonenzymatic replication, as it could increase the chances of mispairing and make the system vulnerable to product inhibition (wherein the just-synthesized copy remains bound to the template sequence, preventing further replication) [44]. However, as described above, we previously proposed that the first nucleobases were precursors to the purine nucleobases [26]—perhaps the same or similar to the intermediates of the modern *de novo* purine biosynthetic pathway—and it would seem reasonable that weaker interactions between the precursors might have offset this high pairing strength. Two examples of possible precursors are inosine and AICAR (Figure 4). Inosine is a purine itself, as well as a precursor to adenosine and guanosine, and in fact plays a key role in genetic coding, occurring in the anticodon wobble position of alanine tRNA [45]. Interestingly, it is utilised due to its ability to form a *purine–purine base pair* (with adenosine) as well as more typical purine–pyrimidine base pairs with cytosine and uracil. When incorporated into RNA oligonucleotides, inosine and 1-(2-deoxy-β-D-ribofuranosyl)-imidazole-4-carboxamide (dICAR) (a 2′-deoxy analogue of AICAR) form base pair interactions with purines as well as pyrimidines [46–48]. The similarity in hydrogen bond-forming potential between adenosine, AICAR and dICAR is shown in Figure 4. However, these pairings between purine precursors are weaker—in some cases significantly—than standard purine–pyrimidine base pairs. In the case of dICAR, this is presumably due to a decrease in base-stacking stabilization due to AICAR possessing an imidazole ring as opposed to a purine double ring structure. Self-complementary RNA oligonucleotide 12-mers containing two inosine- or dICAR base pairs have melting temperatures ~20 °C lower than the comparable RNA duplexes containing standard purine–pyrimidine base pairs only [47], and it is likely that the presence of inosine and AICAR would similarly decrease the stability of pentopyranose duplexes, although this has not been shown experimentally. Nevertheless, we propose that this decrease in stability would have been offset by the greater strength of pentopyranosyl pairing interactions. In addition, it has been shown that pentopyranose duplexes accommodate purine–purine base pairs more easily than RNA duplexes [7], which a number of these interactions would constitute (or at least approximate). In fact, the two opposing effects—inherently more stable duplexes vs. weaker interactions between purine precursors—might have produced a system able to undergo nonenzymatic replication: strong enough for duplex formation and replication, but not so strong as to produce template-product inhibition [44].

The lack of cross pairing between pentopyranose and RNA systems—and the consequent inability to transfer sequence information in the proposed genetic transition—could be seen as a major weakness of our hypothesis. However, genetic transitions will as a rule result in a degree of information loss, due to the alteration in 3D structure (and therefore function) caused by the difference in backbone conformation between the preceding and succeeding informational polymers [7]. As Eschenmoser has pointed out, "Irrespective of its direct communication with the predecessor system, the successor system would have to evolve its phenotype (i.e., chemical catalytic functionality] *de novo*." [7] If, however—as is the case with the non-cross pairing pentopyranose and RNA systems—there is *zero* possibility of information transfer, one might ask what is the point of proposing such a transition, and the need for a pentopyranose system to evolve first? Wouldn't the principle of Occam's razor favour the hypothesis that RNA arose *de novo*? As discussed above, advantages of the hypothesis are:

1. The demonstrated prebiotically plausible formation of the pentose–diphosphate backbone monomers; and, conversely, the absence of this for the RNA backbone subunit.

2. The structural features of pentopyranose systems, for example the large angle of inclination between base pairs and backbone, that make them structurally suited for nonenzymatic replication using plausibly prebiotic 2′,3′-cyclic phosphate-activated monomers.

3. The greater pairing strength of pentopyranose systems—important to counterbalance the initial weaker interactions between purine nucleobase precursors—to allow stable duplex formation and nonenzymatic replication.

4. The fact that a pentopyranose system provides a—structural and possibly also catalytic—stepping stone to RNA, removing some of the difficulties in the RNA backbone arising *de novo*.

Finally, it seems likely that LDH clays might also have played other roles in the origin of life, such as in the formation of the first lipid bilayers. The negatively charged carboxylate anions of $CH_3(CH_2)_{16}COO^-Na^+$, the sodium salt of the long-chain fatty acid octadecanoic (or stearic) acid, adsorb to both sides of the LDH interlayer, forming a tilted version of a lipid bilayer, and increasing the interlayer width to 48 Å (Figure 5) [49]. In conclusion, what we have learned so far regarding LDH clays points to our having only just begun to scratch the surface in relation to this fascinating class of materials.

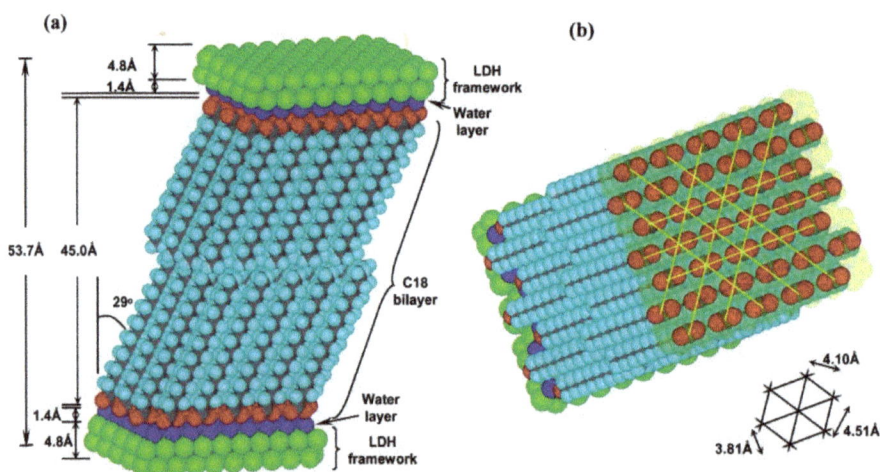

Figure 5. Structural models of the stearate/LDH interaction showing (**a**) tilted stearate bilayer and (**b**) regular packing of stearate anions in the interlayer gallery. Reprinted with permission from [49]. Copyright 2003 American Chemical Society.

Acknowledgments: Thanks to Dave Larsen and Warren Tate for paying the article processing charge (APC), which enabled this article to be published. The author would like to thank Ram Krishnamurthy, Chris Greenwell, Dave Larsen and Jim McQuillan for helpful discussions, and to Ram Krishnamurthy for his comments on an earlier draft of this manuscript.

Conflicts of Interest: The author declares no conflict of interest.

Abbreviations

AICAR	5-Aminoimidazole-4-carboxamide riboside
dICAR	1-(2-Deoxy-β-D-ribofuranosyl)-imidazole-4-carboxamide
G2P	Glyceraldehyde-2-phosphate
GAP	Glycolaldehyde phosphate
GMP	Guanosine-5-monophosphate
LDH	Layered double hydroxide

References

1. Islam, S.; Powner, M.W. Prebiotic systems chemistry: Complexity overcoming clutter. *Chem* **2017**, *2*, 470–501. [CrossRef]
2. Coggins, A.J.; Powner, M.W. Prebiotic synthesis of phosphoenol pyruvate by alpha-phosphorylation-controlled triose glycolysis. *Nat. Chem.* **2017**, *9*, 310–317. [CrossRef]
3. Islam, S.; Bucar, D-K.; Powner, M.W. Prebiotic selection and assembly of proteinogenic amino acids and natural nucleotides from complex mixtures. *Nat. Chem.* **2017**, *9*, 584–589. [CrossRef]
4. Krishnamurthy, R.; Pitsch, S.; Arrhenius, G. Mineral induced formation of pentose-2,4-bisphosphates. *Orig. Life Evol. Biosph.* **1999**, *29*, 139–152. [CrossRef]
5. Pitsch, S.; Wendeborn, S.; Krishnamurthy, R.; Holzner, A.; Minton, M.; Bolli, M.; Miculca, C.; Windhab, N.; Micura, R.; Stanek, M.; et al. Pentopyranosyl oligonucleotide systems. 9th Communication. The β-D-ribopyranosyl-(2′→4′) oligonucleotide system ('pyranosyl-RNA'): Synthesis and resumé of base-pairing properties. *Helv. Chim. Acta* **2003**, *86*, 4270–4363. [CrossRef]
6. Jungmann, O.; Beier, M.; Luther, A.; Huynh, H.K.; Ebert, M.O.; Jaun, B.; Krishnamurthy, R.; Eschenmoser, A. Pentopyranosyl oligonucleotide systems. The α-L-arabinopyranosyl-(4′→2′)-oligonucleotide system: Synthesis and pairing properties. *Helv. Chim. Acta* **2003**, *86*, 1259–1308. [CrossRef]
7. Eschenmoser, A. Etiology of potentially primordial biomolecular structures: From vitamin B12 to the nucleic acids and an inquiry into the chemistry of life's origin: A retrospective. *Angew. Chem. Int. Ed.* **2011**, *50*, 12412–12472. [CrossRef] [PubMed]
8. Mishra, G.; Dash, B.; Pandey, S. Layered double hydroxides: A brief review from fundamentals to application as evolving biomaterials. *Appl. Clay Sci.* **2018**, *153*, 172–186. [CrossRef]
9. Kuma, K.; Paplawsky, W.; Gedulin, B.; Arrhenius, G. Mixed-valence hydroxides as bioorganic host minerals. *Orig. Life Evol. Biosph.* **1989**, *19*, 573–602. [CrossRef]
10. Swadling, J.B.; Suter, J.L.; Greenwell, H.C.; Coveney, P.V. Influence of surface chemistry and charge on mineral–RNA interactions. *Langmuir* **2013**, *29*, 1573–1583. [CrossRef]
11. Xu, Z.P.; Lu, G.Q. Layered double hydroxide nanomaterials as potential cellular drug delivery agents. *Pure Appl. Chem.* **2006**, *78*, 1771–1779. [CrossRef]
12. Arrhenius, G.O. Crystals and life. *Helv. Chim. Acta* **2003**, *86*, 1569–1586. [CrossRef]
13. Boclair, J.W.; Braterman, P.S.; Brister, B.D.; Jiang, J.; Lou, S.; Wang, Z.; Yarberry, F. Cyanide self-addition, controlled adsorption, and other processes at layered double hydroxides. *Orig. Life Evol. Biosph.* **2001**, *31*, 53–69. [CrossRef] [PubMed]
14. Gwak, G.-H.; Kocsis, I.; Legrand, Y.-M.; Barboiu, M.; Oh, J.-M. Controlled supramolecular structure of guanosine monophosphate in the interlayer space of layered double hydroxide. *Beilstein J. Nanotechnol.* **2016**, *7*, 1928–1935. [CrossRef] [PubMed]
15. Hazen, R.M.; Sverjensky, D.A.; Azzolini, D.; Bish, D.L.; Elmore, S.C.; Hinnov, L.; Milliken, R.E. Clay mineral evolution. *Am. Mineral* **2013**, *98*, 2007–2029. [CrossRef]
16. Hazen, R.M. Paleomineralogy of the Hadean eon: A preliminary species list. *Am. J. Sci.* **2013**, *313*, 807–843. [CrossRef]
17. Arrhenius, G.; Sales, B.; Mojzsis, S.; Lee, T. Entropy and charge in molecular evolution – the case of phosphate. *J. Theor. Biol.* **1997**, *187*, 503–522. [CrossRef] [PubMed]
18. Russell, M.J. Green rust: The simple organizing 'seed' of all life? *Life* **2018**, *8*, E35. [CrossRef]
19. Powner, M.W.; Gerland, B.; Sutherland, J.D. Synthesis of activated pyrimidine ribonucleotides in prebiotically plausible conditions. *Nature* **2009**, *459*, 239–242. [CrossRef]
20. Powner, M.W.; Sutherland, J.D.; Szostak, J.W. Chemoselective multicomponent one-pot assembly of purine precursors in water. *J. Am. Chem. Soc.* **2010**, *132*, 16677–16688. [CrossRef]
21. Kim, H.J.; Benner, S.A. Prebiotic stereoselective synthesis of purine and noncanonical pyrimidine nucleotide from nucleobases and phosphorylated carbohydrates. *Proc. Natl Acad. Sci. USA* **2017**, *114*, 11315–11320. [CrossRef] [PubMed]
22. Becker, S.; Thoma, I.; Deutsch, A.; Gehrke, T.; Mayer, P.; Zipse, H.; Carell, T. A high-yielding, strictly regioselective prebiotic purine nucleoside formation pathway. *Science* **2016**, *352*, 833–836. [CrossRef] [PubMed]

23. Becker, S.; Schneider, C.; Okamura, H.; Crisp, A.; Amatov, T.; Dejmek, M.; Carell, T. Wet-dry cycles enable the parallel origin of canonical and non-canonical nucleosides by continuous synthesis. *Nat. Commun.* **2018**, *9*, 163. [CrossRef] [PubMed]

24. Stairs, S.; Nikmal, A.; Bučar, D.K.; Zheng, S.L.; Szostak, J.W.; Powner, M.W. Divergent prebiotic synthesis of pyrimidine and 8-oxo-purine ribonucleotides. *Nat. Commun.* **2017**, *8*, 15270. [CrossRef] [PubMed]

25. Kim, S.C.; O'Flaherty, D.K.; Zhou, L.; Lelyveld, V.S.; Szostak, J.W. Inosine, but none of the 8-oxo-purines, is a plausible component of a primordial version of RNA. *Proc. Natl. Acad. Sci. USA* **2018**, *115*, 13318–13323. [CrossRef] [PubMed]

26. Bernhardt, H.S.; Sandwick, R.K. Purine biosynthetic intermediate-containing ribose-phosphate polymers as evolutionary precursors to RNA. *J. Mol. Evol.* **2014**, *79*, 91–104. [CrossRef] [PubMed]

27. Zhang, Y.; Morar, M.; Ealick, S.E. Structural biology of the purine biosynthetic pathway. *Cell Mol. Life Sci.* **2008**, *65*, 3699–3724. [CrossRef] [PubMed]

28. de Duve, C. *Blueprint for a Cell: The Nature and Origin of Life*; Neil Patterson Publishers, Carolina Biological Supply Company: Burlington, NC, USA, 1991; pp. 133–134.

29. Nierlich, D.P.; Magasanik, B. Phosphoribosylglycinamide synthetase of *Aerobacter aerogenes*. Purification and properties, and nonenzymatic formation of its substrate 5'-phosphoribosylamine. *J. Biol. Chem.* **1965**, *240*, 366–374. [PubMed]

30. Mueller, E.J.; Meyer, E.; Rudolph, J.; Davisson, V.J.; Stubbe, J. N^5-Carboxyaminoimidazole ribonucleotide: Evidence for a new intermediate and two new enzymatic activities in the *de novo* purine biosynthetic pathway of *Escherichia coli*. *Biochemistry* **1994**, *33*, 2269–2278. [CrossRef]

31. Crick, F.H. The origin of the genetic code. *J. Mol. Biol.* **1968**, *38*, 367–379. [CrossRef]

32. Wachtershauser, G. An all-purine precursor of nucleic acids. *Proc. Natl. Acad. Sci. USA* **1988**, *85*, 1134–1135. [CrossRef]

33. Chandrasekhar, K.; Malathi, R. Non-Watson Crick base pairs might stabilize RNA structural motifs in ribozymes – a comparative study of group-I intron structures. *J. Biosci.* **2003**, *28*, 547–555. [CrossRef]

34. Fedor, M.J.; Williamson, J.R. The catalytic diversity of RNAs. *Nat. Rev. Mol. Cell Biol.* **2005**, *6*, 399–412. [CrossRef]

35. Bernhardt, H.S.; Tate, W.P. The transition from noncoded to coded protein synthesis: did coding mRNAs arise from stability-enhancing binding partners to tRNA? *Biol. Direct* **2010**, *5*, 16. [CrossRef]

36. Krishnamurthy, R. On the emergence of RNA. *Isr. J. Chem.* **2015**, *55*, 837–850. [CrossRef]

37. Grégoire, B.; Greenwell, H.C.; Fraser, D.G. Peptide formation on layered mineral surfaces: The key role of brucite-like minerals on the enhanced formation of alanine dipeptides. *ACS Earth Space Chem.* **2018**, *2*, 852–862. [CrossRef]

38. Pitsch, S.; Krishnamurthy, R.; Arrhenius, G. Concentration of simple aldehydes by sulfite-containing double-layer hydroxide minerals: Implications for biopoesis. *Helv. Chim. Acta* **2000**, *83*, 2398–2411. [CrossRef]

39. Ferris, J.P. Montmorillonite catalysis of 30-50 mer oligonucleotides: Laboratory demonstration of potential steps in the origin of the RNA world. *Orig. Life Evol. Biosph.* **2002**, *32*, 311–332. [CrossRef]

40. Choy, J.-H.; Kwak, S.-Y.; Park, J.-S.; Jeong, Y.J.; Portier, J. Intercalative nanohybrids of nucleoside monophosphates and DNA in layered metal hydroxide. *J. Am. Chem. Soc.* **1999**, *121*, 1399–1400. [CrossRef]

41. Aisawa, S.; Ohnuma, Y.; Hirose, K.; Takahashi, S.; Hirahara, H.; Narita, E. Intercalation of nucleotides into layered double hydroxides by ion-exchange reaction. *Appl. Clay Sci.* **2005**, *28*, 137–145. [CrossRef]

42. Swadling, J.B.; Coveney, P.V.; Greenwell, H.C. Stability of free and mineral-protected nucleic acids: Implications for the RNA world. *Geochim. Cosmochim. Acta* **2012**, *1*, 360–378. [CrossRef]

43. Jellicoe, T.C.; Fogg, A.M. Synthesis and characterization of layered double hydroxides intercalated with sugar phosphates. *J. Phys. Chem. Solids* **2012**, *73*, 1496–1499. [CrossRef]

44. Engelhart, A.E.; Powner, M.W.; Szostak, J.W. Functional RNAs exhibit tolerance for non-heritable 2'-5' versus 3'-5' backbone heterogeneity. *Nat. Chem.* **2013**, *5*, 390–394. [CrossRef]

45. Torres, A.G.; Piñeyro, D.; Filonava, L.; Stracker, T.H.; Batlle, E.; Ribas de Pouplana, L. A-to-I editing on tRNAs: Biochemical, biological and evolutionary implications. *FEBS Lett.* **2014**, *588*, 4279–4286. [CrossRef]

46. Carter, R.J.; Baeyens, K.J.; SantaLucia, J.; Turner, D.H.; Holbrook, S.R. The crystal structure of an RNA oligomer incorporating tandem adenosine-inosine mismatches. *Nucleic Acids Res.* **1997**, *25*, 4117–4122. [CrossRef]

47. Johnson, W.T.; Zhang, P.M.; Bergstrom, D.E. The synthesis and stability of oligodeoxyribonucleotides containing the deoxyadenosine mimic 1-(2′-deoxy-β-D-ribofuranosyl)imidazole-4-carboxamide. *Nucleic Acids Res.* **1997**, *25*, 559–567. [CrossRef]

48. Sala, M.; Pezo, V.; Pochet, S.; Wain-Hobson, S. Ambiguous base pairing of the purine analogue 1-(2-deoxy-β-D-ribofuranosyl)-imidazole-4-carboxamide during PCR. *Nucleic Acids Res.* **1996**, *24*, 3302–3306. [CrossRef]

49. Itoh, T.; Ohta, N.; Shichi, T.; Yui, T.; Takagi, K. The self-assembling properties of stearate ions in hydrotalcite clay. *Langmuir* **2003**, *19*, 9120–9126. [CrossRef]

Article

Unevolved De Novo Proteins Have Innate Tendencies to Bind Transition Metals

Michael S. Wang [1,†], **Kenric J. Hoegler** [2,†] and **Michael H. Hecht** [1,*]

1   Department of Chemistry, Princeton University, Princeton, NJ 08540, USA; msw5@princeton.edu
2   Department of Molecular Biology, Princeton University, Princeton, NJ 08540, USA; kenhoegl@gmail.com
*   Correspondence: hecht@princeton.edu; Tel.: +1-609-258-2901
†   These authors contributed equally to this work.

Received: 24 September 2018; Accepted: 4 January 2019; Published: 9 January 2019

Abstract: Life as we know it would not exist without the ability of protein sequences to bind metal ions. Transition metals, in particular, play essential roles in a wide range of structural and catalytic functions. The ubiquitous occurrence of metalloproteins in all organisms leads one to ask whether metal binding is an evolved trait that occurred only rarely in ancestral sequences, or alternatively, whether it is an innate property of amino acid sequences, occurring frequently in unevolved sequence space. To address this question, we studied 52 proteins from a combinatorial library of novel sequences designed to fold into 4-helix bundles. Although these sequences were neither designed nor evolved to bind metals, the majority of them have innate tendencies to bind the transition metals copper, cobalt, and zinc with high nanomolar to low-micromolar affinity.

Keywords: protein design; novel metalloproteins; binary patterned amino acid sequences; prebiotic chemistry

---

1. Introduction

Proteins that bind metals perform many of the essential functions necessary to sustain life [1]. Metalloproteins occur in every organism, and are required for a wide range of catalytic, structural, and signaling functions. Moreover, the amino acid sequences that bind metals and perform these functions are conserved across all three domains of life. These observations suggest that proteins capable of binding metals arose early in the history of life on earth [2–4].

Was metal binding by ancestral sequences a rare occurrence? Or do proteins have an innate tendency to bind metals? It is tempting to address these questions by studying natural proteins and assessing their abilities to bind various metals. However, extant natural proteins differ dramatically from the sequences that arose early in evolution. Current metalloproteins are the products of billions of years of selection for sequences that bind metals with high affinities and specificities. In contrast, the first ancestral sequences that bound metals may have done so with weak affinities, poor specificities, and/or overlapping binding sites.

In an attempt to gain insight into the innate metal binding of unevolved sequences, we focused on a collection of de novo proteins that were designed to fold, but were not explicitly designed to possess any binding or functional activities. The unevolved proteins used in this study were drawn from a combinatorial library of novel sequences, which were designed by binary patterning of polar and nonpolar residues to fold into 4-helix bundles [5–7]. These sequences differ fundamentally both from natural proteins and from de novo designed metalloproteins: In contrast to natural proteins, the binary patterned sequences are not biased by billions of years of selection for life-sustaining functions, and in contrast to rationally designed metalloproteins [8–12], the binary patterned sequences are not biased by explicit designs for metal binding sites.

To assess the abilities of these proteins to bind transition metals, 52 different sequences were incubated with transition metal ions immobilized on beads [13,14]. The metal binding ability of several of these proteins was further characterized by equilibrium dialysis and isothermal titration calorimetry. Although the sequences in this collection were neither designed nor evolved to bind metals, the majority of the de novo proteins exhibited some level of metal binding. Several of the novel sequences bound with affinities comparable to some natural metallo-peptides and proteins. The nanomolar affinity of some proteins for zinc is roughly comparable to human serum albumin (HSA), which has $K_d \sim 3 \times 10^{-8}$ [15]. These findings suggest that—at least in some cases—unevolved proteins have an innate tendency to bind transition metals.

2. Materials and Methods

Gene Libraries, Strains, and Growth Conditions. Genes for 3GL proteins were previously cloned into and expressed from a modified pCA24N vector containing a chloramphenicol (CAM) resistance cassette [16]. Protein expression is controlled by a T5 promoter and lac operator.

Electrocompetent BW25113 cells were made according to standard procedures. Luria broth was supplemented with 30 µg/mL CAM. The working concentration of isopropyl β-D-1-thiogalactopyranoside (IPTG) was 100 µM.

The 3GL plasmids were prepared in DNase free H_2O to a final concentration of 50 ng/µL. Thus, 2 µL of the library were transformed into 100 µL of electrocompetent cells. Cells were then diluted 1:100 and 50 µL of cells were plated on LB/agar containing 30 µg/mL CAM. Plates were incubated at 37 °C for 12–16 h, after which single colonies became visible.

Screen for Soluble Expressers. Overnight cultures were started from single colonies and grown for 12 h at 37 °C in LB containing 30 µg/mL CAM and 100 µM IPTG. Cell lysates were normalized to an OD_{600} of 1.0. Then, 1.5 mL of overnight culture was spun-down and re-suspended in 250 µL of BugBuster Protein Extract Reagent (Novagen, Madison, WI, USA). After 20 min incubation, samples were spun for 30 min at 13,000 rpm. 20 µL of the supernatants were diluted 1:1 with 2× Laemmli sample buffer and run on SDS-PAGE (Bio-Rad, Hercules, CA, USA). Proteins were visualized by staining with Coomassie blue.

Colonies that expressed high levels of protein were re-grown overnight in 5 mL LB containing 30 µg/mL CAM. The plasmid was purified from the overnight culture using a QIAprep Spin Mini kit (Qiagen). Sequencing of the 3GL gene was done by Sanger sequencing (Genewiz S., Plainsfield, NJ, USA).

Preparation of Immobilized Metal Beads. Here, 80 µL of Sepharose beads chemically attached to iminodiacetic Acid (IDA) (Sigma-Aldrich, St. Lousi, MO, USA) were incubated with 100 µL of 250 mM $CoCl_2$, $CuCl_2$, or $ZnCl_2$ for 20 min at room temperature. These were washed several times with wash buffer (50 mM Tris HCl, 500 mM NaCl, pH 7.6). Separation of beads and supernatants was done at low rpm (<3000) to ensure beads were not damaged.

Assays for Metal Binding. For this, 1.5 mL of overnight culture were spun-down and re-suspended in 250 µL of BugBuster Protein Extract Reagent (Novagen). After incubation for 20 min, samples were spun for 30 min at 13,000 rpm. Supernatants were incubated with iminodiacetic acid Sepharose Beads (Sigma-Aldrich) charged with Co^{2+}, Cu^{2+}, or Zn^{2+} for 1 h at 4 °C. The beads were washed 4 times for 4 min with stringent wash buffer containing 50 mM Tris HCl pH 7.6, 500 mM NaCl and the following concentrations of imidazole: 10 mM for cobalt, 50 mM for copper, and 15 mM for zinc. Beads were then suspended in 2× Laemmli sample buffer (Bio-Rad), heated to 95 °C for 5 min, and spun for 2 min at 13,000 rpm. Supernatants were normalized to the initially-loaded lysate volume and both were run on SDS-PAGE (Bio-Rad). Densitometry was analyzed using Image-Quant TL.

Protein Expression and Purification. BL21 *E. coli* cells expressing the protein of interest were grown overnight at 37 °C on LB plates containing 30 µg/mL CAM. Single colonies were picked and grown in liquid cultures (LB with 30 µg/mL CAM) for 12 h at 37 °C. Expression was induced at OD 600 ~0.5 by addition of 100 µM IPTG, and cells were allowed to grow for 16 h at 37 °C. Cells were lysed using an EmulsiFlex French press (Avestin, Ottawa, ON, Canada). Protein was purified in two steps.

First, lysates were run over a His Trap HP nickel column (GE Healthcare, Chicago, IL, USA) washed with wash buffer (200 mM Na_2HPO_4, 500 mM NaCl, pH 7.4) and eluted with elution buffer (200 mM Na_2HPO_4, 500 mM NaCl, 250 mM imidazole, pH 7.4). Even without a His-tag, the proteins bind the nickel column, presumably because of the high number of histidine residues on their surfaces. HisZero was the only protein that did not stick to the nickel column, and was instead first purified by Anion Exchange on a HiTrap Q HP column (GE Healthcare) with start buffer (200 mM Na_2HPO_4, 100 mM NaCl, pH 7.6) and elution buffer (200 mM Na_2HPO_4, 1.5 M NaCl, pH 7.6) in a linear gradient. Second, fractions containing the desired protein were run over a HiLoad 26/60 superdex 75 size exclusion column (GE Healthcare) in buffer (10 mM Tris, 100 mM NaCl, pH 7.6). Protein purity and identity were verified by HPLC (Agilent 100 system, Agilent Zorbax 300SB-C18 column) followed by ESI-MS (Agilent 6210 TOF LC/MS).

Equilibrium Dialysis. A Pierce™ Rapid Equilibrium Dialysis (RED) Device, 8 kDa MWCO (ThermoFisher, Waltham, MA, USA) was used to analyze metal binding. Here, 5 µM of pure protein in standard buffer (10 mM Tris, 100 mM NaCl, pH 7.6) were placed in chamber A of the apparatus, and increasing amounts of $CoCl_2$, $CuCl_2$, or $ZnCl_2$ diluted in standard buffer were placed in chamber B. The dialysis apparatus was shaken at 800 rpm for 2 h at room temperature to reach equilibrium. An aliquot of the metal solution was then removed from chamber B and an equal volume of higher-concentration metal solution was added. From a series of such measurements, the curve of bound vs. free metal was determined. Measurements were not taken above concentrations where the protein precipitated, as determined by spinning down separate solutions of proteins incubated with metals. All equilibrium dialysis experiments were performed in triplicate.

The concentration of metal in chamber B was measured using 500 µM 4-(2-pyridylazo)resorcinol (PAR) (Thermo Fischer Scientific) in standard buffer. This was done by external calibration using dilutions of TraceCERT® metal standards (SigmaAldrich) in standard buffer. The linear relationship between binding and absorbance (495 nm) enabled determination of the concentration of metal in chamber B. The limit of detection (LoD = 3σ) was 60 nM for Co^{2+}, 120 nM for Cu^{2+}, and 75 nM for Zn^{2+}. Bound metal was identified as the difference between total metal present in the dialysis and the free metal measured. All data were baselined against a buffer blank run without protein. Calculations were performed using standard protocols, and binding curves were fit using least-squares fitting.

Isothermal Titration Calorimetry (ITC). A MicroCal PEAQ-ITC calorimeter (Malvern Panalytical, Malvern, UK) was used to verify several of the binding curves. Each titration was performed at 25.0 °C, with the reaction cell containing 250 µL of protein (10 or 20 µM) in standard buffer (10 mM Tris, 100 mM NaCl, pH 7.6), and the injection syringe filled with Co(II), Cu(II), or Zn(II) metal solutions (100, 250, or 500 µM depending on stoichiometry) in standard buffer. The enthalpy evolved upon metal binding was monitored by proxy of the energy needed to hold the temperature constant. Each titration experiment was performed using 18 injections of 2 µL with 8-s duration and a 120 to 160-s interval between injections. Reference power of 41.9 µW, high-feedback mode, and a stirring speed of 1000 rpm were used for all experiments. All data were analyzed by using the MicroCal PEAQ-ITC analysis software. To obtain the binding enthalpies, the observed enthalpy values were corrected for the enthalpy of dilution observed when the protein was saturated with metal.

3. Results

3.1. A Collection of Unevolved De Novo Proteins

We previously described a strategy that uses binary patterning of polar and nonpolar residues to design libraries of novel α-helical proteins [5–7]. In this approach, the sequence locations of polar and nonpolar residues are specified explicitly; however, the exact identities of the residues at each position are not specified, and are varied combinatorially. Combinatorial diversity is enabled by using degenerate DNA codons to construct a collection of synthetic genes encoding the library of novel proteins. The codon NTN (N = A,T,C,G) is used to encode five nonpolar amino acids (Phe, Leu, Ile,

Met, Val), and the alternate codon VAN (V = A,C,G) is used to encode six polar amino acids (His, Glu, Gln, Asp, Asn, Lys). We have created several libraries of binary patterned proteins designed to fold into 4-helix bundles. Proteins from these libraries were characterized biophysically, and several 4-helix bundle structures were determined at high resolution [17–20]. An example protein structure, S-824, is shown in Figure 1.

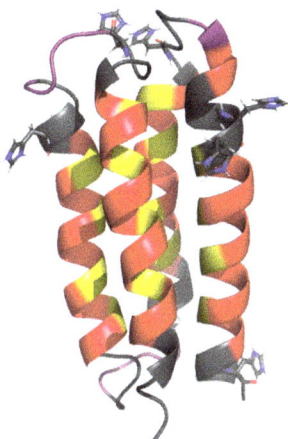

Figure 1. NMR structure of the 4-helix bundle protein, S-824 (PDB code 1P68) [18]. Polar residues are colored red, nonpolar residues are yellow, and nonhelical residues are purple. Residues that were conserved in the design of this library are in grey, with conserved histidines shown explicitly.

From these collections, we isolated a novel protein (called ConK) that rescues *E. coli* from lethal concentrations of copper [21]. This last finding suggests that novel proteins, which have not undergone evolutionary selection, can nonetheless have significant impacts on metal binding and/or homeostasis

For the current study, we wished to assess the ability of proteins—chosen arbitrarily from this library of unevolved sequences—to bind transition metals. To facilitate rapid screening, we chose a subset of proteins from our third generation library (3GL) [7] that expressed high levels of soluble protein.

To prepare this collection, we transformed the 3GL plasmid library into *E. coli* BW25113 and plated on LB agar supplemented with chloramphenicol, as described previously [16]. Single colonies were chosen arbitrarily and grown overnight in the presence of 100 µM IPTG to induce expression. Cell numbers were normalized, and clarified lysates were run on SDS-PAGE and stained with Coomassie Brilliant Blue. If a dark band was seen at approximately 12 kDa (the mass of a 3GL protein of 102 residues), the clone was included in the sub-library. Each member was provided the name naïve metal binder (NMB), followed by a number. Thus, NMB 5 is the fifth well-expressed soluble protein isolated. Overall, approximately two-thirds of the clones expressed high levels of soluble protein. In this manner, 52 proteins were selected for further characterization.

3.2. Construction of Weak Metal Binders as Negative Controls

Preliminary screens suggested the majority of our de novo proteins bind transition metals. Therefore, in order to establish a baseline for proteins that do not bind transition metals, we designed two new sequences based on sequence S-824, a protein from a binary patterned library, which was shown previously to form the stable well-ordered 4-helix bundle structure shown in Figure 1 [16,17].

The first putative non-binder was designed by mutating all six histidines in the combinatorial regions of S-824 to other polar residues. Histidine is a known metal-binding residue, and we wished to use a sequence devoid of histidines in the combinatorial regions as one negative control. This variant was named S-824-HC (HisConserved). This sequence still contains the six conserved histidines in the turn and helical regions that were held constant in the library design.

The second negative control was made using the Rosetta Sequence Tolerance protocol [22] to replace the six remaining histidines in the S-824-HC variant with residues that do not bind metals. This variant was named S-824-HZ (HisZero). These new proteins allowed us to establish two lower limits for metal binding from de novo proteins: the HisConserved shows the lower limit available to the library that was tested, while the HisZero shows the binding available if histidines are removed entirely from the design. Sequences of all proteins used in this study are shown in Supplementary Material Table S1.

3.3. Screening Novel Proteins for Binding to Transition Metals

Immobilized metal affinity chromatography (IMAC) resins were used to screen the sub-library of 52 well-expressed soluble de novo proteins for binding to transition metals (Figure 2). When using IMAC resins, several metal and chelator combinations are possible. We chose to test three transition metals known to play roles in a range of biological systems: cobalt, copper, and zinc. We immobilized these metals on the resin using the tridentate chelator, iminodiacetic acid (IDA). We chose this chelator because it leaves the most metal ion coordination sites available, but still binds transition metals strongly enough to ensure the metal would remain bound to the bead during stringent washes. IDA Sepharose beads (Sigma Aldrich) were charged with Co^{2+}, Cu^{2+}, or Zn^{2+}.

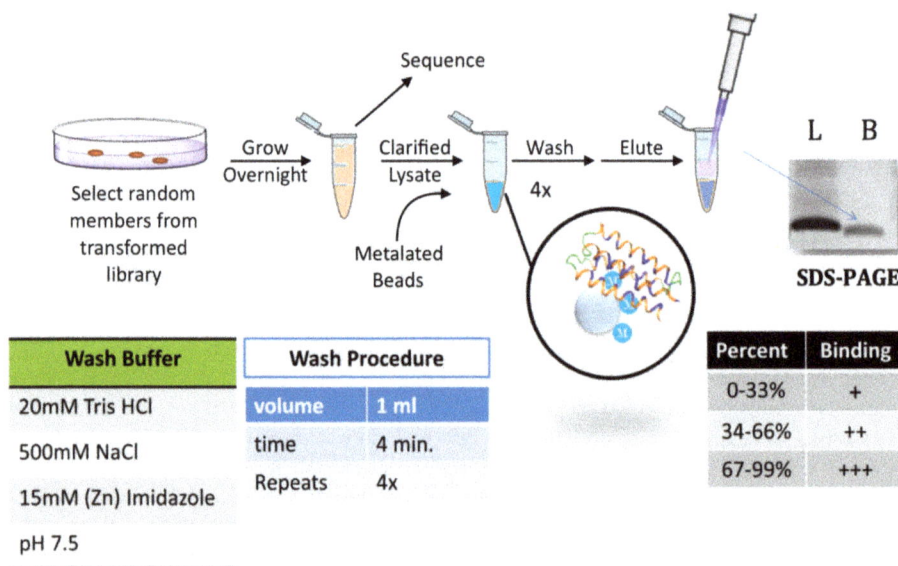

Figure 2. Screen for metal binding. Cells expressing a de novo protein were grown overnight in the presence of isopropyl β-D-1-thiogalactopyranoside (IPTG). Plasmids were then sequenced and cells were lysed. Clarified lysates were incubated with iminodiacetic acid (IDA) Sepharose beads that had been charged with a metal cation. The bound proteins underwent several stringent washes (see "wash buffer" and "wash procedure" panels). Samples were then eluted by boiling in SDS sample buffer, run on SDS-PAGE, stained with Coomassie, and quantified by densitometry. The amount of protein loaded (L) was compared to the amount that remained bound (B) to the beads. Based on the percentage bound, proteins were binned into three groups (bottom right panel).

Single clones from the sub-library were grown overnight in the presence of 30 µg/mL IPTG to induce expression. Clarified lysates from these cultures were incubated with the metalated beads, and then washed with a buffer containing salt and imidazole. A high concentration of salt (500 mM NaCl) was used to shield non-specific electrostatic interactions with the resin. Imidazole concentrations for the washes were established by control experiments with the HisConserved protein described above. HisConserved was removed from the beads by washes with imidazole concentrations of 10 mM for cobalt, 50 mM for copper, and 15 mM zinc. In contrast to HisConserved, the HisZero protein did not stick to the metalated beads.

To assess metal binding, we compared the amount of de novo protein that remained bound to the metalated bead (after stringent washing) to the amount initially loaded. Samples were run on SDS-PAGE and bands were quantified by densitometry. Examples are shown in Figure 3. Based on the fraction of protein that remained bound to the metalated resin, the 52 sequences were grouped into three classes: If <33% of the protein remained bound to the bead, the protein was designated a non-binder or a weak binder (+); if 33–66% of the protein remained, it was classified as a moderate metal binder (++); and if 66–100% remained bound, it was classified as a strong binder (+++). The majority of proteins bound cobalt weakly, zinc moderately, and copper strongly (Figure 4). Despite the stringency of imidazole washes being varied to match the Irving-Williams series (10 mM for cobalt, 15 mM for zinc, and 50 mM for copper), the identified ordering still corresponds with the affinities expected from this series [23]. A full list of proteins and their screened relative binding affinities for the three metals is shown in Supplementary Material Table S2.

To compare our findings with these de novo proteins to natural *E. coli* proteins, we repeated the procedure using *E. coli* cells not containing an expression vector. As shown in Figure 3, under these conditions, the vast majority of endogenous *E. coli* proteins do not bind divalent metals on the resin.

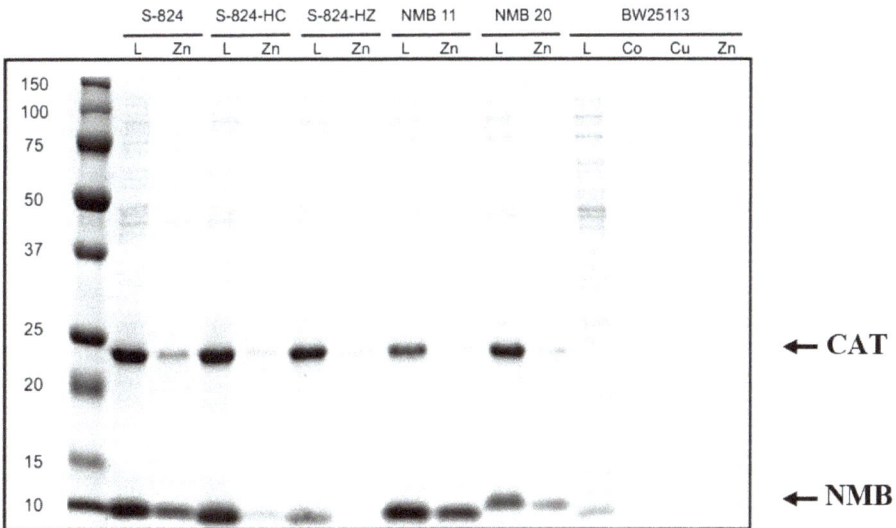

Figure 3. SDS-PAGE comparing bound protein to loaded protein. A representative gel showing protein binding to metalated beads. Each gel was run with protein S-824 as a positive binding control and with S-824-HC as a negative binding control. A sample of the clarified lysate that had been loaded onto the bead (L) was run next to the sample of protein that remained bound to the bead following several stringent washes (metal). Our novel metal binding proteins (labeled "NMB") run near 12 kDa. Densitometry designated NMB 11 as a moderate zinc binder (++) while NMB 20 was designated a weak zinc binder (+). The four rightmost lanes correspond to the *E. coli* strain without the expression plasmid. The band running near 25 kDa is chloramphenicol acetyl transferase (CAT), which is also encoded on the expression plasmid.

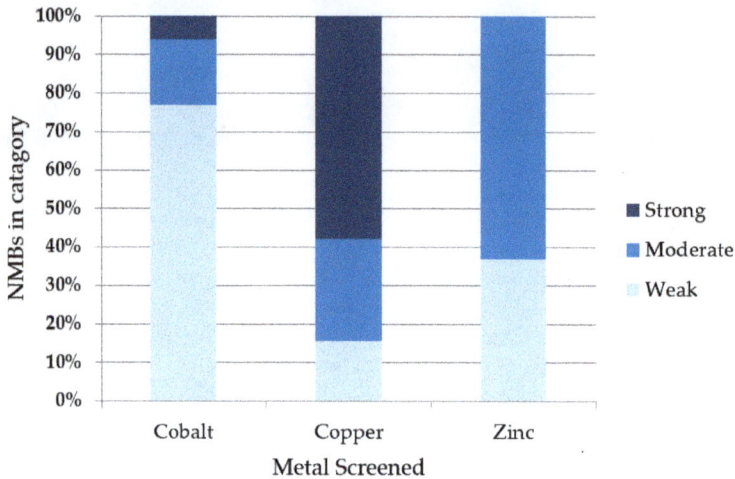

Figure 4. Summary of binding by 52 de novo proteins to three transition metals. Co^{2+}, Cu^{2+}, or Zn^{2+} ions were immobilized on IMAC beads and incubated with cell lysates expressing 52 different de novo proteins. Binders were binned into three groups based on the fraction of protein that remained bound to the metalated bead after stringent washes with imidazole. The majority of NMBs bound copper strongly, zinc moderately, and cobalt weakly.

3.4. Estimation of Binding Affinities

Binding to metalated beads showed that many of our unevolved proteins have some affinity for the tested transition metals. However, binding to metalated beads provides little information about affinity constants, and no information about binding stoichiometry.

To estimate the metal binding affinity and stoichiometry of the novel proteins, we used equilibrium dialysis. In this technique, a selectively permeable membrane divides two chambers. The purified protein is added to chamber A, and increasing concentrations of a metal salt (e.g., $ZnCl_2$) are added chamber B. The membrane allows metal ions (but not protein) to flow between chambers until equilibrium is established. By measuring the final concentration of metal in chamber B, one can determine how much metal bound to the protein in chamber A.

Because our proteins were neither evolved nor designed to bind metals, we anticipated that binding might not be limited to single well-defined sites; and multiple sites with similar affinities might occur on a single protein. Therefore, we analyzed our data to estimate the stoichiometry of binding, and to calculate an "apparent" dissociation constant ($K_{d,app}$), which represents the ensemble binding affinity for multiple sites on a protein, rather than an exact K_d for each individual site. The procedure for calculating stoichiometry and $K_{d,app}$ from the equilibrium dialysis measurements is summarized in Supplementary Material Equation (1)–(8).

We measured metal binding for six NMB proteins and three control proteins for cobalt, copper, and zinc (Figure 5). These NMB proteins were chosen because they represented a range of apparent metal affinities in the initial screen. For example, NMB 11 seemed to prefer zinc over other metals, whereas NMB 20 had the opposite trend and disfavored zinc binding. The three controls, S-824, S-824-HisConserved, and S-824-HisZero, were also assayed. Binding data for all proteins are shown in Table 1, with binding curves in Supplementary Material Figure S1.

The equilibrium dialysis experiments identified affinities ranging from 200 nM (NMB 37 binding copper) to 2.3 μM (NMB 24 binding cobalt). They also ranged in stoichiometry from one binding event (NMB 39 binding copper) to four (NMB 39 binding zinc). Thus, the stoichiometry can differ for a given protein binding different metals. This suggests certain sites bind metals selectively.

Table 1. Equilibrium dialysis results for metal binding for six NMB proteins and three controls. The apparent binding constants ($K_{d,app}$), and the stoichiometry of metal ions per protein (max bound equivalents) are shown. "-" indicates very weak electrostatic binding, as discussed in the text.

Protein	Metal	$K_{d,app}$ (nM)	Max Bound Equivalents	Protein	Metal	$K_{d,app}$ (nM)	Max Bound Equivalents
S-824	Co^{2+}	700	1.5	NMB 24	Co^{2+}	2300	3
	Cu^{2+}	700	1.5		Cu^{2+}	1000	3
	Zn^{2+}	1000	3		Zn^{2+}	600	2
S-824-HC	Co^{2+}	900	1.5	NMB 25	Co^{2+}	1200	3
	Cu^{2+}	300	1		Cu^{2+}	1200	2
	Zn^{2+}	600	2		Zn^{2+}	500	4
S-824-HZ	Co^{2+}	–	–	NMB 37	Co^{2+}	1600	3
	Cu^{2+}	–	–		Cu^{2+}	200	2
	Zn^{2+}	1500	1		Zn^{2+}	1200	4
NMB 11	Co^{2+}	600	3	NMB 39	Co^{2+}	700	2
	Cu^{2+}	800	3		Cu^{2+}	600	1
	Zn^{2+}	300	4		Zn^{2+}	1000	4
NMB 20	Co^{2+}	300	1.5				
	Cu^{2+}	600	2				
	Zn^{2+}	800	3				

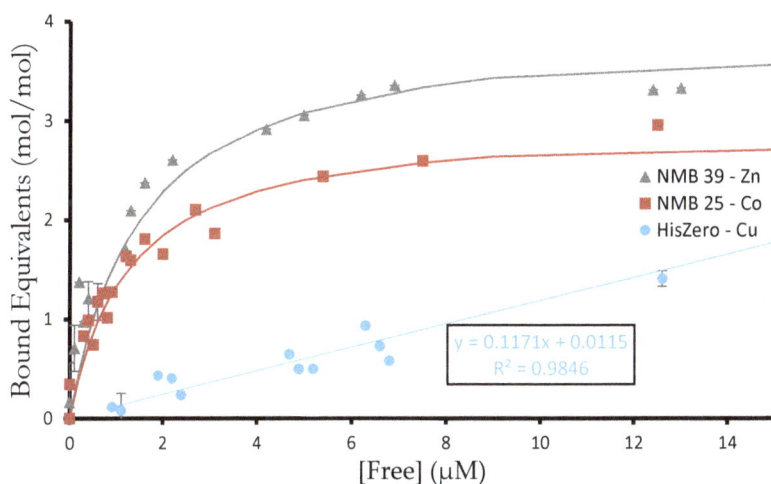

Figure 5. Binding of de novo proteins to different metals. Equilibrium dialysis was used to determine the binding coefficient, [Metal bound]/[Protein total]. As protein is titrated with Co^{2+}, Cu^{2+}, or Zn^{2+}, the $K_{d,app}$ is the concentration of metal ion at which half of the metal binding sites on the protein are occupied. Shown are three representative binding curves from the 27 experiments. NMB 39 binds 4 equivalents of zinc with $K_{d,app} \cong 1$ μM. NMB 25 binds three equivalents of cobalt with $K_{d,app} \cong 1.25$ μM. The binding of HisZero to copper shows linear behavior proportional to free metal, which is characteristic of weak nonspecific electrostatic interactions.

Several of the de novo proteins bind non-integer equivalents of metal. This suggest that in some cases, a metal ion may be shared between two protein monomers. We also note that partial occupancy of the metal coordination sphere by the de novo protein does not preclude simultaneous binding by solvent and/or buffer molecules. In particular, Tris is known to complex Cu(II) with millimolar affinity [24]. Buffer interactions in direct competition with protein and/or in the formation of ternary complexes with metal and protein may cause the calculated $K_{d,app}$ to underestimate the true metal binding affinity.

The two negative controls reveal further nuances of binding. The HisConserved variant bound at least one equivalent of each metal ion, showing that the conserved histidine residues are sufficient for some of the binding observed in the NMB proteins. In contrast, the HisZero variant did not bind Co^{2+} or Cu^{2+} with significant affinity, and bound one equivalent of Zn^{2+} with affinity comparable to other NMB proteins. As HisZero shows the binding possible from combinatorial regions without histidine, it is likely that carboxylic acid residues in the binary pattern are sufficient to bind Zn^{2+}. Overall, both negative controls exhibited poor metal binding and established the minimum binding for library proteins.

To confirm the affinities and stoichiometries measured by equilibrium dialysis, we also measured these properties for NMB 39 S-824 and S-824-HisZero by Isothermal Titration Calorimetry (ITC). The ITC results (Supplementary Materials Table S3 and Figure S2) corroborate the equilibrium dialysis measurements in order of magnitude, but not always stoichiometry. For example, both techniques showed that S824 binds three equivalents of Zn(II), and HisZero binds Zn(II), but not Co(II) or Cu(II). However, equilibrium dialysis differed from ITC by identifying 4 binding events for NMB 39 binding to Zn(II), while ITC identified only 1.5 binding events.

The difference in apparent stoichiometry between equilibrium dialysis and ITC probably reflects the different properties measured by these two techniques: Equilibrium dialysis relies on a colorimetric indicator to assay metal concentrations. In contrast, ITC records the ΔH of binding. For NMB 39, the measured ΔH of binding was on average -44.5 kJ/mol. Smaller enthalpies associated with additional binding sites might be dwarfed by this signal. Consequently, these sites may not be seen, thereby diminishing the apparent stoichiometry observed by ITC.

A surprising counterpoint is that for S-824 binding to Co(II) and NMB39 binding to Cu(II), ITC detects *more* binding events than does equilibrium dialysis. The commonality between these cases is that they are the least enthalpically-favored metal binding of those tested (i.e., lowest ΔH relative to other metals). While ITC may detect these binding sites, they may reach equilibrium more slowly in the dialysis experiments, and so be undetected. Despite occasional differences, these two orthogonal techniques both detect multiple binding sites with low micromolar affinities.

In general, separate binding events are observed only when these events have distinctly different affinities. Consequently, two distinct binding events were detected by ITC only in the case of S-824 binding Co. For this reason, S-824 binding Co had a higher affinity measured by ITC than suggested by the IMAC screen or by equilibrium dialysis. Because these separate events can be difficult to observe, we chose to report the binding values as the "apparent" average of multiple events.

Though it is difficult to correlate the solution experiments with the bead experiments, both suggest ready metal binding. A possible cause of discrepancies between the techniques is that the beads rely on binding to a tridentate IDA, which occupies half the coordination sites of the metal, while the equilibrium dialysis experiments present hydrated metals with all sites potentially available. Another difference is that the stringent washes in the IMAC screen are calibrated to wash away the HisConserved protein and may remove weakly bound proteins.

4. Discussion

Metal binding proteins are ubiquitous in nature. Nearly half of all proteins contain a metal [25], and a survey of 1371 different enzymes with known 3-D structures estimated that 47% required a metal for function and 41% contained a metal at their catalytic centers [26]. Because they play essential roles in biology and are conserved across all forms of life [1], it has been suggested that metals—and cofactors containing metals—played key roles in the first replicating entities that gave rise to life on earth. These considerations suggest that the earliest ancestral amino acid sequences may have been selected, in part, for their intrinsic abilities to bind metal ions.

As a proxy for ancient proto-proteins, we studied a library of synthetic de novo proteins with no evolutionary history; and assessed their abilities to bind transition metals. Because design of a metal-binding site was not a part of the library design, and so must arise by chance, we considered the possibility that binding would be rare among our unevolved sequences. Consequently, when planning this project, we expected to require high-throughput methods, such as phage display [27], to find rare needles in a large haystack. However, as described above, our experimental result showed that artificial metalloproteins are not rare, and occur frequently in our collection of de novo proteins.

The observed frequency of metal binding suggests that—at least in some cases—unevolved proteins possess an innate ability to bind transition metals. This supports the suggestion that the early selection of the 20 proteinogenic amino acid side chains may have been influenced by a biological requirement for polymers that bind transition metals [28].

The de novo proteins in our collection have no evolutionary history. Yet they are likely to differ from ancestral natural proteins in several ways: First, the library proteins are all α-helical, while natural proteins contain α-helices, β-strands, and non-repetitive structures. Second, because the combinatorial library was constructed using the degenerate codons VAN and NTN to encode polar and nonpolar residues, respectively, cysteine, which is encoded by TGC and TGT, does not occur in our proteins. Yet, cysteine is an important metal-binding residue in nature. Third, the de novo proteins are rich in exposed His- Glu-, and Asp- residues. Thus, although unbiased by evolutionary history, our binary pattern proteins are biased to contain an abundance of the metal binding residues His, Asp, and Glu. In particular, the de novo proteins in our collection have an abundance of solvent-exposed histidines; an average of 12 histidines per 102-residue sequence. The HisConserved control protein contains six histidines, while some proteins in the collection contain as many as 18. The abundance of histidine residues results from the binary code strategy, which specifies histidine as one of only six residues that can occupy surface exposed positions on all four α-helices [5–7].

While the abundance of histidines in our proteins makes them different from most proteins in the current biosphere (which contain ~2.3% histidine), ancestral sequences may have also been rich in histidine. It has been proposed that histidine was one of the earliest genetically encoded amino acids, and was originally encoded by all four CAN codons. According to this proposal, glutamine arose later, and adopted the CAA and CAG codons by "codon capture" [29]. If this was indeed the case, then ancestral proteins may have contained abundant histidines, and like our unevolved de novo proteins, may have bound transition metals more readily than their current descendants.

Glutamine may have evolved as an alternative to histidine, in part, to limit the tendency of primordial sequences to bind non-specifically (and promiscuously) to metals and metal-containing cofactors. This is consistent with our finding that current *E. coli* proteins—which have a lower percentage of histidines than our de novo proteins—also have a lower tendency to bind metalated resins (Figure 3).

Much as codon capture by glutamine may have decreased the abundance of histidines in the *sequences* of modern proteins, the scarcity of solvent-exposed histidines in the *structures* of modern proteins may result from negative selection against the type of non-specific metal binding that occurs so readily among the binary patterned de novo structures.

The IMAC resin screens for sites that allow a metal to simultaneously bind both the resin and the protein. In modern proteins, however, most histidines occur in buried or partially buried active sites [30,31]. They are not abundant on protein surfaces, and rarely occur as exposed clusters that would lead to promiscuous metal binding. Consequently, our IMAC screen may isolate natural metalloproteins only if they (1) are in their *apo* form, (2) have binding sites that are not deeply buried, and (3) do not bind so tightly as to abstract the metal from the resin. Likewise for the de novo proteins, the IMAC screen would fail to identify NMBs with buried sites or with a tightly bound metal that co-purifies with the protein following cell lysis.

Removing solvent-exposed histidines in the negative controls, HisConserved and HisZero, decreased binding to the metalated resin (Figure 2). Therefore, we considered the possibility that binding to metalated resin might be a simple function of the number of histidines in each of the de novo sequences. To test this possibility, we compared the number histidines (and the metal binding carboxylates, Glu and Asp) per 102-residue sequence against the fraction of each protein retained by the metalated resin. No correlation was observed (Supplementary Material Figures S3–S5). Nonetheless, the failure of the histidine-free control HisZero to bind the metalated resin suggests, broadly, that the binding observed in the library relies on histidine residues more than Asp or Glu.

The lack of correlation between the number of histidines per sequence and binding to the metalated beads indicates that binding of transition metals by unevolved de novo proteins—like binding by natural proteins—depends on the relative positioning (not just the frequency) of metal binding side chains in the three-dimensional structure of the protein. Assuming that many of these sites are the same sites that bind the free metal, we modelled the relative locations of His, Asp, and Glu side chains on the predicted structures of our NMB proteins. Assuming that a binding site requires two or more metal binding residues to be close in space, we used QUARK ab initio structure prediction [32] to visualize which residues might be positioned in ways that could lead to a primitive metal binding site (Figure 6a). The manual search considered both ligand strength (H > D,E) and the distance between residues. A list of possible binding sites is shown in Table 2. In those cases where a putative site has several nearby metal-binding residues, it is not possible to predict which will actually be involved in binding. For example, in the sequence $H_1XXH_4H_5$ the metal could be coordinated between residues 1 and 4, 1 and 5, or all three. The possible binding motifs for protein S-824 and the NMB proteins assayed by equilibrium dialysis are plotted by frequency of occurrence in Figure 6b.

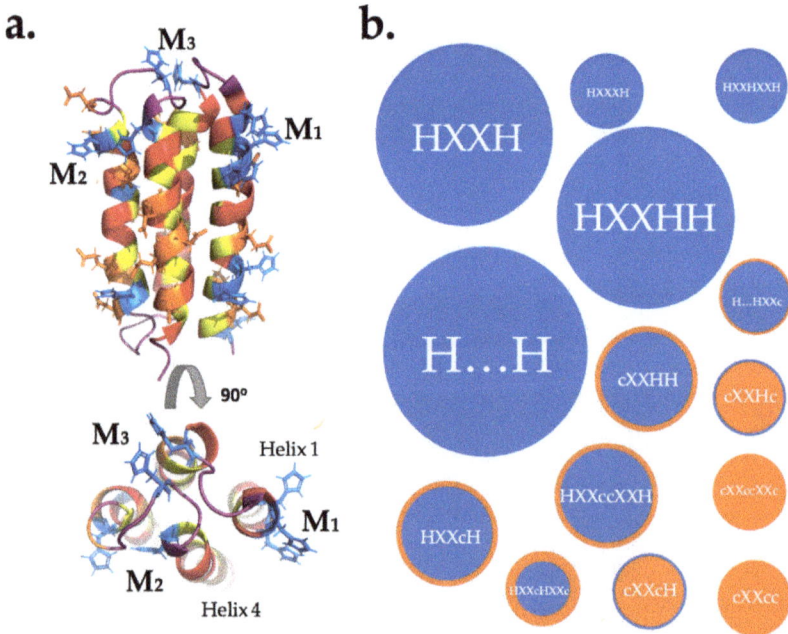

Figure 6. Possible binding sites in a binary-patterned 4-helix bundle. (**a**) NMR structure of S824 (PDB code 1P68). Polar residues are in red, nonpolar residues are in yellow, potential metal-coordinating residues are colored blue (His) or orange (Asp, Glu). The figure shows three ways metal ions (M) could be coordinated: by residues in one helix (M_1) between two helices (M_2) or by loop residues (M_3). (**b**) Frequency of possible binding motifs in S-824 and all NMB proteins, where H is His and c is either carboxylic acid Glu or Asp. Mutants of S-824 were not included as their binding sites are a subset of those in S-824. The size of a bubble is determined by the abundance of that motif, while color is dictated by abundance of H (blue) against c (orange). Ellipsis (...) indicate a large separation of residues on different helices.

Table 2. Possible binding sites on the assayed proteins. Conserved residues are in blue text. Binding sites involving a conserved residue are placed in a "C" row while independent binding sites are in an "I" row. Stoichiometry is from equilibrium dialysis.

Protein		Binding Motifs				Stoichiometry		
						Co^{2+}	Cu^{2+}	Zn^{2+}
S-824	C	$H_{24} \dots H_{74}$	$H_{20} \dots H_{30}XXD_{33}$	$H_{80}XXH_{83}H_{84}$	$H_{94}XXE_{97}E_{98}XXH_{101}$	1.5	1.5	3
	I	$D_7XXE_{10}D_{11}XXE$	$E_{42}XXH_{45}D_{46}$					
S-824-HC	C	$H_{24} \dots H_{74}$	$E_{97}E_{98}XXH_{101}$			1.5	1	2
	I	$D_7XXE_{10}D_{11}XXE$						
S-824-HZ	I	$D_7XXE_{10}D_{11}XXE$				0	0	1
NMB 11	C	$H_{24} \dots H_{74}$	$H_{30}D_{32}XXH_{35}$	$H_{83}H_{84}XXH_{87}$	$H_{98}XXH_{101}$	3	3	4
	I	$H_{59}XXH_{62}$	$E_{66}XXD_{69}E_{70}$					
NMB 20	C	$H_{24} \dots H_{74}$	$H_{97}H_{98}XXH_{101}$			1.5	2	3
	I	$H_{14}XXH_{17}H_{18}$	$E_{42}XXD_{45}H_{46}$					
NMB 24	C	$H_{24} \dots H_{74}$	$H_{32}XXE_{34}E_{35}$	$H_{84}H_{85}XXH_{87}$	$H_{97}H_{98}XXH_{101}$	3	3	2
	I	$H_{39}XXD_{42}E_{43}XXH_{46}$						
NMB 25	C	$H_{24} \dots H_{74}$	$H_{80}XXH_{83}H_{84}$			3	2	4
	I	$H_{11} \dots H_{43}$	$H_{31}D_{32}XXH_{35}$	$H_{43}XXH_{46}$	$H_{59}XXH_{62}$			
NMB 37	C	$H_{24} \dots H_{74}$	$H_{83}H_{84}XXD_{87}$	$H_{98}XXH_{101}$		3	2	4
	I	$D_7XXD_{10}H_{11}$	$H_{31}XXE_{34}H_{35}XXE_{38}$					
NMB 39	C	$H_{24} \dots H_{74}$	$H_{30}D_{32}XXE_{35}$	$D_{80}XXH_{83}H_{84}$	$H_{95}XXH_{98}XXH_{101}$	2	1	4
	I							

Mutagenesis of S-824 to remove histidines revealed two important sites. The first is a conserved binding site between residues H_{24} and H_{74}, shown as M_3 in the Figure 6. Based on the HisConserved construct, this should bind 1 equivalent of each metal. This conserved sequence is responsible for the large abundance of H . . . H motifs in Figure 6. The second is that the observed binding of zinc to HisZero could arise from a DXXED motif. This uses three carboxylic acid residues, which are hard bases, and would be expected to bind Zn^{2+}, which is a harder metal than Co^{2+} or Cu^{2+}. Both these binding insights should hold true for all the NMB proteins: The $H_{24} \dots H_{74}$ motif is conserved and motifs with only carboxylic acid residues, and no histidines, might selectively bind zinc but not the other metals. Further studies involving mutagenesis of each NMB will be required to confirm which of the predicted sites are responsible for the observed binding.

It is also noteworthy that several of the conserved histidines participate in approximately half the proposed binding. Thus, H_{83} and H_{84} are responsible for the common motif HXXHH, shown in Figure 6. Likewise, H_{101} is a participant in three of the six HXXH motifs. Because these conserved histidines appear in all of the sequences in our library, these findings can also be viewed as analogous to the evolutionary improvement of ancestral metal binding sequences.

As noted in the preceding section, many of the putative binding sites include conserved residues. Yet, 12 of the 27 proposed binding sites do *not* involve any conserved residues. They result from a semi-random combinatorial exploration of sequence space, and thereby highlight the high frequency at which metal binding can occur among sequences that were neither designed nor evolved to bind metals.

The simple model for binding shown in Figure 6 does not lead to occupancy of the entire coordination sphere of the metal. Therefore, the proposed binding sites are incomplete and binding sites that bridge two protein molecules are possible. This is consistent with our experimental results, which indicate that some NMB proteins may form ternary complexes, and in some cases, precipitate proteins that are brought together. Other studies have suggested an important role for metals in protein-protein interfaces [33]. Our work supports such a role for metals in the early evolution of protein-metal complexes. This also emphasizes the evolutionary counterpoint: ternary protein-metal complexes can promote amyloid formation [34].

The apparent dissociation constants ($K_{d,app}$) for the six characterized NMB proteins reveal high-nanomolar to low-micromolar affinities. This range is similar to natural proteins like human serum albumin (HSA), but dramatically different from the tight binding measured for many proteins; superoxide dismutase, for example, has sub-picomolar affinity for zinc [15]. This affinity is, however, comparable to some de novo proteins explicitly designed to bind metals [11].

The tight metal binding affinities of many modern proteins is necessary in the metal-limited environment of the modern cell, which requires that evolved natural proteins must have high affinities and specificities to be metalated correctly in vivo. In contrast, our unevolved de novo proteins, with their moderate affinities and promiscuous specificities, would not be metalated in the highly evolved cytoplasm of *E. coli* [25].

Most of the proteins in our collection bound copper stronger than zinc, which they bound stronger than cobalt (Figure 4). This suggests, not surprisingly, that innate unevolved binding correlates with the Irving-Williams series. While some metal preference is indicated by both affinity and stoichiometry, the apparent binding is both promiscuous and ubiquitous; quite different from modern natural proteins. Because binding different metals can allow natural proteins to exhibit new functions [35], the permissive binding observed in these unevolved proteins might serve as a source from which selectivity and function can arise following natural or artificial selection.

In summary, the findings reported here suggest that some level of metal binding is not difficult to achieve. Instead, binding occurs frequently in collections of sequences containing abundant His, Asp, and Glu residues, even though the positions and relative geometries of these residues were not specified a priori. These results show that the ability of proteins to bind transition metals requires neither eons of natural selection or years of rational/computational design. Instead, it appears that unevolved de novo proteins have innate tendencies to bind transition metals.

Supplementary Materials: The following are available online at http://www.mdpi.com/2075-1729/9/1/8/s1, Figure S1: Binding Curves.

Author Contributions: Conceptualization, K.J.H. and M.H.H.; Methodology, K.J.H. and M.S.W.; Validation, K.J.H. and M.S.W.; Formal Analysis, K.J.H. and M.S.W.; Investigation, K.J.H. and M.S.W.; Resources, M.H.H.; Data Curation, K.H.H. and M.S.W.; Writing—Original Draft Preparation, K.J.H., M.S.W., and M.H.H.; Writing—Review and Editing, M.S.W. and M.H.H.; Visualization, K.J.H. and M.S.W.; Supervision, M.H.H.; Project Administration, M.H.H.; Funding Acquisition, M.H.H.

Funding: This research was funded by NSF grant number MCB-1409402.

Conflicts of Interest: The authors declare no conflict of interest.

References

1. Nies, D.H.; Silver, S. *Molecular Microbiology of Heavy Metals*; Springer: Berlin/Heidelberg, Germany, 2007.
2. Goldman, A.D.; Samudrala, R.; Baross, J.A. The evolution and functional repertoire of translation proteins following the origin of life. *Biol. Direct* **2010**, *5*, 15. [CrossRef] [PubMed]
3. Weiss, M.C.; Sousa, F.L.; Mrnjavac, N.; Neukirchen, S.; Roettger, M.; Nelson-Sathi, S.; Martin, W.F. The physiology and habitat of the last universal common ancestor. *Nat. Microbiol.* **2016**, *1*, 16116. [CrossRef] [PubMed]
4. Raanan, H.; Pike, D.H.; Moore, E.K.; Falkowski, P.G.; Nanda, V. Modular origins of biological electron transfer chains. *PNAS* **2018**, *115*, 1280–1285. [CrossRef] [PubMed]
5. Kamtekar, S.; Schiffer, J.; Xiong, H.; Babik, J.; Hecht, M. Protein design by binary patterning of polar and nonpolar amino acids. *Science* **1993**, *262*, 1680–1685. [CrossRef] [PubMed]
6. Hecht, M.H.; Das, A.; Go, A.; Bradley, L.H.; Wei, Y. De novo proteins from designed combinatorial libraries. *Protein Sci.* **2004**, *13*, 1711–1723. [CrossRef] [PubMed]
7. Bradley, L.H.; Kleiner, R.E.; Wang, A.F.; Hecht, M.H.; Wood, D.W. An intein-based genetic selection allows the construction of a high-quality library of binary patterned de novo protein sequences. *Protein Eng. Des. Sel.* **2005**, *18*, 201–207. [CrossRef] [PubMed]
8. Regan, L. The design of metal-binding sites in proteins. *Annu. Rev. Biophys. Biomol. Struct.* **1993**, *22*, 257–281. [CrossRef]
9. Regan, L. Protein design: Novel metal-binding sites. *Trends Biochem. Sci.* **1995**, *20*, 280–285. [CrossRef]

10. Calhoun, J.R.; Kono, H.; Lahr, S.; Wang, W.; DeGrado, W.F.; Saven, J.G. Computational design and characterization of a monomeric helical dinuclear metalloprotein. *J. Mol. Biol.* **2003**, *334*, 1101–1115. [CrossRef]

11. Chakraborty, S.; Kravitz, J.Y.; Thulstrup, P.W.; Hemmingsen, L.; DeGrado, W.F.; Pecoraro, V.L. Design of a three-helix bundle capable of binding heavy metals in a triscysteine environment. *Angew. Chem. Int. Ed. Engl.* **2011**, *50*, 2049–2053. [CrossRef]

12. Mills, J.H.; Sheffler, W.; Ener, M.E.; Almhjell, P.J.; Oberdorfer, G.; Pereira, J.H.; Parmeggiani, F.; Sankaran, B.; Zwart, P.H.; Baker, D. Computational design of a homotrimeric metalloprotein with a trisbipyridyl core. *PNAS* **2016**, *113*, 15012–15017. [CrossRef] [PubMed]

13. Block, H.; Maertens, B.; Spriestersbach, A.; Brinker, N.; Kubicek, J.; Fabis, R.; Labahn, J.; Schäfer, F. Immobilized-metal affinity chromatography (IMAC): A review. *Methods Enzymol.* **2009**, *463*, 439–473. [CrossRef] [PubMed]

14. Sun, X.; Chiu, J.; He, Q. 2D PAGE: Sample Preparation and Fractionation. In *Fractionation of Proteins by Immobilized Metal Affinity Chromatography*; Posch, A., Ed.; Humana Press: Totowa, NJ, USA, 2008; pp. 205–212.

15. Kochańczyk, T.; Drozd, A.; Krężel, A. Relationship between the architecture of zinc coordination and zinc binding affinity in proteins–insights into zinc regulation. *Metallomics* **2015**, *7*, 244–257. [CrossRef]

16. Fisher, M.A.; McKinley, K.L.; Bradley, L.H.; Viola, S.R.; Hecht, M.H. De Novo Designed Proteins from a Library of Artificial Sequences Function in Escherichia Coli and Enable Cell Growth. *PLoS ONE* **2011**, *6*, e15364. [CrossRef] [PubMed]

17. Wei, Y.; Liu, T.; Sazinsky, S.; Moffet, D.; Pelczer, I.; Hecht, M. Selecting Folded Proteins from a Library of Secondary Structural Elements. *Protein Sci.* **2003**, *12*, 92–102. [CrossRef] [PubMed]

18. Wei, Y.; Kim, S.; Fela, D.; Baum, J.; Hecht, M.H. Solution structure of a de novo protein from a designed combinatorial library. *PNAS* **2003**, *100*, 13270–13273. [CrossRef] [PubMed]

19. Go, A.; Kim, S.; Baum, J.; Hecht, M.H. Structure and dynamics of de novo proteins from a designed superfamily of 4-helix bundles. *Protein Sci.* **2008**, *17*, 821–832. [CrossRef]

20. Arai, R.; Kobayashi, N.; Kimura, A.; Sato, T.; Matsuo, K.; Wang, A.F.; Platt, J.M.; Bradley, L.H.; Hecht, M.H. Domain-Swapped Dimeric Structure of a Stable and Functional De Novo Four-Helix Bundle Protein, WA20. *J. Phys. Chem. B* **2012**, *116*, 6789–6797. [CrossRef]

21. Hoegler, K.J.; Hecht, M.H. A de novo protein confers copper resistance in Escherichia coli. *Protein Sci.* **2016**, *25*, 1249–1259. [CrossRef] [PubMed]

22. Smith, C.A.; Kortemme, T. Predicting the Tolerated Sequences for Proteins and Protein Interfaces Using Rosetta Backrub Flexible Backbone Design. *PLoS ONE* **2011**, *6*, e20451. [CrossRef] [PubMed]

23. Irving, H.M.N.H.; Williams, R. 637. The stability of transition-metal complexes. *J. Chem. Soc.* **1953**, *637*, 3192–3210. [CrossRef]

24. Hanlon, D.P.; Watt, D.S.; Westhead, E.W. The Interaction of divalent metal ions with tris buffer in dilute solution. *Anal. Biochem.* **1966**, *16*, 225–233. [CrossRef]

25. Waldron, K.J.; Rutherford, J.C.; Ford, D.; Robinson, N.J. Metalloproteins and metal sensing. *Nature* **2009**, *460*, 823–830. [CrossRef] [PubMed]

26. Andreini, C.; Bertini, I.; Cavallaro, G.; Holliday, G.L.; Thornton, J.M. Metal ions in biological catalysis: From enzyme databases to general principles. *JBIC* **2008**, *13*, 1205–1218. [CrossRef] [PubMed]

27. Smith, G.P. Filamentous fusion phage: Novel expression vectors that display cloned antigens on the virion surface. *Science* **1985**, *228*, 1315–1317. [CrossRef] [PubMed]

28. Van der Gulik, P.; Massar, S.; Gilis, D.; Buhrman, H.; Rooman, M. The first peptides: The evolutionary transition between prebiotic amino acids and early proteins. *J. Theor. Biol.* **2009**, *261*, 531–539. [CrossRef] [PubMed]

29. Brooks, D.J.; Fresco, J.R.; Lesk, A.M.; Singh, M. Evolution of Amino Acid Frequencies in Proteins Over Deep Time: Inferred Order of Introduction of Amino Acids into the Genetic Code. *Mol. Biol. Evol.* **2002**, *19*, 1645–1655. [CrossRef] [PubMed]

30. Cun, S.; Lai, Y.-T.; Sun, H. Structure-oriented bioinformatic approach exploring histidine-rich clusters in proteins. *Metallomics* **2013**, *5*, 904–912. [CrossRef] [PubMed]

31. Baud, F.; Karlin, S. Measures of residue density in protein structures. *PNAS* **1999**, *96*, 12494–12499. [CrossRef]

32. Xu, D.; Zhang, Y. Ab initio protein structure assembly using continuous structure fragments and optimized knowledge-based force field. *Proteins* **2012**, *80*, 1715–1735. [CrossRef]

33. Kozin, S.A.; Mezentsev, Y.V.; Kulikova, A.A.; Indeykina, M.I.; Golovin, A.V.; Ivanov, A.S.; Tsvetkov, P.O.; Makarov, A.A. Zinc-induced dimerization of the amyloid-β metal-binding domain 1–16 is mediated by residues 11–14. *Mol. BioSyst.* **2011**, *7*, 1053–1055. [CrossRef] [PubMed]
34. Salgado, E.; Ambroggio, X.I.; Brodin, J.D.; Lewis, R.A.; Kuhlman, B.; Tezcan, F.A. Metal templated design of protein interfaces. *PNAS* **2010**, *107*, 1827–1832. [CrossRef] [PubMed]
35. Baier, F.; Chen, J.; Solomonson, M.; Strynadka, N.C.; Tokuriki, N. Distinct metal isoforms underlie promiscuous activity profiles of metalloenzymes. *ACS Chem. Biol.* **2015**, *10*, 1684–1693. [CrossRef] [PubMed]

Article

Production of Carbamic Acid Dimer from Ammonia-Carbon Dioxide Ices: Matching Observed and Computed IR Spectra

Zikri Altun [1], Erdi Bleda [1] and Carl Trindle [2],*

[1] Physics Department, Marmara University, 34722 Istanbul, Turkey; altunzikri@gmail.com (Z.A.); erdia@g.clemson.edu (E.B.)
[2] Chemistry Department, University of Virginia, Charlottesville, VA 22902, USA
* Correspondence: cot@virginia.edu; Tel.: +1-434-770-9197

Received: 20 February 2019; Accepted: 19 April 2019; Published: 23 April 2019

Abstract: The production of complex molecules in ammonia–carbon dioxide ices is presumed to pass through species of formula $H_3N:CO_2$ with further addition of ammonia and carbon dioxide. One possible landmark, carbamic acid, H_2NCOOH, has been implicated among the products of warming and irradiation of such ices. Experimental study of the IR spectra of residues has suggested the presence of related species, including weakly bound 1:1 and 2:1 complexes of ammonia with carbon dioxide, zwitterionic carbamic acid, ammonium carbamate, and the dimer of carbamic acid. We computed the energetics and vibrational spectra of these species as well as the complex between ammonia and carbamic acid for gas and condensed phases. By means of a new spectrum-matching scoring between computed and observed vibrational spectra, we infer species that are most probably present. The leading candidates are ammonium carbamate, the carbamic acid–ammonia complex, and the carbamic acid dimer.

Keywords: prebiotic chemistry; carbamic acid; carbon dioxide-ammonia ices; infrared spectra; anharmonicity

1. Experimental Evidence for Carbamic Acid

The evolution of complex and perhaps biologically relevant molecules from the simple molecules well established to exist in the interstellar medium is a central issue in astrochemistry [1]. Among the species of small molecules containing C, H, N, and O that may be formed on grains in ices are variants of carbamic acid, H_2NCOOH. Such species can be precursors of biologically significant species including amino acids. Existence of a T-shaped 1:1 complex of ammonia with carbon dioxide was inferred from molecular beam studies by Fraser et al. [2]. Terlouw and Schwarz [3] showed by mass spectrometric means that carbamic acid could be produced in the gas phase. Frasco [4] and Hisatsune [5] inferred from IR spectroscopy of the residue from VUV-irradiated and warmed ammonia–carbon dioxide ices that ammonium carbamate $NH_4(+)H_2NCOO(-)$ is formed in the solid. Chen et al. [6] irradiated ammonia–carbon dioxide–water ice with 4–20 eV photons, and monitored the 250 K residue by IR. They assigned certain features of the spectra to non-zwitterionic H_2NCOOH.

Bossa et al. [7] interpreted the IR spectrum of low-temperature ammonia–carbon dioxide ices as suggesting the presence of the 1:1 and 2:1 complexes of ammonia with carbon dioxide, and modeled the structures by DFT calculations. In a separate investigation, Bossa et al. [8] warmed 1:1 ammonia–carbon dioxide ice from 10 K to 260 K and monitored the residue by FTIR and mass spectrometry. A species Bossa et al. [8] identified as ammonium carbamate decomposes to ammonia and carbon dioxide above 220 K. Other signals emerged in their study that they associate with carbamic acid, carbamic acid dimer, and ammonium carbamate.

Rodriguez-Lazcano et al. [9] studied solid mixtures of ammonia and carbon monoxide by vapor deposition and also hyperquenching on a cold plate with and without water, over a T range of 120–240 K. They report that ammonium carbamate is the primary product but that carbamic acid species are formed as well; in the presence of water, ammonium bicarbonate is also detected.

Noble and co-workers [10] studied the kinetics of production of ammonium carbamate over a temperature range of 70–90 K. Carbamic acid was proposed as an intermediate; several IR features were attributed to carbamic acid and its dimer.

Irradiation of model ices with high energy electrons or ions is of long-standing interest. Berit and co-workers [11] irradiated carbon dioxide–ammonia ices with 30 keV beams of He cation, but identification of species produced by the beam was difficult owing to intense absorption by the abundant ammonia, carbon dioxide, and (synthesized) water. Khanna and Moore [12] irradiated a composite solid composed of a layer of ammonia below a layer of carbon dioxide–water ice with 1 MeV protons; IR analysis of the 250 K sublimate suggested the presence of zwitterionic $H_3N(+)$-$COO(-)$ carbamic acid. Jheeta et al. [13] irradiated ammonia–carbon dioxide ices with 1 keV electrons, and inferred the production of ammonium carbamate from FTIR spectra.

Lv et al. [14] irradiated ice mixtures including ammonia and carbon dioxide with 144 keV $S(+9)$ cations. While simple warming seemed to produce ammonium carbamate and its dimer, irradiation produced N_2O, OCN anion, and CO. Munoz-Caro and co-workers [15] directed 8.8 eV photons and also beams of 620 meV $Zn(+26)$ and 19.6 meV $Ne(+9)$ ions (emulating cosmic rays) towards methanol-ammonia ice. IR monitoring suggested production of carbonyl groups, perhaps from aldehydes, carboxylic acids, and esters.

Some experimental modeling investigations employ ices with three components. Vinogradoff et al. [16] employed an interstellar ice analog composed of water, CO_2, ammonia, and formaldehyde. Bombardment with UV and ions produced a species identified as ammonium carbamate; carbamic acid was proposed as a catalyst. Noble and co-workers [17] studied several ices including HCN. Ammonium cyanide was identified as a predominant product. Esmaili and co-workers [18] irradiated carbon dioxide–methane–ammonia ices with electron beams with energy up to 70 eV, and attributed the production of glycine to low energy electrons

We note that all these studies rely on a few observed vibrational frequencies for the identification of specific molecules or at least functional groups, often in a condensed phase. A variety of computational models guided these assignments, but in no case is the reliability of the identification established.

2. Computational Modeling of Species Emerging from Water–Ammonia–Carbon Dioxide Ices

Modeling of species and processes occurring in ices presents considerable challenges, even beyond the study of orderly condensed phases. Woon [19] has discussed the special requirements of treating the chemistry and spectra of condensed phases. Since the experimental evidence summarized above for small molecules proposed to be formed in water–ammonia–carbon dioxide ices is derived from IR spectra of residues, computational modeling of those molecules without appreciation of the effects of the medium seems problematic. Modeling of carbamic acid and related species in their gas phase has therefore attracted considerable attention as a preliminary to more demanding studies of condensed phase mixtures. The complex of ammonia with carbon dioxide was first modeled by Amos et al. [20] by SCF methods, and later by Tsipis and Karipidis by DFT methods [21]. Remko and co-workers [22] have shown that carbamic acid is thermodynamically unstable relative to constituent NH_3 and CO_2 using SCF and MP2 theory as well as the CBS-QB3 thermochemical scheme. Ramachandran et al. [23] estimated a considerable (>50 kcal/mol) activation barrier to carbamic acid dissociation. Wen and Brooker [24] report SCF calculations on the zwitterionic form $H_3N(+)$-$CO_2(-)$, finding it slightly less stable than the H_2NCOOH form of carbamic acid in gas. The vibrational spectra for gas phase carbamic acid and the N,N–dimethyl variant have been evaluated in DFT models by Remko [25] and Jamróz and Dobrowski [26]. The 2:1 ammonia:carbon dioxide complex and the ammonia–carbamic acid complex have been similarly described [27]. Dell'Amico et al. [28] suggest that carbamic acid is implicated as a transient species accompanying the condensation of ammonia and carbon dioxide. Noble and co-workers [10] used B3LYP/6311G(d,p) to model the production of carbamate anion and carbamic acid in a cluster composed of a single CO_2 and six ammonia molecules, concluding that those two products are comparably stable.

3. Plan of This Report

We first describe our methods for evaluation of structures and energies of candidate species in gas and condensed phases including ammonia, carbon dioxide, carbamic acid and its dimer, and 1:1 and 2:1 complexes of ammonia and carbon dioxide. Cations $NH_4^{(+)}$, the zwitterion of carbamic acid, and the salt ammonium carbamate are treated with special assumptions detailed below. All species reported so far as possibly present in CO_2–NH_3 ices are treated.

We establish a figure of merit that is intended to judge the quality of matching between observed and computed vibrational frequencies. A "match" is counted if an observed frequency falls within a frequency interval (FI) defined by the MP2 harmonic frequency and the frequency produced by anharmonic correction. We chose to define a FI for the following reason: harmonic frequencies computed by MP2 systematically over-estimate experimental values that include intrinsic anharmonicities. Anharmonic frequencies, however, typically underestimate experimental values. We suggest that the two computed frequencies establish a range (the FI) that should capture the physical system's value. When an observed frequency falls within the FI, we award a point. If an observed value falls outside the computed FI by a few wave numbers, we award a half point. We use the figure of merit to judge reported frequencies from several experimental investigations.

4. Computational Methods for Structure and Energetics

For conventionally bound neutral species, we have chosen the thermochemical scheme W1BD [29] as implemented in the Gaussian 09 suite [30]. W1BD employs density functional models to obtain molecular geometry and vibrational frequencies and many-body corrections (Breuckner doubles) for correlation energy. The salt ammonium carbamate $(NH_4+)(H_2NCOO-)$ and the zwitterionic form of carbamic acid $H_3N(+)COO(-)$ require special treatment. We first conducted W1BD calculations on separate ammonium cations and carbamate anions with no allowance for solvent, but of course medium effects are necessary for proper description of charged species in a condensed phase. For all systems, we estimated polarizable medium stabilization energies by ωB97XD/aug-cc-pVDZ calculations [31,32] with and without Tomasi's model of a polarized continuum [33] with parameters for water using the Gaussian 09 suite. No specific interactions such as H–bonding were addressed. We estimated frequency shifts arising from the medium by CBSQB3 calculations using Gaussian 09 [30], again, with and without inclusion of the Tomasi."water" medium.

5. Species Addressed

Figure 1 shows the structures studied here. For later convenience, we introduce abbreviated names, including C for carbon dioxide and A for ammonia. Carbamic acid is called CA. Analogous abbreviated labels for other species are employed. The energy reference point is defined by the W1BD gas phase energy summed for sufficient carbon dioxide and ammonia molecules to account for all atoms in each structure. Thus, the reported energy is the gas phase binding energy of a species relative to those stable dissociation fragments. We see that conventionally represented carbamic acid (that is, not the zwitterion) in the gas phase is 0.75 kcal/mol less stable than $CO_2 + NH_3$. The zwitterionic form of carbamic acid (CA–Z) is considerably less stable without the stabilization of the condensed phase (ice), but the special assumptions made in the calculation prevent us from assigning a meaningful value for its relative energy. The same issue arises for the ammonium carbamate NH_4+–H_2NCO_2- salt, called A(+):C(−). The W1BD energies put the unsolvated and separate ammonium and carbamate ions both at about 145 kcal/mol and the unsolvated zwitterion at about 36 kcal/mol above the unsolvated ammonia and carbon dioxide.

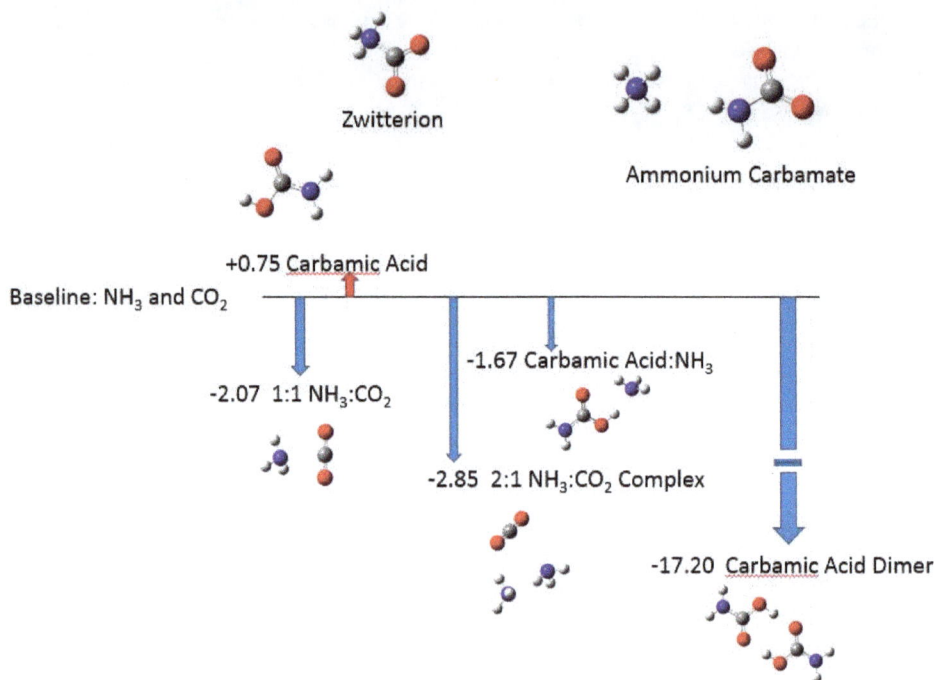

Figure 1. Species addressed in this report. Values are W1BD enthalpies in kcal/mol relative to equivalent numbers of fragments of NH_3 and CO_2.

6. Results and Discussion

The reference fragments A and C can form a 1:1 A:C complex in vacuum, stabilized by 2.07 kcal/mol, or a 2:1 A:A:C complex stabilized by 2.85 kcal/mol. These structures feature a defining interaction between ammonia's N and carbon dioxide's C; the N ... C interaction is comparable with a hydrogen bond in strength. Carbamic acid H_2NCOOH is unstable by 0.75 kcal/mol. The complexation between carbamic acid CA and ammonia A forms CA:A. Forming the carbamic acid–ammonia species is stabilizing by 2.42 kcal/mol; the CA:A complex is 1.67 kcal/mol lower in energy than the stable fragments of carbon dioxide and two ammonias. The striking feature of our calculations is that the carbamic acid dimer $(CA)_2$ is so strongly stabilized, lying 17.2 kcal/mol below the stable fragments of two carbon dioxide molecules and two ammonia molecules. These relative energies are displayed in Figure 1 (no medium correction).

7. Appreciation of Medium Effects: Stabilization Energies

Solvent corrections are not implemented in the W1BD suite, but we can get a first appreciation of such effects from simpler models. The medium stabilization energies shown in Table 1 are obtained from ωB97XD/aug-cc-pVDZ calculations with and without Tomasi's model of a polarized continuum with parameters for water. Solvation has a large effect on charged species (the ammonium and carbamate ions) and highly dipolar species (the zwitterionic form of carbamic acid). However, at least in our approximate estimate, these species do not complete in stability with the dimer. Apart from the charged species, ammonium and carbamate and the zwitterionic form of carbamic acid, all effects are less than 10 kcal/mol. This, however, is sufficient to revise the profile of energies substantially.

Table 1. Solvation effects on relative energies of carbamate species.

Species	CA	$(CA)_2$	Z	$NH_4(+)$ and Carbamate(-)	$NH_3{:}CO_2$	$2NH_3{:}CO_2$	Carbamic Acid:NH_3	NH_3	CO_2
Medium E (kcal/mol)	−6.68	−8.16	−23.01	−139.03	−3.11	−3.58	−6.77	−3.04	−1.50
Relative to solvated NH_3 and CO_2	−1.39	−16.30	17.80	13.24	−0.65	+1.14	−0.87	R-0-	-0-

Medium-adjusted enthalpies are shown in Figure 2. The carbamic acid monomer becomes more stable than fragments of ammonia and carbon dioxide. In contrast, the 2:1 ammonia:carbon dioxide complex is less stable than those dissociation products. The 1:1 ammonia:carbon dioxide complex is still stable, but the complexation of ammonia with carbamic acid is endoergic (requires energy input). The dimer is still much more stable than any alternative. Of course, formation of the dimer is disfavored by entropic factors if the abundance of carbamic acid is low.

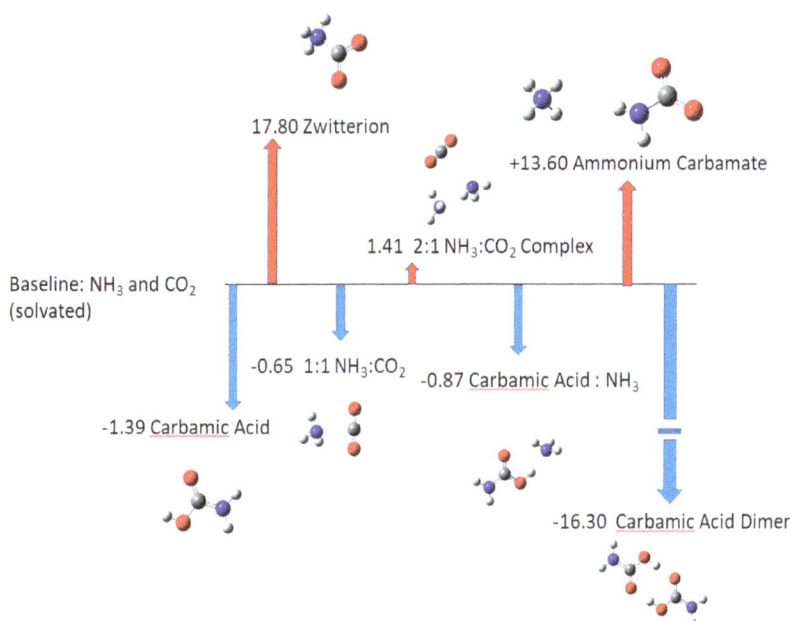

Figure 2. Relative enthalpies with estimated medium effect. Values in kcal/mol relative to equivalent numbers of ammonia and carbon dioxide molecules in medium.

8. Does the Medium Shift Frequencies Substantially?

We used the composite thermochemical scheme CBS-QB3 to judge the impact of the medium on computed harmonic frequencies. This method permits reoptimization of each structure in response to the medium, and estimates frequency shifts arising from both structure changes and the effective potential as corrected for the presence of the polarizable medium. These calculations show that, most often, frequency shifts are minor (less than 20 cm^{-1}). Some corrections are more serious: the leading example is the carbamic acid–ammonia complex, for which the $H_3N \ldots$ H–O hydrogen bond is strengthened and shortened (and the OH stretch is red shifted) and the $H_2NH \ldots$ O= bond is weakened. The vibrational frequencies of the zwitterionic form of carbamic acid are systematically red-shifted, by about 3–5%. The carbamic acid dimer shows medium-induced red shifting of about 50 cm^{-1} for the out-of-plane motion of OH bonds near 950 cm^{-1} and also for the composite motion combining the OCO asymmetric stretches and the COH bends near 1750 cm^{-1}. More typically, the response to the medium is minor for frequencies above 500 cm^{-1}; this is the case for even the weakly bound 1:1 A:C and 2:1 A:A:C$_2$ complexes, for which shifts are no more than 20 cm^{-1}. Details are compiled in the supplementary information.

9. Are Carbamic Species Distinguishable by Computed Vibrational Spectra?

We use the independently calculated MP2/aug-cc-pVDZ//W1BD (no medium correction) harmonic frequencies with anharmonic corrections obtained by the Gaussian 09 suite to judge the assignments to structures in experimental reports. Our comparison is between "frequency intervals" defined as the range between harmonic and anharmonic computed frequencies. Specifically, as shown in Table 2, the highest frequency FI is 3563 to 3721 cm^{-1} for the monomer, while the highest frequency FI is 3474 to 3631 cm^{-1} for the 1:1 complex A:C or H$_3$N:CO$_2$. The two FIs overlap, so we count a match (shown as "YES" in Table 2.) We define a figure of merit, or matching fraction, as the ratio of experimental frequencies matched in the range MP2 (blue extreme) to MP2 + anharmonic corrections (red extreme). Our first exercise is to see how similar the MP2 computed frequency ranges are among the several species. Seemingly, structurally distinct molecules with high similarity would be hard to distinguish by their vibrational spectra. Entries are percentages of FIs common to two molecules. Of course, the diagonal values in this table must be 100%, but totally disjointed sets of FIs could produce a score of zero.

Table 2. Computed frequency intervals (frequencies in wave numbers) for two carbamic system species and the match index indicating that of the 15 frequency intervals (FIs) calculated for the carbamic acid monomer; counterparts for six observed frequencies can be found in the set of FIs calculated for the 1:1 ammonia:carbon dioxide complex.

Monomer M:H$_2$N–COOH		Match to M Found in 1:1 Spectrum?	1:1 Complex H$_3$N:CO$_2$	
Harmonic	Anharmonic		Harmonic	Anharmonic
3721	3563	Yes (FIs overlap)	3631	3474
3823	3648	Maybe (FIs almost touch)	3629	3464
3599	3456	Yes (FIs overlap)	3477	3328
1854	1814	No		
1647	1602	Yes (FIs almost coincide)	1648	1602
1373	1322	Maybe (FIs almost touch)	1308	1248
1239	1210	Maybe (FIs almost touch)		
1094	1066	Yes (FIs overlap)	1076	1005
945	921	No		
776	745	No		
609	615	Maybe (FIs almost touch)	634	652
566	429	No		
521	513	No		
446	401	No		
341	260	No		
Total FI	15	6 (40%) accounted		

Here is how we arrive at a value for a similarity index. Consider the computed MP2 frequencies for the carbamic acid monomer and for the 1:1 ammonia–carbon dioxide complex. Computed MP2 frequencies in harmonic approximation and anharmonic corrections are reported in Table 2. Consider the Monomer's first frequency interval (FI), from 3721 to 3563 cm^{-1}. This FI overlaps with the first FI for the 1:1 complex, so we could recognize one point of agreement. The second FI for the monomer

(3823 to 3648 cm^{-1}) lies slightly outside any FI for the 1:1 complex. We could perhaps award half a point for the near miss. The FI 3599 to 3456 cm^{-1} finds a match in the 1:1 complex set, so it is given one full point. Continuing to the FI of the monomer 1854 to 1814 cm^{-1}, we find no counterpart in the 1:1 complex set of FIs. The FI 1647 to 1602 cm^{-1} does overlap with the FIs of both complexes 1648 to 1602 cm^{-1} and 1648 to 1589 cm^{-1}.

Continuing in this fashion, we find 6.0 points of agreement in the 15 FIs for the carbamic acid monomer, so the similarity index for the 1:1 complex to the monomer is 6.0/15 or 40%. If we set aside the near misses, we find 4/15 points of agreement (27%).

Comparisons among computed FIs are collected in Table 3. A sample entry can be read as the percent of all FIs computed for [row label] found in the set computed for [column label]. More specifically, of all 28 FIs computed for $(CA)_2$, 60% are found in the FIs computed for the complex CA:A. In contrast, of the FIs computed for $(CA)_2$, only 27% are to be found in the computed FIs for the 1:1 complex A:CO_2, while 88% are to be found in the 1:1 carbamic acid–ammonia complex CA:A. (These entries are bold in Table 3.) Unsurprisingly, the greatest degree of similarity is between the 1:1 complex of ammonia with carbon dioxide and the 2:1 complex of ammonia with carbon dioxide. These results illustrate a high degree of similarity among all proposed species, and serve as a precaution against overconfident attribution of observed frequencies to particular species.

Table 3. Similarity indices for computed vibrational frequencies; maximum value is 100. These values include half point values for near-misses. Entries in bold face are discussed in the text.

Similarity	CA	CA–Z	A:CO_2	A:A:CO_2	A+:C–	CA:A	$(CA)2$
CA (15)	-	43	40	43	63	60	57
CA–Z (12)	42	-	33	33	50	71	58
A:CO_2 (10)	60	50	-	95	80	70	60
A:A:CO_2	79	38	94	-	59	76	62
A+:C– (15)	63	40	43	70	-	77	43
CA:A	64	59	52	67	59	-	88
$(CA)_2$ (28)	**57**	52	**27**	**27**	45	**88**	-

Intensities and response of frequencies to isotopic substitution could enhance discrimination. Experimental traces allow at least semiquantitative evaluation of intensities. Inferences would be complicated by the fact that observed intensities are affected by both the intrinsic properties of the mode and the abundance of the species in question.

While to our knowledge no data on isotope effects have been included in reports of experimental studies of synthetic ices, nothing in principle prevents such refinements. To provide a first impression of possible results, we report computed harmonic frequencies for the most intense vibrational transitions for the set of candidates (supplementary information). Those modes likely to show substantial shifts upon perdeuteration are identified. For species incorporating the carboxyl group and a coordinated NH_3, we expect major effects on the OH and NH stretching regions, with shifts approaching the limit of $1/\sqrt{(2)}$. Lesser effects are to be observed for bends involving H(D) atoms. Perdeuteration may allow discrimination between closely related species, notably the complexes of carbon dioxide with one and two ammonia molecules. Perdeuteration seems to have a notable impact only on the 1100 cm^{-1} NH_3 pyramidalization band of the carbon dioxide–ammonia complex, while many more of the modes of the $CO_2(NH_3)_2$ complex are affected.

10. Fingerprint Regions?

Of course our method of frequency interval matching is not how IR spectra are ordinarily used to discriminate between or among species in a mixture. One looks for distinct absorption frequencies by which specific structures can be recognized. Ideally. such descriptive features should be intense and easily detected. Table 4 shows computed FIs for a number of strong and moderately intense IR features. For example, among the computed FIs for the monomer, we find one FI (1854 to 1814 cm^{-1}) intensely absorbing and unique to the monomer. Another FI of the monomer (1239 to 1210 cm^{-1}) is close to FIs for the ammonium carbamate salt and the A:CO_2 1:1 complex. As the detailed tables in the supplementary information show, it is often the case that the most distinctive FI for a particular structure is matched by at least one FI for another species in the set. For example, there are no distinctive FIs uniquely present for the 1:1 complex. The FI 2380 to 2341 cm^{-1} for the 1:1 complex is the most characteristic, held in common only by the 2:1 A:A:CO_2 complex. This is to be expected from the very similar structure of the 1:1 and 2:1 complexes. In this case, we offer no prospect of distinguishing the 1:1 from the 2:1 complex by their IR spectra.

Table 4. Most prominent features of the computed vibrational spectra (frequencies in wave numbers; frequency intervals computed by MP2/aug-cc-pVTZ).

Species	Strong Transitions (FI)	Medium-Strength Transitions (FI)
NH_3 CO_2 complex	None	2380–2341
$(NH_3)_2$ CO_2 complex	None	2381–2343
Monomer	1854–1814	1647–1602; 1373–1322; 1239–1210
Dimer	3163–2783; 1763–1714; 1626–1584; 1385–1355	3786–3514; 976–951
Zwitterion	1807–1788; 1443–1389	
Ammonium carbamate	1670–1639	3646–3346; 1308–1268; 827–797; 665–630
Carbamic acid–NH_3 complex	3108–2757; 1779–1738; 1361–1317	1612–1557; 1115–1059

Still, the information contained in the computed spectra may be sufficient to exclude certain structures from the list, and perhaps assign probabilities to the presence of specific forms. Consider the CA dimer: its computed FI 1498 to 1456 finds a counterpart in the FIs for the carbamic acid complex with ammonia, 1495 to 1432. Two frequencies of the dimer ca. 1380–1350 cm^{-1} are distinguishable from any FI in the complex, as is another FI from 550–564 cm^{-1}. These FIs can serve as fingerprint regions of the spectra.

11. Is It Feasible to Attribute Observed Frequencies to Specific Species?

It is certainly tempting to associate observed frequencies with plausible structures, and investigators have not resisted the temptation. Here, we review several attributions and establish similarity scores for those proposals. Full details and instructions for scoring are in the supplementary information.

Khanna and Moore [12] report IR frequencies for species appearing in the condensed residue after irradiation. They assign a number of IR bands to ammonium carbamate and others to the zwitterionic form of carbamic acid. Detailed compilations of the frequencies observed by Khanna and Moore [12] and the frequency intervals (FIs) computed for each of our seven species are provided in the supplementary data. We consider all 31 frequencies observed and reported by Khanna and Moore [12]. Scoring in the same way as we evaluated matches among computed spectra, we find that the computed fundamental frequencies score 14.5 for the carbamic acid–ammonia complex; 11.5 for carbamic acid dimer, 11.0 for ammonium carbamate, and 9.5 for each of the three other structures, i.e., the zwitterionic carbamic acid and the 1:1 and 2:1 ammonia–carbon dioxide complexes. No single species accounts for as much as half of the observed frequencies. If we include the leading candidate and add the carbamic acid–ammonia complex, and also ammonium carbamate, the score rises to 19, or 61% of the reported observed frequencies. None of the remaining species add so substantially to the match score. For example, the score for the carbamic acid–ammonia complex plus the carbamic acid dimer rises only to 17 (55%). The implication is that one should include the carbamic acid–ammonia complex in the discussion of ammonia–carbon dioxide ices.

Why does it seem impossible to match all reported vibrational frequencies with computed FIs? We compute only fundamental frequencies, while experimental reports may include overtones and combination bands in the set of observed frequencies. There are many low-frequency modes in these species, and indeed our calculations of anharmonicity predict overtones and combinations, some of which have reasonably large intensity. We have not included these frequencies in our match scores, thinking that the general outcome would be indicated in the first stage of the analysis.

The 15 frequencies reported by Hisatsune [5] also suggest that the carbamic acid–ammonia complex is significant. It receives the highest score (7.5), while the ammonium carbamate salt earns 7.0; other structures lag. The combination of the complex and the salt earns a 10.0 score.

Bossa, et al. [7,8] observe 15 frequencies, which they assign to ammonium carbamate and the dimer of carbamic acid. The leading score of 10.5 is earned by the carbamic acid–ammonia complex, while ammonium carbamate is second with 7.5 points. Considering both the complex and the salt increases the total score to 12.0 points. The dimer by itself scores 6.5 points while the combination of dimer and salt totals 12.0 points. The set of three species complex, salt, and dimer does not increase the score further. Deciding between the presence of CA:A complex and $(CA)_2$ dimer requires closer attention.

Chen et al. [6] report five frequencies, and conclude that the carbamic acid monomer is present owing to the observed absorption at 1720 cm^{-1}, assigned to COO asymmetric stretching. Calculations conflict with this assignment, however. DFT calculations (scaled by a factor of 0.98) and MP2 calculations with anharmonicity corrections put this motion at about 1800 cm^{-1}. Lower frequencies near 1720 cm^{-1} are calculated for ammonium carbamate, the dimer, and the complex of carbamic acid with ammonia. The best match (score 4.0) with the five frequencies reported by Chen et al. [6] is found for the carbamic acid:ammonia complex.

Rodriguez-Lazcano et al. [9] report 17 frequencies at 140 K and 20 frequencies at 240 K from their vapor deposition studies (we do not include the "shoulders"). They infer the presence of ammonium carbamate and offer evidence for a carbamic acid species. They suggest that the species could be isolated carbamate anion, without ruling out the possibility of any or all of neutral, zwitterionic, and dimeric carbamic acid species. Our matching exercise for the high temperature data shows the carbamic acid–ammonia complex to be the leading single candidate at high temperature with a score of 12, followed by the dimer (10) and ammonium carbamate (9.5). The best score for two species is 16.5 for the dimer and ammonium carbamate, followed by 15.5 for ammonium carbamate and the carbamic acid–ammonia complex. The low-temperature data also suggest the presence of ammonium carbamate and allow the possibility of either or both the dimer and the ammonia–carbamic acid complex.

Noble et al. [10] report 14 frequencies in an ammonia–carbon dioxide ice between 70 and 90 K. They suggest that ammonium carbamate is a major product of the warming, and assign some frequencies to carbamic acid and to its dimer. In our scoring the carbamic acid–ammonia complex achieves a score of 10, followed by ammonium carbamate (8.5) and then the carbamic acid dimer and the 2:1 ammonia–carbon dioxide complex. The best score for a two-component system includes the carbamic acid–ammonia complex and ammonium carbamate, totaling 12.

12. Conclusions

Ammonium carbamate and the carbamic acid complex with ammonia seem to be the most likely products of irradiation of ammonia–carbon dioxide ices. The carbamic acid dimer is also a candidate, but the abundance of ammonia in most of the experimentally studied solid environments also points to production of the ammonia complex. The pathway for conversion of ammonium carbamate to the ammonia–carbamic acid complex in the solid state is worth exploring. We recommend that investigations of models of prebiotic synthesis of small molecules and astronomical data include consideration of the possibility of the ammonia–carbamic acid complex.

Supplementary Materials: The following are available online at http://www.mdpi.com/2075-1729/9/2/34/s1, Coordinates of all species, detailed description of evaluation of figure of merit MATCH for all comparisons.

Author Contributions: Formal analysis, C.T.; Investigation, Z.A., E.B., and C.T.

Funding: The authors are grateful to BAPKO, the research office of Marmara University, for financial support of this study. Thanks also to the Body Foundation for travel expenses.

Conflicts of Interest: The authors declare no conflict of interest.

References

1. Herbst, E.; Yates, J.T. Introduction: Astrochemistry. *Chem. Rev.* **2013**, *113*, 8707–8709. [CrossRef] [PubMed]

2. Fraser, G.T.; Leopold, R.K.; Klemperer, W. The rotational spectrum, internal rotation, and structure of NH3–CO2. *J. Chem. Phys.* **1984**, *81*, 2577. [CrossRef]

3. Terlouw, J.K.; Schwarz, H. The Generation and Characterization of Molecules by Neutralization-Reionization Mass Spectrometry (NRMS). *Angew. Chem. Int. Ed. Engl.* **1987**, *26*, 805–808. [CrossRef]

4. Frasco, D.L. Infrared Spectra of Ammonium Carbamate and Deuteroammonium Carbamate. *J. Chem. Phys.* **1964**, *41*, 2134. [CrossRef]

5. Hisatsune, I.C. Low-temperature infrared study of ammonium carbamate formation. *Can. J. Chem.* **1984**, *62*, 945. [CrossRef]

6. Chen, Y.J.; Nuevo, M.; Hsieh, J.M.; Yih, T.S.; Sun, W.H.; Ip, W.H.; Fung, H.S.; Chiang, S.Y.; Lee, Y.Y.; Chen, J.M.; et al. Carbamic acid produced by the UV/EUV irradiation of interstellar ice analogs. *A&A* **2007**, *464*, 253–257.

7. Bossa, J.B.; Duvernay, F.; Theulé, P.; Borget, F.; Chiavassa, T. VUV irradiation of carbon dioxide (CO2) and ammonia (NH3) complexes in argon matrix. *Chem. Phys.* **2008**, *354*, 211–221. [CrossRef]

8. Bossa, J.B.; Theulé, P.; Duvernay, F.; Borget, F.; Chiavassa, T. Carbamic acid and carbamate formation in NH3:CO2 ices – UV irradiation versus thermal processes. *A&A* **2008**, *492*, 719–724.

9. Rodriguez-Lazcano, Y.; Mate´, B.; Herrero, V.J.; Escribano, R.; Galvez, O. The formation of carbamate ions in interstellar ice analogues. *Phys. Chem. Chem. Phys.* **2014**, *16*, 3371–3380.

10. Noble, J.A.; Theule, P.; Duvernay, F.; Danger, G.; Chiavassa, T.; Ghesquiere, P.; Minerva, T.; Talbi, D. Kinetics of the production of ammonium carbamate. *Phys. Chem. Chem. Phys.* **2014**, *16*, 23604–23615. [CrossRef] [PubMed]

11. Benit, J.; Bibring, J.-P.; Rocard, F. Chemical Irradiation Effects in Ices. *Nucl. Instrum. Methods Phys. Rese.* **1988**, *32*, 349–353. [CrossRef]

12. Khanna, R.K.; Moore, M.H. Carbamic acid: molecular structure and IR spectra. *Spectrochim. Acta A* **1999**, *55*, 961–967. [CrossRef]

13. Jheeta, S.; Ptasinskan, S.; Sivaraman, B.; Mason, N.J. The Irradiation of ammonia: carbon dioxide ice at 30 K. *Chem. Phys. Lett.* **2012**, *543*, 208–212. [CrossRef]

14. Lv, X.Y.; Boduch, P.; Ding, J.I.; Domaracka, A.; Langlinay, T.; Palumbo, M.E.; Rotard, H.; Strazzulla, G. Thermal and energetic processing of ammonia and carbon dioxide bearing solid mixtures. *Phys. Chem. Chem. Phys.* **2014**, *16*, 3433–3441. [CrossRef] [PubMed]

15. Munoz Caro, G.M.; Dartois, E.; Boduch, P.; Rothard, H.; Domaracka, A.; Jimenez-Escobar, A. Comparison of UV and high energy ion irradiation of methanol: ammonia ice. *A&A* **2014**, *556 Pt 2*, A93.

16. Vinogradoff, V.; Duvernay, F.; Fray, N.; Bouilloud, M.; Chiavassa, T.; Cottin, H. Carbon dioxide influence on the thermal formation of complex organic molecules in interstellar ice analog. *Astrophy.s J. Lett.* **2015**, *809*, L18. [CrossRef]

17. Noble, J.A.; Theule, P.; Borget, F.; Danger, G.; Chovat, M.; Duvernay, F.; Mispelaer, F.; Chiavassa, T. The thermal reactivity of HCN and NH3 in interstellar ice analogues. *Mon. Notices Royal Astron. Soc. (MNRAS)* **2013**, *428*, 3262–3273. [CrossRef]

18. Esmaili, S.; Bass, A.D.; Clotier, P.; Sanche, L.; Huels, M.A. Glycine formation in CO2:CH4:NH3 ices induced by 0-70 eV electrons. *J. Chem. Phys.* **2018**, *148*, 164702. [CrossRef]

19. Woon, D.E. Quantum chemical protocols for modeling reactions and spectra in astrophysical ice analogs: the challenging case of the C+ + H2O reaction in icy grain mantles. *Phys. Chem. Chem. Phys.* **2015**, *17*, 28705. [CrossRef]

20. Amos, R.D.; Handy, N.C.; Knowles, P.J.; Rice, J.E.; Stone, A.J. Ab-Initio Prediction of Properties of CO2, NH3, and CO2 … NH3. *J. Phys. Chem.* **1985**, *89*, 2186–2192. [CrossRef]

21. Tsipis Constantinos, A.; Karipidis Paraskevas, A. Mechanistic Insights into the Bazarov Synthesis of Urea from NH3 and CO2 Using Electronic Structure Calculation Methods. *J. Phys. Chem. A* **2005**, *109*, 8560–8567. [CrossRef]

22. Remko, M.; Rodeb, B.M. Ab initio study of decomposition of carbamic acid and its thio and sila derivatives. *J. Mol. Struct. Theochem* **1995**, *339*, 125–131. [CrossRef]

23. Ramachandran, B.R.; Halpern, A.M.; Glendening, E.D. Kinetics and Mechanism of the Reversible Dissociation of Ammonium Carbamate: Involvement of Carbamic Acid. *J. Phys. Chem. A* **1998**, *102*, 3934–3941. [CrossRef]

24. Wen, N.; Brooker, M.H. Ammonium Carbonate, Ammonium Bicarbonate, and Ammonium Carbamate Equilibria: A Raman Study. *J. Phys. Chem.* **1995**, *99*, 359–368. [CrossRef]

25. Remko, M. The gas-phase acidities of substituted carbonic acids. *J. Mol. Struct. Theochem* **1999**, *492*, 203–208. [CrossRef]

26. Jamróz, M.H.; Dobrowolski, J. Theoretical IR spectra and stability of carbamic acid complexes. *Vibrat. Spectrosc.* **2002**, *29*, 217–221. [CrossRef]

27. Jamróz, M.H.; Dobrowolski, J.C.; Borowiak, M.A. Borowiak Theoretical IR spectra of the (2:1) ammonia–carbon dioxide system. *Vibrat. Spectrosc.* **2000**, *22*, 157–161. [CrossRef]

28. Dell'Amico, D.B.; Calderazzo, F.; Labella, L.; Marchetti, F.; Pampaloni, G. Converting Carbon Dioxide into Carbamato Derivatives. *Chem. Rev.* **2003**, *103*, 3857–3897. [CrossRef]

29. Barnes, E.C.; Petersson, G.A.; Montgomery, J.A.; Frisch, M.J.; Martin, J.M.L. Unrestricted Coupled Cluster and Brueckner Doubles Variations of W1 Theory. *J. Chem. Theory Comput.* **2009**, *5*, 2687–2693. [CrossRef]

30. Frisch, M.J.; Trucks, G.W.; Schlegel, H.B.; Scuseria, G.E.; Robb, M.A.; Cheeseman, J.R.; Scalmani, G.; Barone, V.; Mennucci, B.; Petersson, G.A.; et al. *Gaussian 09, Revision E.01*; Gaussian, Inc.: Wallingford, CT, UK, 2013.

31. Chai, J.D.; Head-Gordon, M. Long-range corrected hybrid density functionals with damped atom-atom dispersion corrections. *Phys. Chem. Chem. Phys.* **2008**, *10*, 6615–6620. [CrossRef]

32. Dunning, T.H., Jr. Gaussian basis sets for use in correlated molecular calculations. I. The atoms boron through neon and hydrogen. *J. Chem. Phys.* **1989**, *90*, 1007–1023. [CrossRef]

33. Tomasi, J.; Persico, M. Molecular Interactions in Solution: An Overview of Methods Based on Continuous Distributions of the Solvent. *Chem. Rev.* **1994**, *94*, 2027–2094. [CrossRef]

Article

Prebiotic Soup Components Trapped in Montmorillonite Nanoclay Form New Molecules: Car-Parrinello Ab Initio Simulations

Juan Francisco Carrascoza Mayén [1,2], Jakub Rydzewski [3], Natalia Szostak [1,2,4], Jacek Blazewicz [1,2,4] and Wieslaw Nowak [3,*

[1]   Institute of Computer Science, Poznan University of Technology, 60-965 Poznan, Poland;
      Francisco.Carrascoza@cs.put.poznan.pl (J.F.C.M.); Natalia.Szostak@cs.put.poznan.pl (N.S.);
      jblazewicz@cs.put.poznan.pl (J.B.)
[2]   European Center for Bioinformatics and Genomics, Poznan University of Technology, 60-965 Poznan, Poland
[3]   Institute of Physics, Faculty of Physics, Astronomy and Informatics, N. Copernicus University, 87-100 Torun,
      Poland; jr@fizyka.umk.pl
[4]   Institute of Bioorganic Chemistry, Poznan Academy of Sciences, 61-704 Poznan, Poland
[*]   Correspondence: wiesiek@fizyka.umk.pl; Tel.: +48-56611-3204

Received: 16 April 2019; Accepted: 30 May 2019; Published: 4 June 2019

Abstract: The catalytic effects of complex minerals or meteorites are often mentioned as important factors for the origins of life. To assess the possible role of nanoconfinement within a catalyst consisting of montmorillonite (MMT) and the impact of local electric field on the formation efficiency of the simple hypothetical precursors of nucleic acid bases or amino acids, we performed ab initio Car–Parrinello molecular dynamics simulations. We prepared four condensed-phase systems corresponding to previously suggested prototypes of a primordial soup. We monitored possible chemical reactions occurring within gas-like bulk and MMT-confined four simulation boxes on a 20-ps time scale at 1 atm and 300 K, 400 K, and 600 K. Elevated temperatures did not affect the reactivity of the elementary components of the gas-like boxes considerably; however, the presence of the MMT nanoclay substantially increased the formation probability of new molecules. Approximately 20 different new compounds were found in boxes containing carbon monoxide or formaldehyde molecules. This observation and an analysis of the atom–atom radial distribution functions indicated that the presence of Ca^{2+} ions at the surface of the internal MMT cavities may be an important factor in the initial steps of the formation of complex molecules at the early stages of the Earth's history.

Keywords: ab initio molecular dynamics; catalysis; prebiotic soup; origins of life

1. Introduction

The origins of life theories are based on hypothetical chemical scenarios that lead to the formation of biomolecules, starting with substances that could be found in a given proto-earth-like system [1–3]. These models share the assumption that there should be a way to explain the synthesis of complex biomolecules, starting from simpler molecular elements, and the notion that such a construction should happen in a scaled manner [4–6]. Despite still debated particular conditions present in the early Earth [7], formation of building blocks of life was possibly facilitated by appropriate physical factors such as reducing atmosphere, strong electric field, UV radiation, mineral catalytic surfaces, cometary impact, or high temperature [8]. Thus, most of those theories meet at a point where simple substances, such as ammonia, carbon monoxide/dioxide, molecular oxygen, and water, form a molecular intermediate prior to the formation of nucleotides or amino acids [9]. Studies of these elementary reactions and physical factors leading to more complex precursors of biosystems are important in understanding origins of life.

Numerous scenarios of prebiotic chemistry have been proposed so far, and many reviews on this topic have been published [6,9–14]. Probably the soundest test complementing such hypotheses was the Miller–Urey experiment [15]. In this experiment, for the first time, a series of biomolecular elementary building blocks, such as glycine and formamide, was obtained from a strongly reducing mixture of simple gases, such as methane, ammonia, hydrogen, and water. The substrates were sealed in a vessel in which an electric discharge was applied for a prolonged period of time. After the Miller–Urey test, similar experiments were performed [1,2,16] showing richer chemistry, and even formation of nucleobases [17].

Among other factors, the catalytic role of solid surfaces has been widely studied in the context of the origins of life. In this paper, we ask what may be the role of the nanoconfinement of simple molecules in montmorillonite clay in facilitating formation of the biomolecule precursors. We tackle this question using ab initio simulations which are able to indicate possible formation of new chemical species. As a part of an extended introduction, we give below selected accounts on some experimentally supported hypotheses, especially those restoring to catalytic interactions with solid surfaces and/or nanometric size compartments. More details may be found in Reference [18].

1.1. Possible Origins of Life Scenarios—A Short Review

In XXth century, an iron–sulphur hypothesis for the origins of life was proposed in a series of 1988–92 articles by G. Wächtershäuser, which attracted considerable attention [19,20]. In this scenario, not only high temperature (400 K) and pressure present in hydrothermal vents were important, but also the catalytic properties of iron sulfide minerals played a major role. This original hypothesis and a postulate of primitive autocatalytic metabolism were recently criticized as assumptions regarding concentrations of required reagents were unrealistic [21].

Another scenario was developed by Sutherland, Powner, Szostak, and coworkers, in which a HCN molecule was a key point to prebiotic chemistry. Notably, formaldehyde, studied here, and 2-aminooxazole were also involved in this scheme [5,10,12,22–25]. It was postulated that a linkage between amino acids, RNA, and lipids was possible through cyanosulfidic chemistry [12]. In an important paper [10], a hypothesis that all basic precursors of biological molecules were formed in cyanosulfidic reactions, partially catalyzed by minerals, was elaborated and supported by experimental data. An interesting geochemical scenario was dependent on schreibersite ((Fe,Ni)$_3$P), HCN, hydrosulfide (HS$^-$), copper, and ultraviolet light under postimpact conditions [3]. All of these experiments suggested that solid surfaces are potential places of catalytic reactions crucial for the origins of life.

Another group pursued a scenario based on formamide synthetic chemistry, stressing advantages and simplicity of one pot synthesis [1,26–31]. In a review by Saladino et al. [32], reactions of HCN with formamide catalyzed by various meteorites were described. The condensation of formamide on the surface of 15 minerals was analyzed as well [33]. Measuring the stability of the obtained products [34] provided useful insights into the ribonucleic acid (RNA) oligomer degradation processes. Clearly, the type of catalytic surface is important; for example, studies of iron–sulfur minerals [35] showed that basalts provide better stability for RNA oligomers than the other surfaces.

Costanzo et al. described procedures for obtaining a set of biomolecules using UV–Vis light, electricity, heat, and high-energy proton bombardment [28]. An analysis of meteorite samples found glycine and formamide among other biomolecules on the surface [36,37].

It is worth to mention experiments that have shown reactions between formamide (NH$_2$CHO) and thermal water (358 K) in the presence of meteorites, in the environment mimicking a plausible and "natural" prebiotic scenario. The results indicate that meteorites from classes: stony iron, chondrite, achondrite, effectively catalyze the synthesis of numerous organic biological compounds including carboxylic acids, nucleobases, amino acids, and sugars [38].

1.2. Catalytic Surfaces, Nanoconfinement, and Biogenesis

Reactions between elementary chemicals discussed in our paper (NH_3, H_2O, CO, formaldehyde, and HCN) have proven to be possible spontaneously [39] in condensed phase as well as in a liquid state [40]. By providing sufficient time under favorable conditions, these simple compounds are capable of forming more complex structures leading finally to nucleotides or amino acids, and polymers [41]. Nevertheless, the reactions are low-yield, and a long period of time is required to achieve substantial amounts of the products. Some natural aids can be incorporated in the physical environment, such as additional sources of energy, catalytic surfaces or nanoconfinement. These, first, increase the probability of a reaction occurrence, and second, result in a considerably faster development of various products and higher yields of reactions.

Chemical reactions in a limited space were considered in the past in the context of biogenesis. For review of this problem, we refer to a recent article by Dass et al. [42].

The properties of chemicals in a very constrained space are often much different than those in bulk. Molecular crowding increases the probability of reactive contacts, and strong electric field gradients help to polarize and direct molecules. Thus, chemistry in the compartments of nanometer dimensions is expected to have some peculiarities. One may expect that the reactivity of relatively inert molecules may be affected, to the extent that new species are created in the nanoconfinement conditions. This hypothesis is tested in this paper. Since in prebiotic times the Earth's volcanic activity was strong, the environment might contain minerals formed from ashes and having porous structures, similar to our model of smectite MMT. Systems ready to accommodate various mixtures of elementary compounds might be quite abundant, thus we postulate that the computational studies of chemistry in nano-reactors are paramount in completing the full set of physical factors governing the formation of complex biomolecules. It is known, that theoretical studies of nanoconfinement are difficult to perform and rare. Therefore, we applied an advanced computational methodology to study the effects of trapping compounds, which were probably present in the primordial soup, on their reactivity while locked in minerals.

MMT is a mineral, belonging to a subclass of smectites, a representative of nanoclays, named after its discovery at Montmorillon (France) in the XIXth century. It is formed by weathering of volcanic ash under poor drainage conditions or in saline environment. It has a unique structure with a layer of loosely bound positive ions (Na^+ or Ca^{2+}) located between negatively charged aluminosilicate surfaces. Because of its ability to absorb water and its catalytic properties, MMT has many applications in oil drilling industry, paper production, and dog food enrichment. MMT was considered in the past for its possible role in the origins of life. Namely, in 2003, Szostak et al. reported that the special electrical properties of MMT particles aid phospholipid vesicle formation. The formation rates were accelerated 100-fold after the addition of MMT to the solution of phospholipids [43]. These authors also hypothesized that at the same time, MMT nanopores could hold RNA molecules. Further experiments focused on MMT as a catalyst were reported by Ferris [44] , Joshi et al. [45], and Jheeta and Joshi [46]. Interestingly, clay minerals similar to MMT were found on Mars by the Opportunity probe. The current interest in the properties of this system is high mainly because of the possible carbon dioxide sequestration [47].

1.3. Theoretical Chemistry in Origins of Life Research

We use theoretical chemistry which is a useful and well-established approach for studies on surface catalytic effects. The adsorption of nucleobases over surface of several clays was often described the at density functional theory (DFT) level [48]. The modeling of kaolinite by the Leszczynski team [49,50] underscored the role of calcium for the adsorption of formamide over the clay's surface. Bhushan et al. reported manganese oxides as a possible catalyst for the nucleobase synthesis [1].

The roles of TiO_2 and the UV light in the formation of adenine and thymine from formamide were reported also [51,52]. More recently, Ferus et al. suggested that life started during the late heavy bombardment period of the Earth by meteorites [30]. This team modeled machinery capable of simulating shock waves in the laboratory by high-pressure effects. Their study was supported by DFT calculations and metadynamics free energy profiles calculations as well [17]. Important insights came from papers published by Goldman and coworkers, who showed (i.a.) that impact-induced shock compression of cometary ices, followed by expansion to ambient conditions, can produce complexes that resemble glycine [53,54]. DFT ab initio molecular dynamics (MD) on picosecond timescale showed that shock waves may drive the synthesis of transient C–N bonded oligomers.

Classical "static" quantum chemical studies of reaction profile energies required "ad hoc" assumptions on reaction coordinates and fixed products. A much better (but more expensive) approach is to use ab initio MD, which does not depend on a force field. In MD, interacting substrates, usually located in a box with periodic boundary conditions, move in time, collide, and under favorable arrangements make products.

Important contribution of ab initio MD (AIMD) to prebiotic chemistry was work by D. Marx et al. [55] The findings from this group work were summarized in an excellent review on chemistry occurring in nanoconfined water [56]. The authors modeled water in 'moderate nanoconfinement' between mackinawite mineral sheets. The cage consisted of two $Fe_{32}S_{32}$ parallel layers situated at the top and bottom of a supercell preserving the spacing of 5.03 Angstroms. Prebiotic peptide cycle was studied and, among others, free energy profiles of glycine reactions with small molecules were calculated using the Car–Parrinello MD method (CPMD) [57]. It was found that nanoconfined water exerts charge stabilizing effects. In comparison with ambient water some reaction barriers are strongly affected by such conditions.

Stirling et al. (2016) used AIMD to monitor the reaction leading from NO_2 to NH_3 and catalyzed by the presence of iron minerals [58]. More advanced CPMD simulations were performed while studying pyrite (FeS_2) as the catalyst [2,29–31,33]. Extensive CPMD modeling was also employed by Ferus et al. in a recent study of nucleic acid components formation in Miller–Urey atmosphere [17].

Several in silico studies of more elementary reactions leading to the formation of biomolecular fragments were published in the recent years. The Miller–Urey experiment was modeled by Saitta and Saija [39], who performed an illuminating theoretical analysis of elementary gases present in primordial soup, mimicking this experiment using CPMD. An external electric field was found to be a crucial factor leading to glycine formation via formamide. The same group performed ab initio MD simulations, and successfully described the reversible formation of formamide from very simple precursors NH_3 + CO, both in gas phase and in solution [59] Notably, a new methodology for studies of elementary reactions channels was proposed in that paper. The same approach, i.e., AIMD simulations, was applied to monitor synthesis of nucleotides from nucleobases and 5-phospho-α-D-ribose-1-diphosphate [60]. Simulations showed that this reaction may happen in mildly basic pH and 400 K, a temperature postulated for prebiotic hydrothermal conditions, with a free-energy cost estimated as 1.2 and 3.3 kcal/mol for uracil and adenine, respectively.

Classical MD simulations for MMT clay were performed in the past as well (see [47,61] and the references therein), but only nonreactive force fields were applied, and MMT effects on the reactivity of biomolecule precursors have not been studied computationally yet.

1.4. Our Aim

Here we test the hypothesis that MMT (or minerals with a similar composition/structure) might contribute to the formation of complex organic molecules during the prebiotic period of the Earth's history. We do not coin any particular scenario with MMT as a key component. We do not try to reproduce full reactions paths leading from primordial soup components to known amino acids or nucleobases. Instead, we rather point out that unique catalytic properties of this nanoclay mineral might facilitate formation of variety of complex organic molecules even without extremely high temperatures, electric discharge, UV radiation, or high impact physical factors. We bring attention of the chemist community to a possible role and significance of metal ions adsorbed in minerals for origin of life studies.

In this work, we exploit the same modeling methodology as that used by Saitta and Saija [39], but we extended their approach to elucidate the hypothetical role of MMT on the path to the elementary building blocks of RNA or amino acids. We used AIMD simulations to evaluate a potential effect of calcium ions present in MMT, a catalytic role of the two-dimensional nanoconfinement of primordial soup components in the MMT nanopores, and the elevated temperature of 400 K on the probability of new compounds formation. To the best of our knowledge, this is the first CPMD study on the role of the MMT nanoconfinement in hypothetical processes related to early stages of biomolecular evolution. We have found that, in our model, MMT nanoclay alone facilitates formation of nearly 20 new organic compounds just from water, ammonia, methane, nitrogen, and carbon monoxide mixture at 20 ps time scale, even in the absence of an external electric field.

2. Methods

2.1. Systems

Following the protocol presented by Saitta and Saija [39], we modeled two types of "primordial soup" systems: g and m. The g systems were virtually identical with those studied by Saitta and Saija [39] and were used here as a reference. The m systems had g mixtures confined in the slab of MMT, as shown in Figure 1 and Table 1. The g and m systems (called later "boxes") contained components corresponding to the four stages of the glycine formation process called Miller–Strecker reaction, which were coded as follows: 0—the original Miller–Urey substrates, 1—reactants, 2—intermediates, and 3—products. The compositions of boxes were carefully selected in work [39], in order to reproduce the intermediate and end products of the Strecker reaction, and to allow each box to have a compatible number of each atom types, thus we used exactly the same systems. The boxes were electrically neutral. The preliminary initial positions of the molecules in the starting boxes were generated by PACKMOL [62] at the density of 1 g/mL. The steepest descent and simulated annealing methods were used to optimize the geometry of the initial structures. In the first step the ions' positions were relaxed using the steepest descent algorithm, the electrons where kept on relaxation using a 0.5 ps damped electron dynamics, afterwards ions' positions were further optimized with a damped 1 ps dynamics, and finally both, electrons and ions were relaxed once more time using the steepest descent algorithm. Then molecular dynamics was started using the Verlet algorithm for both types of particles, i.e., ions and electrons, with the increasing temperature controlled by the Nose–Hoover [63] thermostat at frequency of 13.5 Thz. Before collecting data, 2 picoseconds of equilibration in an appropriate temperature (300 K, 400 K, or 600 K) was applied in each box.

(a)   (b)

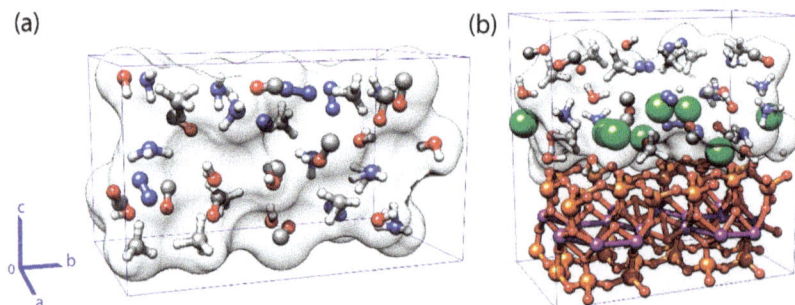

Figure 1. Models of g "condensed phase" (**a**) and m for "condensed phase" + montmorillonite (MMT) boxes (**b**). MMT is depicted using a "balls and sticks" representation, calcium atoms are shown as green spheres, and components of the g box are depicted using sticks and transparent surfaces. The borders of the PBC boxes and the definitions of axes are also shown.

Table 1. Systems modeled, dimensions: $a = 10.30$ Å, $b = 17.96$ Å.

Box	Temperature	No. Atoms	Chemical Species	Dimension c [Å]
g0300	300	176	32 H2O; 8 NH3; 8 CH4; 4 H2	10.43
g0400	400			
g0600	600			
g1300	300	126	8 H2O; 8 NH3; 8 CH4; 5 N2; 10 CO	7.41
g1400	400			
g1600	600			
g2300	300	126	9 H2O; 9 NH3; 9 Formaldehyde; 9 HCN	7.41
g2400	400			
g2600	600			
g3300	300	126	9 NH3; 9 Glycine	7.41
g3400	400			
g3600	600			
m0300	300	328	4 MMT; 32 H2O; 8 NH3; 8 CH4; 4 H2	18.07
m0400	400			
m0600	600			
m1300	300	278	4 MMT; 8 H2O; 8 NH3; 8 CH4; 5 N2; 10 CO	17.90
m1400	400			
m1600	600			
m2300	300	278	4 MMT; 9 H2O; 9 NH3; 9 Formaldehyde; 9 HCN	17.90
m2400	400			
m2600	600			
m3300	300	278	4 MMT; 9 NH3; 9 Glycine	17.90
m3400	400			
m3600	600			

Because of the periodic boundary conditions (PBC), the mixtures in the m-type boxes were effectively confined in nanocages (~1.0 nm × 1.8 nm × 0.8 nm, see Figure 1b). The pressure in the model cavity is difficult to control and, due to the low compressibility of water, may be high. The dimensions of the boxes are presented in Table 1.

2.2. CPMD Simulations

CPMD simulations relay on the classical motion of heavy ions interacting with each other and experiencing a potential from fast moving electrons. In contrast to the classical MD, CPMD explicitly includes the electrons as active degrees of freedom via fictitious dynamical variables [57]. To reduce the number of electrons and computational time, atomic inner core electrons are replaced by plane–wave pseudopotentials, and electronic correlation effects are included in specially designed exchange–correlation functionals adopted from the DFT methods. Therefore, chemical bonding may be studied using CPMD.

CPMD simulations were performed using Quantum Espresso 5.3.0 [64] with the Perdew–Burke–Ernzerhov exchange and correlation functional [65] and the softcore pseudo-potentials by del Corso [66] with a kinetic energy cutoff of 35 Ry and a charge density cutoff of 280 Ry. The fictitious electronic mass was set to 500 a.u. The ion dynamics was performed in the NVT ensemble using the Verlet algorithm [67] and the Nose–Hoover [63] thermostat at a frequency of 13.5 THz. Each system was simulated at 300 K, 400 K, and 600 K, at 1 atm external pressure, for 20 ps with a time step of 0.1 fs. As already mentioned, before production runs a gradual relaxation of minimized geometry was adopted together with an increasing temperature heating phase and 2 ps equilibration runs.

The trajectories were analyzed using PLUMED 2.4 [68] and VMD [69]. A typical 20 ps CPMD run took about 12 days on a 10 processor (28 cores each) cluster at the PCSS Computing Center (Poznan, Poland). Twenty-four runs were performed.

We monitored the distributions of all of the heavy atoms forming our simple molecules from all the steps of the Miller–Urey experiment (boxes 0–3 in the g and m systems) by calculating the atom–atom radial distribution functions $g_{AB}(r)$. The radial distribution functions for atoms B around atoms A were calculated as follows

$$g_{AB}(r) = \frac{1}{4\pi\rho_B r^2} \frac{\Delta N_{A-B}}{\Delta r}$$

where ρ_B is the number density of atoms B and ΔN_{A-B} is the average number of atoms B lying in the region r to $r + \Delta r$ from a type-A atom.

3. Results and Discussion

We aimed at monitoring the chemical reactions possibly occurring in "gaseous" g-type boxes and the same molecular systems but confined in the MMT slab nanopore, i.e., m-type boxes. Further, we monitored the effects of the increasing temperature on the chemical reactivity of these mixtures as well. Twenty-four computational boxes were modeled in total (Table 1). The 20-ps timescale of the sampling ab initio mechanical dynamics is typical for such studies [39], given the high demand of computational time required for the AIMD simulations of such large systems as those studied here (126–328 atoms). The catalytic effects observed in this relatively short time window should be even more pronounced on geological timescales when not only a limited set of new compounds observed here might be formed, but formation of other complex molecules should be reasonably expected. Some of our newly formed molecules may be short lived compounds. Much longer simulations are required to sample all possible chemistry in our model systems.

Using homemade scripts and computer graphics, we searched for the formation of new chemical species. We evaluated the probabilities of reactions in each box by measuring the number of molecule–molecule close contacts, effective clashes leading to new species, and the atom–atom radial distribution functions g(r). The definition of effective clashes is given in Section 3.1.

The dynamics of the MMT slab was also monitored by its root-mean-square deviation (RMSD), radial distribution functions, and distortions of the slab geometry by inspecting the atomic position plots for each type of metal ion. We identified the newly formed species (see Table 2 and Figure 2), and evaluated their lifetimes in the course of the 20-ps CPMD trajectories (Table 3).

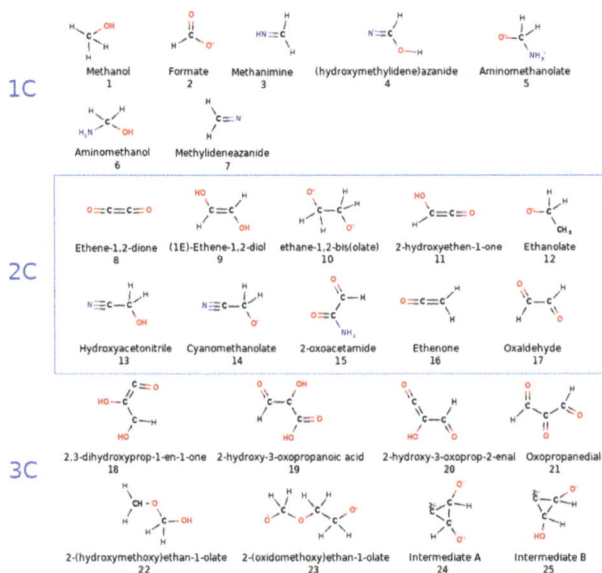

Figure 2. All new products noticed during 20 ps time of CPMD simulations in 24 boxes.

Table 2. Number of newly formed effective atomic contacts (clashes) in each 20-ps Car–Parrinello MD method (CPMD) trajectory and a list of the observed new transient products in each box. For the chemical formulas of the products, see Figure 2.

Box	C C	C N	C O	Products
g0300	0	0	0	0
g0400	0	0	0	0
g0600	0	0	0	0
g1300	0	0	0	0
g1400	0	0	0	0
g1600	0	0	0	0
g2300	0	15	0	6
g2400	0	1	43	6
g2600	6	0	0	14
g3300	0	0	0	0
g3400	0	0	0	0
g3600	0	0	0	0
m0300	0	0	0	0
m0400	0	0	0	0
m0600	0	0	0	0
m1300	30	0	5	24, 25
m1400	24	0	12	8, 9, 11, 18, 19, 20, 21
m1600	4	0	3	9, 11, 12, 15, 18, 21
m2300	9	7	2	1, 3, 4, 5, 13, 22
m2400	17	2	9	1, 5, 6, 7, 10, 14, 22, 23
m2600	17	1	6	1, 2, 3, 6, 7, 10, 12, 13
m3300	0	7	16	0
m3400	0	20	15	0
m3600	0	7	4	0

Table 3. The average number of atomic Ca–X short contacts (clashes) per picosecond during 20 ps CPMD trajectories. The contact is defined as a Ca–X distance smaller than 2.5 Å.

Box	C	H	N	O
m0300	0	126	30	905
m0400	0	108	30	964
m0600	0	140	32	1083
m1300	10	41	37	649
m1400	10	19	56	666
m1600	8	47	50	882
m2300	3	13	70	518
m2400	4	28	90	603
m2600	9	8	181	568
m3300	0	1	9	825
m3400	0	3	30	915
m3600	0	4	20	985

3.1. Effects of MMT on Chemical Reactivity

We assumed that changes in the chemical reactivity or the possible catalytic effects of confinement and the presence of the MMT mineral could be monitored using the statistics of heavy atoms contacts. We defined such contact as an effective collision (clash) if two atoms A and B remained closer than a given threshold R(A–B), for a time period longer than 100 frames (725 fs). The following thresholds were adopted; R(C–C) = 1.64 Å, R(C–N) = 1.57 Å, and R(C–O) = 1.53 Å. The values were based on standard bond lengths and 0.1 Å was added to account for vibrational effects.

The results of scanning all 24 trajectories are presented in Table 2. We performed analysis of the convergence of a number of reactive clashes (where present, data not shown) and found that this number is almost constant in the 10 to 20 ps range.

First, we observed that the boxes g0 and m0 were not reactive in our simulations. Miller's primordial soup components (water, ammonia, methane, and hydrogen) present in this box did not react spontaneously on our 20-ps timescale neither in the pure condensed phase (g) nor after the confinement and presence of MMT (m). Moderate or elevated temperatures (400 K or 600 K) did not affect this observation. This result was not surprising, as it is known that g0 shows no spontaneous reactivity and an electric field (an electric discharge) was required in the experiment to initiate the production of an amino acid with a reasonable yield. Saitta and Saija did not observe any reactivity in the identical box in their 20-ps CPMD simulations [39]. Neither the presence of the mineral confinement nor a local electric field from Ca^{2+} ions increased reactivity in this mixture. It is worth mentioning that the chemical conditions assumed in the Miller–Urey experiment are sometimes disputed, since the early Earth atmosphere had probably lower concentrations of ammonia and hydrogen [7]. The lack of reactivity observed in the CPMD modeling indicates that additional physical factors were required to trigger complex organic molecules formation.

Identical results (no reactive clashes) were observed for boxes g3 and m3. At all of the considered temperatures, the chemical composition of these boxes was constant; the boxes contained a 1:1 glycine and ammonia mixture. Interestingly, even the presence of eight Ca^{2+} ions on the surface of the MMT crystal did not activate glycine or ammonia toward making a new compound, at least within our simulation timescale. We observed numerous collisions in m3 box (Table 2) but these were transient ammonia–glycine encounters that did not result in any new stable molecules. This result is encouraging with respect to the problem of the origin of life: once formed, a simple amino acid (such as glycine here) has a good chance to remain stable even under the harsh conditions of confinement and the presence of the clay with strong local electric fields.

In contrast, boxes 1 and 2 (reactants and intermediates in model Miller's experiment, see Table 1 for the composition) exhibited reactivity under both isolated and confined settings, leading to 25 new chemical species (see Figure 2). We divided them into group 1C (compounds **1–7**), 2C (**8–17**), and 3C+ (**18–25**) according to the number of carbon atoms present.

The components of g1 remained inert at all of the three temperatures; the same outcome as that reported in the Saitta and Saija simulations [39]. However, the same molecules from box 1 (water, ammonia, methane, nitrogen, and carbon monoxide) underwent frequent reactive collisions in our MMT pore model, leading to the formation of 11 different products (see Table 2 and Figure 2). There were four compounds having a new short C–C chain (ethane-1,2-dione, (E)-ethene-1,2-diol, 2-hydroxyethen-1-on, and 2-oxoacetamide; nos. **8**, **9**, **11**, and **15** in Figure 2) and four compounds (nos. **18**, **19**, **20**, and **21**) with a three-carbon atom long chain. Among them, the notable one was 2-hydroxy-3-oxopropanoic acid (no. **19**), which contained a newly formed carboxylic group—a fundamental part of all of the known elementary amino acids. Further, the observed formation of 2-oxoacetamide (**15**) was considerably important, albeit it was formed only at 600 K in the m1 box.

We found that the most numerous and diverse products were generated in the box m2400, which corresponded to the simulations at 400 K. Frequencies of reactive clashes, summarized in Table 3, illustrate this finding. Eight new compounds were observed (**1**, **5**, **6**, **7**, **10**, **14**, **23**, and **22**; Table 2 and Figure 2). There are molecules having up to six heavy atoms connected in one chain (**22** and **23**). We observed the formation of methanol (**1**). As in box 2, we initially had formaldehyde and HCN, and we observed the formation of critically important new carbon–nitrogen bonds in azaniumylmethanolate and aminomethanol (**5**, **6**). Interestingly, aminomethanol (**6**) was formed in purely g boxes g2300 and g2400, but almost all of the other complex products, except (**14**) required the presence of MMT. The snapshots of representative reactive collisions are presented in Figure 3, and the typical time evolution of the distances between the reacting fragments are shown in Figure 4.

Once the new bond is formed it typically last for many picoseconds (see Figure 4), though one should note that not all discovered species survive till the end of our 20 ps CPMD runs. Several of newly formed species thus they may have a transient nature and not necessarily lead to stable products. The formation of a new bond was clearly facilitated by the activation of polar molecules containing oxygen through interactions with the Ca^{2+} ions. In our model system, we did not introduce any additional water molecules, usually present in such a clay, except those already assigned to the reaction mixture, to mimic, or rather to enhance, the strong local electric field present on the surface (or in the pores) of the MMT mineral. These calcium ions were therefore considerably mobile and exerted strong activating catalytic effects on water, formaldehyde, and carbon monoxide. To a lesser extent, ammonia was activated by Ca^{2+}. These observations were based on the visual inspection of all of the trajectories and data presented in Table 3, where statistics of the close contacts of different heavy atoms with calcium ions is presented. We observed that electronegative oxygen atoms were far often more coordinated to Ca^{2+} than neutral carbon atoms. As expected, the number of clashes increases with an increase in the temperature (Table 3).

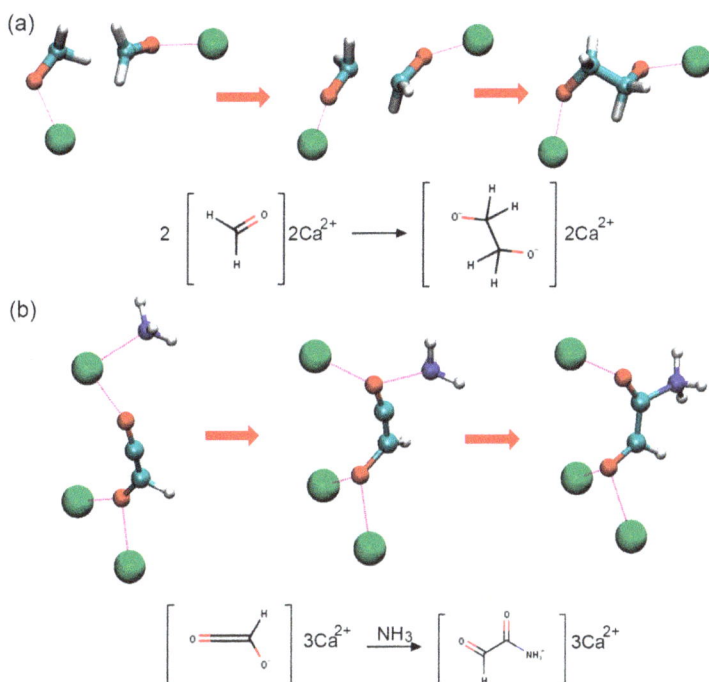

Figure 3. (**a**) Snapshots from a m2400 CPMD trajectory illustrating a new C–C bond formation. The formaldehyde ion is activated by Ca^{2+} (green spheres) and reacts with carbon monoxide forming oxaldehyde ion (Figure 2, **17**). (**b**) Snapshots from m1600 CPMD trajectory show a new C–N bond formation. The oxaldehyde ion (**17**) coordinated by two Ca^{2+} ions react with ammonia forming 2-oxoacamide (Figure 2, **15**).

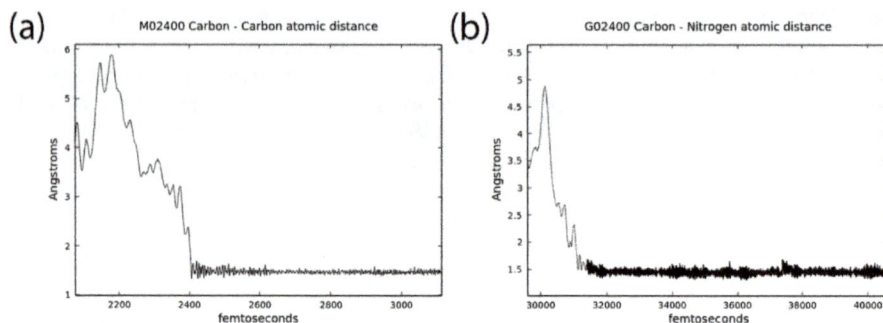

Figure 4. Distances between selected atoms in a fragment of CPMD trajectory showing C–C bond formation (**a**) and N–C bond formation (**b**) in m2400 box.

One should note that in the regions close to the surface of minerals, say 3–5 Å, quite often strong electric fields are present. For example, in a recent paper by Laporte et al. [70], in the vicinity of a hydrated MgO surface, an electric field of 1–3 V/Å was calculated, and the electrostatic potential goes up to 5 eV. We estimated that in our MMT slab system electric field is of the same order of magnitude far from ions and surface, but the main catalytic effect comes from the presence of a very high field introduced by Ca^{2+} ions (see Supplementary Materials Figure S6).

The MMT slab remained considerably stable through all the simulations. This stability was confirmed by the low values of RMSD calculated along the 20-ps trajectories: the highest average RMSD values were calculated to be 0.38 Å, 0.57 Å, and 0.72 Å for the Si, Al, and Ca ions, respectively (see Table 4). Clearly, the Ca^{2+} ions were mobile in our models as expected; this observation was confirmed by the trace plots shown in Figure S3 (in Supplementary Materials). In real MMT clays, the alkali atoms (Ca^{2+} and Na^+) have variable stoichiometry, are mobile, and coordinate labile water molecules [61].

Table 4. Root-mean-square deviation (RMSD) (in Å) for the MMT atoms. Data from 20-ps trajectories of m type boxes. Each analyzed trajectory contains 5000 frames.

Box	Al			Si			Ca		
	Avg.	Min.	Max.	Avg.	Min.	Max.	Avg.	Min.	Max.
m0300	0.17	0.03	0.24	0.18	0.04	0.24	0.63	0.03	0.86
m0400	0.20	0.04	0.27	0.21	0.08	0.24	0.43	0.03	0.75
m0600	0.28	0.06	0.39	0.29	0.06	0.39	0.62	0.05	1.00
m1300	0.38	0.01	0.58	0.36	0.01	0.51	0.66	0.05	1.00
m1400	0.21	0.07	0.34	0.25	0.04	0.46	0.51	0.04	0.82
m1600	0.27	0.11	0.39	0.54	0.06	1.00	0.58	0.04	0.99
m2300	0.16	0.04	0.38	0.46	0.03	0.55	0.66	0.05	1.00
m2400	0.21	0.04	0.34	0.43	0.03	0.59	0.72	0.03	1.00
m2600	0.28	0.04	0.45	0.57	0.04	0.73	0.71	0.04	0.99
m3300	0.25	0.04	0.45	0.22	0.04	0.30	0.28	0.03	1.00
m3400	0.18	0.05	0.50	0.25	0.05	0.42	0.36	0.03	0.52
m3600	0.29	0.05	0.28	0.37	0.06	0.48	0.56	0.05	1.00

Abbreviators: Avg. = average; Min. = minimum; Max. = maximum.

Two types of reactive collision mechanisms in the m systems were qualitatively distinguished: (i) triggered by the catalytic role of calcium ions (i.e., strong local electric field) and (ii) nucleophilic bimolecular substitution in which one bond is broken and another bond is formed synchronously (a S_N2 mechanism). In our simulations, in m type boxes, both types were represented, while in g type boxes, water-assisted polarization of ammonia and formaldehyde preceded S_N2 type new bond formation.

It would be interesting to discriminate between possible catalytic role of confinement of small molecules in a limited space in a clay cage (m0–m3 boxes), and the role of strong local electric fields possibly exerted by Ca^{2+} ions. A more systematic computational study of this problem requires collecting extensive statistics, using other model systems and calls for a separate study.

3.2. Primordial Soup Ingredient Dynamics

One may expect, inspired by the observation made by Szostak group for phospholipids [43], that the MMT surface effects may lead to the preferential adsorption and ordering of components of the primordial soup. We monitored the distributions of all of the heavy atoms—C, N, and O—from basic steps of the Miller–Urey experiment (boxes 0–3 in the g and m systems) by calculating the atom–atom radial distribution functions $g_{AB}(r)$ (for definition cf. Methods).

Data are presented in Figure 5, Figures S1 and S2.

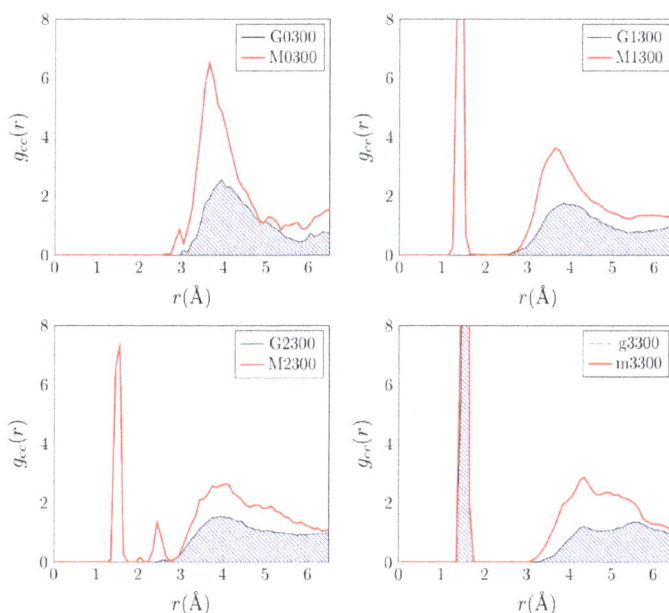

Figure 5. Radial carbon-carbon distance distribution functions, $g_{CC}(r)$ (in a.u.), calculated along 20-ps CPMD trajectories at 300 K.

In Figure 5, we compare $g_{CC}(r)$ calculated for the g and m boxes at 300 K. The presence of MMT changed the distributions of the carbon–carbon distances. The most notable effect was a substantial increase, by a factor of 2, in the population of the carbon pairs observed at the distance of 4 Å. The narrow maxima in $g_{CC}(r)$ at 1.5–2.0 Å were attributed to the fact that the Ca^{2+} ions and the MMT surface tended to coordinate CO, formaldehyde, and glycine. The lack of such a maximum in $g_{CC}(r)$ for the g0300 box might be explained by the lower number of carbon atoms in this mixture than in the other ones (8 vs. 18) and the fact that in g0, carbon atoms were present only in CH_4, i.e., a nonpolar molecule not coordinated by the clay ions. A similar ordering effect of MMT was also observed at 400 K and 600 K (see Figure S1). The plots of g(r) for the calcium ion–any heavy atom distances varied from box to box but depended only slightly on temperature (see Figure S2). The Ca^{2+} ions exhibited (as expected) considerable mobility as they were loosely coupled to the Si–Al mineral core (see Figure S3). In contrast, the positions of the Al and Si ions did not change considerably during the CPMD trajectory and the vibrations of the crystal were within a reasonable range (Figures S4 and S5).

The simplified model of MMT cage adopted is not perfect. Clays have variable stoichiometry, interlayer distances, defects, smaller, and variable mobile ions (Ca^{2+} and Na^+) density. However, these theoretical data clearly showed that the nanoconfinement in MMT changed the dynamics of all of the elementary mixtures 0–3, mimicking to some extent the primordial soup. In general, carbon atoms were localized closer to each other and this effect alone increased the probability of the formation of more complex molecules. This was particularly observed in the elevated temperature simulations (m1400 and m2400). The shorter C–C distances in MMT may be only partially attributed to possible higher pressure present in these boxes. We would rather explain this reactivity by the strong polarizing effects of Ca^{2+} ions present in our model. We packed as many as eight ions in a small volume just to maintain the stoichiometry of the MMT nanoclay, and to increase the probability of reactions (if any) in our short time scale AIMD simulations.

The presence of sulfidic anions (HS^-, HSO_3^-, and SO_3^{2-}) in certain areas of shallow water was proposed to be critical for formation of biomolecular systems [71]. Such mixtures are worth studying using the theoretical framework presented here. It is also worth to explore possible effects of internal cavity pressure and temperature-induced changes in density of the reacting mixtures neglected in our study. Reach chemistry observed upon the nanoconfinement opens also a possibility for further computational tests of alternative scenarios leading to elementary precursors of biomolecules relevant for emergence of life. Calculations of free energy profiles along reaction pathways, not only classical [72], but similar to those proposed in [17] would be also desirable but are beyond the scope of this exploratory work. Since we have found many quickly formed molecules from the intermediate Miller–Urey test boxes m1 and m2, this indicates that some complex, but not necessarily useful ("waste"), compounds might be formed in the early Earth conditions discussed here as well [73].

4. Conclusions

Life is based on complex molecules formed from simpler components. In the discovering of the very first steps in the origins of life, various scenarios of the formation of such elementary building blocks have to be considered. In this paper, we have addressed an intriguing question: To what extent does the confinement of the components of the hypothetical primordial soup affect the synthesis of new, more complex chemical compounds? We placed several (discussed in the literature) test mixtures in the nanopores of the MMT mineral model, frequently considered as a catalyst in the formation of biology-related compounds. Using advanced CPMD simulations, we have compared the propensity to reactivity of four standard chemical mixtures localized in a condensed phase environment (modeled by applying PBC) and an MMT nanoclay slab. The structural model of the mineral was based on crystallography data, except for the presence of a grid of eight nonhydrated Ca^{2+} ions, which were introduced to mimic the effect of a strongly localized electric field. The ions were located in typical crystallographic positions of their hydrated counterparts, and were hydrated by the water molecules present in the mixture studied. The system was therefore relatively crowded but within physical limits. The effects of nanoconfinement were dependent on a chemical composition of the prebiotic soup mixture.

Boxes m0 and m3 remained nonreactive despite presence of the MMT model slab and Ca^{2+} ions. We found that even within a relatively short timescale of 20 ps, the MMT cavity substantially increased the reactivity of boxes 1 (water, ammonia, methane, nitrogen, and carbon monoxide) and 2 (water, ammonia, formaldehyde, and cyanide), which were composed of the intermediates of the Strecker reaction as discussed in the Miller–Urey experiment. As expected, at the elevated temperatures (400 K and, added for a reference to earlier paper, 600 K), the catalytic effectivity of MMT was higher and the largest number of more than 20 diverse products/intermediates was observed at 400 K. The elevated temperature, especially 400 K, could have been easily achieved locally in the Earth Hadean Era hydrothermal conditions. Among other species, we observed the formation of important carboxylic group and 2-oxoacetamide. Therefore, we have concluded that both the presence of Ca^{2+} and the confinement led to a higher probability of reactive collisions in some of the mixtures studied. The detailed discrimination what factor, namely Ca^{2+} ions or the nanoconfinement, plays a major role in this increased reactivity requires additional extensive and statistically sound tests. Such research requires large computational resources and was out of the scope of present study. Notably, these effects were present only if the chemical composition of the boxes was adequate; for example, for both the Miller–Urey experiment substrates (methane, hydrogen, ammonia, and water; box 0) and the products (ammonia and glycine; box 3), the MMT mineral did not exhibit any catalytic activity on 20 ps simulations time scale. Thus, our study adds new arguments supporting the popular notion that mineral surfaces and compartmentalization have to be considered as important factors in the origin of complex organic molecules. We think that such molecules may be critical for biological systems formation, both in terrestrial and extraterrestrial settings.

Supplementary Materials: The following are available online at http://www.mdpi.com/2075-1729/9/2/46/s1, Figure S1: Radial distribution functions g(r) (in a.u.) for carbon-to-carbon distances r (in Å) at 300 K, 400 K, and 600 K in all the simulated systems, Figure S2: Radial distribution functions g(r) (in a.u.) for distances between a heavy atom and the calcium ions in all the simulated systems, Figure S3: Traces of calcium ions projected on the plane axc. Figure S4: Traces of aluminum ion positions projected on the plane axb, Figure S5: Mobility of silicon atoms represented by traces projected on the axb plane' Figure S6: Model total electric potential (in V) calculated for selected frames from m1 and g1 CPMD trajectories.

Author Contributions: Conceptualization, J.R., J.B., and W.N.; Data curation, J.F.C.M. and J.R.; Funding acquisition, J.B.; Investigation, J.F.C.M., N.S., and W.N.; Methodology, J.F.C.M., J.R., and W.N.; Project administration, J.B.; Resources, N.S.; Software, J.F.C.M. and J.R.; Supervision, J.B. and W.N.; Visualization, J.F.C.M.; Writing—original draft, J.F.C.M., J.R., N.S., and W.N.; Writing—review & editing, J.F.C.M., J.R., N.S., J.B., and W.N.

Funding: The Poznan team was supported by DS grants 09/91/DSPB/0649 and POIR.04.02.00-30-A004/16-00, "ECBIG - Europejskie Centrum Bioinformatyki i Genomiki" (The European Centre of Bioinformatics and Genomics), IV Priority Axis of the Operational Program Smart Growth 2014-2020.

Acknowledgments: The infrastructures of the Interdisciplinary Centre for Modern Technologies of Nicolaus Copernicus University in Torun (Poland) and the European Center for Bioinformatics and Genomics affiliated to Poznan University of Technology (Poznan, Poland) were used in this work. The most demanding computations were performed using Poznan Super Computing and Networking Center.

Conflicts of Interest: No competing financial interests exist.

Abbreviations

a.u.	atomic units
Å	Angstroms
AIMD	ab initio molecular dynamics
Approx.	approximately
atm	atmospheres
CCL2	CC chemokine ligand 2
CCR2	CC chemokine receptor 2
CCR5	CC chemokine receptor 5
CPMD	Car-Parrinello molecular dynamics
DFT	density functional theory
Dimension c	refers to the size of the vector c
fs	femtoseconds
g	gases only box
g/mL	grams per milliliter
g(r)	refers to the atom-atom radial distribution function
HCN	hydrogen cyanide
i.a.	Latin meaning "among other things"
i.e.,	Latin meaning "in essence/in other words"
K	Kelvin
kcal/mol	kilocalories per mol
m	montmorillonite and gases box
MD	molecular dynamics
MMT	montmorillonite

nm   nanometer
NVT   refers to the canonical enssemble at
PBC   periodic boundary conditions
pH   potential hydrogen
ps   picoseconds
RMSD   root mean square deviation
RNA   ribonucleic acid
Ry   Rydberg atomic units
THz   terahertz
TLC   thin layer chromatography
UV   ultraviolet
UV–Vis   ultraviolet-visible

References

1. Bhushan, B.; Nayak, A.; Kamaluddin. Catalytic Role of Manganese Oxides in Prebiotic Nucleobases Synthesis from Formamide. *Orig. Life Evol. Biosph.* **2016**, *46*, 203–213. [CrossRef] [PubMed]
2. Pollet, R.; Boehme, C.; Marx, D. Ab Initio Simulations of Desorption and Reactivity of Glycine at a Water-Pyrite Interface at "Iron-Sulfur World" Prebiotic Conditions. *Orig. Life Evol. Biosph.* **2006**, *36*, 363–379. [CrossRef] [PubMed]
3. Ritson, D.J.; Battilocchio, C.; Ley, S.V.; Sutherland, J.D. Mimicking the Surface and Prebiotic Chemistry of Early Earth Using Flow Chemistry. *Nat. Commun.* **2018**, *9*, 1821. [CrossRef] [PubMed]
4. Szostak, N.; Synak, J.; Borowski, M.; Wasik, S.; Blazewicz, J. Simulating the Origins of Life: The Dual Role of RNA Replicases as an Obstacle to Evolution. *PLoS ONE* **2017**, *12*, e0180827. [CrossRef] [PubMed]
5. Szostak, J.W. The Origin of Life on Earth and the Design of Alternative Life Forms. *Mol. Front. J.* **2017**, *01*, 121–131. [CrossRef]
6. Kitadai, N.; Maruyama, S. Origins of Building Blocks of Life: A Review. *Geosci. Front.* **2018**, *9*, 1117–1153. [CrossRef]
7. Ranjan, S.; Sasselov, D.D. Influence of the UV Environment on the Synthesis of Prebiotic Molecules. *Astrobiology* **2016**, *16*, 68–88. [CrossRef] [PubMed]
8. Lambert, J.-F. Adsorption and Polymerization of Amino Acids on Mineral Surfaces: A Review. *Orig. Life Evol. Biosph.* **2008**, *38*, 211–242. [CrossRef]
9. Ruiz-Mirazo, K.; Briones, C.; de la Escosura, A. Chemical Roots of Biological Evolution: The Origins of Life as a Process of Development of Autonomous Functional Systems. *Open Biol.* **2017**, *7*, 170050. [CrossRef]
10. Patel, B.H.; Percivalle, C.; Ritson, D.J.; Duffy, C.D.; Sutherland, J.D. Common Origins of RNA, Protein and Lipid Precursors in a Cyanosulfidic Protometabolism. *Nat. Chem.* **2015**, *7*, 301–307. [CrossRef]
11. Saladino, R.; Crestini, C.; Pino, S.; Costanzo, G.; Di Mauro, E. Formamide and the Origin of Life. *Phys. Life Rev.* **2012**, *9*, 84–104. [CrossRef] [PubMed]
12. Sutherland, J.D. The Origin of Life-Out of the Blue. *Angew. Chem. Int. Ed.* **2016**, *55*, 104–121. [CrossRef] [PubMed]
13. Trevors, J.T. Hypothesized Origin of Microbial Life in a Prebiotic Gel and the Transition to a Living Biofilm and Microbial Mats. *C. R. Biol.* **2011**, *334*, 269–272. [CrossRef] [PubMed]
14. Wächtershäuser, G. Groundworks for an Evolutionary Biochemistry: The Iron-Sulphur World. *Prog. Biophys. Mol. Biol.* **1992**, *58*, 85–201. [CrossRef]
15. Miller, S.L. A Production of Amino Acids Under Possible Primitive Earth Conditions. *Science* **1953**, *117*, 528–529. [CrossRef] [PubMed]
16. Bartley, B.A.; Kim, K.; Medley, J.K.; Sauro, H.M. Synthetic Biology: Engineering Living Systems from Biophysical Principles. *Biophys. J.* **2017**, *112*, 1050–1058. [CrossRef]
17. Ferus, M.; Pietrucci, F.; Saitta, A.M.; Knížek, A.; Kubelík, P.; Ivanek, O.; Shestivska, V.; Civiš, S. Formation of Nucleobases in a Miller-Urey Reducing Atmosphere. *Proc. Natl. Acad. Sci. USA* **2017**, *114*, 4306–4311. [CrossRef]
18. Leyton, P. The Role of Minerals on Prebiotic Synthesis: Comment on "Formamide and the Origin of Life" by R. Saladino et al. *Phys. Life Rev.* **2012**, *9*, 116–117. [CrossRef]

19. Huber, C.; Eisenreich, W.; Hecht, S.; Wächtershäuser, G. A Possible Primordial Peptide Cycle. *Science* **2003**, *301*, 938–940. [CrossRef]

20. Huber, C.; Wächtershäuser, G. Peptides by Activation of Amino Acids with CO on (Ni,Fe)S Surfaces: Implications for the Origin of Life. *Science* **1998**, *281*, 670–672. [CrossRef]

21. Chandru, K.; Gilbert, A.; Butch, C.; Aono, M.; Cleaves, H.J. The Abiotic Chemistry of Thiolated Acetate Derivatives and the Origin of Life. *Sci. Rep.* **2016**, *6*, 29883. [CrossRef] [PubMed]

22. Powner, M.W.; Gerland, B.; Sutherland, J.D. Synthesis of Activated Pyrimidine Ribonucleotides in Prebiotically Plausible Conditions. *Nature* **2009**, *459*, 239–242. [CrossRef] [PubMed]

23. Stairs, S.; Nikmal, A.; Bučar, D.-K.; Zheng, S.-L.; Szostak, J.W.; Powner, M.W. Divergent Prebiotic Synthesis of Pyrimidine and 8-Oxo-Purine Ribonucleotides. *Nat. Commun.* **2017**, *8*, 15270. [CrossRef] [PubMed]

24. Sutherland, J.D. Opinion: Studies on the Origin of Life—The End of the Beginning. *Nat. Rev. Chem.* **2017**, *1*, 0012. [CrossRef]

25. Xu, J.; Tsanakopoulou, M.; Magnani, C.J.; Szabla, R.; Šponer, J.E.; Šponer, J.; Góra, R.W.; Sutherland, J.D. A Prebiotically Plausible Synthesis of Pyrimidine β-Ribonucleosides and Their Phosphate Derivatives Involving Photoanomerization. *Nat. Chem.* **2017**, *9*, 303–309. [CrossRef] [PubMed]

26. Carota, E.; Botta, G.; Rotelli, L.; Di Mauro, E.; Saladino, R. Current Advances in Prebiotic Chemistry Under Space Conditions. *Curr. Org. Chem.* **2015**, *19*, 1963–1979. [CrossRef]

27. Costanzo, G.; Saladino, R.; Crestini, C.; Ciciriello, F.; Di Mauro, E. Formamide as the Main Building Block in the Origin of Nucleic Acids. *BMC Evol. Biol.* **2007**, *7* (Suppl. 2), S1. [CrossRef]

28. Costanzo, G.; Giorgi, A.; Scipioni, A.; Timperio, A.M.; Mancone, C.; Tripodi, M.; Kapralov, M.; Krasavin, E.; Kruse, H.; Sponer, J.E.J.; et al. Non-Enzymatic Oligomerization of 3′,5′ Cyclic CMP Induced by Proton- and UV-Irradiation Hints at a Non-Fastidious Origin of RNA. *ChemBioChem* **2017**. [CrossRef]

29. Ferus, M.; Knizek, A.; Sponer, J.; Sponer, J.E.; Civis, S. RADICAL SYNTHESIS OF NUCLEIC ACID BASES FROM FORMAMIDE IN IMPACT PLASMA. *Chem. List.* **2015**, *109*, 406–414.

30. Ferus, M.; Nesvorný, D.; Šponer, J.; Kubelík, P.; Michalčíková, R.; Shestivská, V.; Šponer, J.E.; Civiš, S. High-Energy Chemistry of Formamide: A Unified Mechanism of Nucleobase Formation. *Proc. Natl. Acad. Sci. USA* **2015**, *112*, 657–662. [CrossRef]

31. Jeilani, Y.A.; Nguyen, H.T.; Newallo, D.; Dimandja, J.-M.D.; Nguyen, M.T. Free Radical Routes for Prebiotic Formation of DNA Nucleobases from Formamide. *Phys. Chem. Chem. Phys.* **2013**, *15*, 21084. [CrossRef] [PubMed]

32. Saladino, R.; Botta, G.; Pino, S.; Costanzo, G.; Di Mauro, E. Materials for the Onset. A Story of Necessity and Chance. *Front. Biosci. (Landmark Ed.)* **2013**, *18*, 1275–1289. [PubMed]

33. Nguyen, H.T.; Jeilani, Y.A.; Hung, H.M.; Nguyen, M.T. Radical Pathways for the Prebiotic Formation of Pyrimidine Bases from Formamide. *J. Phys. Chem. A* **2015**, *119*, 8871–8883. [CrossRef] [PubMed]

34. Cossetti, C.; Crestini, C.; Saladino, R.; di Mauro, E. Borate Minerals and RNA Stability. *Polymers* **2010**, *2*, 211–228. [CrossRef]

35. Saladino, R.; Neri, V.; Crestini, C.; Costanzo, G.; Graciotti, M.; Di Mauro, E. Synthesis and Degradation of Nucleic Acid Components by Formamide and Iron Sulfur Minerals. *J. Am. Chem. Soc.* **2008**, *130*, 15512–15518. [CrossRef] [PubMed]

36. Martins, Z.; Botta, O.; Fogel, M.L.; Sephton, M.A.; Glavin, D.P.; Watson, J.S.; Dworkin, J.P.; Schwartz, A.W.; Ehrenfreund, P. Extraterrestrial Nucleobases in the Murchison Meteorite. *Earth Planet. Sci. Lett.* **2008**, *270*, 130–136. [CrossRef]

37. Saladino, R.; Botta, G.; Delfino, M.; Di Mauro, E. Meteorites as Catalysts for Prebiotic Chemistry. *Chem. Eur. J.* **2013**, *19*, 16916–16922. [CrossRef] [PubMed]

38. Botta, L.; Saladino, R.; Bizzarri, B.M.; Cobucci-Ponzano, B.; Iacono, R.; Avino, R.; Caliro, S.; Carandente, A.; Lorenzini, F.; Tortora, A.; et al. Formamide-Based Prebiotic Chemistry in the Phlegrean Fields. *Adv. Sp. Res.* **2018**, *62*, 2372–2379. [CrossRef]

39. Saitta, A.M.; Saija, F. Miller Experiments in Atomistic Computer Simulations. *Proc. Natl. Acad. Sci. USA* **2014**, *111*, 13768–13773. [CrossRef] [PubMed]

40. Cafferty, B.J.; Fialho, D.M.; Khanam, J.; Krishnamurthy, R.; Hud, N.V. Spontaneous Formation and Base Pairing of Plausible Prebiotic Nucleotides in Water. *Nat. Commun.* **2016**, *7*, 11328. [CrossRef] [PubMed]

41. Attwater, J.; Holliger, P. Origins of Life: The Cooperative Gene. *Nature* **2012**, *491*, 48–49. [CrossRef] [PubMed]

42. Dass, A.; Jaber, M.; Brack, A.; Foucher, F.; Kee, T.; Georgelin, T.; Westall, F. Potential Role of Inorganic Confined Environments in Prebiotic Phosphorylation. *Life* **2018**, *8*, 7. [CrossRef] [PubMed]

43. Hanczyc, M.M.; Fujikawa, S.M.; Szostak, J.W. Experimental Models of Primitive Cellular Compartments: Encapsulation, Growth, and Division. *Science* **2003**, *302*, 618–622. [CrossRef] [PubMed]

44. Ferris, J.P. Catalysis and Prebiotic Synthesis. *Rev. Mineral. Geochem.* **2005**, *59*, 187–210. [CrossRef]

45. Joshi, P.C.; Pitsch, S.; Ferris, J.P. Selectivity of Montmorillonite Catalyzed Prebiotic Reactions of D, L-Nucleotides. *Orig. Life Evol. Biosph.* **2007**, *37*, 3–26. [CrossRef] [PubMed]

46. Jheeta, S.; Joshi, P. Prebiotic RNA Synthesis by Montmorillonite Catalysis. *Life* **2014**, *4*, 318–330. [CrossRef] [PubMed]

47. Cygan, R.T.; Romanov, V.N.; Myshakin, E.M. Molecular Simulation of Carbon Dioxide Capture by Montmorillonite Using an Accurate and Flexible Force Field. *J. Phys. Chem. C* **2012**, *116*, 13079–13091. [CrossRef]

48. Michalkova, A.; Robinson, T.L.; Leszczynski, J. Adsorption of Thymine and Uracil on 1:1 Clay Mineral Surfaces: Comprehensive Ab Initio Study on Influence of Sodium Cation and Water. *Phys. Chem. Chem. Phys.* **2011**, *13*, 7862–7881. [CrossRef] [PubMed]

49. Dawley, M.M.; Scott, A.M.; Hill, F.C.; Leszczynski, J.; Orlando, T.M. Adsorption of Formamide on Kaolinite Surfaces: A Combined Infrared Experimental and Theoretical Study. *J. Phys. Chem. C* **2012**, *116*, 23981–23991. [CrossRef]

50. Michalkova Scott, A.; Dawley, M.M.; Orlando, T.M.; Hill, F.C.; Leszczynski, J. Theoretical Study of the Roles of Na^+ and Water on the Adsorption of Formamide on Kaolinite Surfaces. *J. Phys. Chem. C* **2012**, *116*, 23992–24005. [CrossRef]

51. Saladino, R.; Ciambecchini, U.; Crestini, C.; Costanzo, G.; Negri, R.; Di Mauro, E. One-Pot TiO_2-Catalyzed Synthesis of Nucleic Bases and Acyclonucleosides from Formamide: Implications for the Origin of Life. *Chembiochem* **2003**, *4*, 514–521. [CrossRef] [PubMed]

52. Senanayake, S.D.; Idriss, H. Photocatalysis and the Origin of Life: Synthesis of Nucleoside Bases from Formamide on TiO2(001) Single Surfaces. *Proc. Natl. Acad. Sci. USA* **2006**, *103*, 1194–1198. [CrossRef] [PubMed]

53. Goldman, N.; Reed, E.J.; Fried, L.E.; William Kuo, I.-F.; Maiti, A. Synthesis of Glycine-Containing Complexes in Impacts of Comets on Early Earth. *Nat. Chem.* **2010**, *2*, 949–954. [CrossRef] [PubMed]

54. Martins, Z.; Price, M.C.; Goldman, N.; Sephton, M.A.; Burchell, M.J. Shock Synthesis of Amino Acids from Impacting Cometary and Icy Planet Surface Analogues. *Nat. Geosci.* **2013**, *6*, 1045–1049. [CrossRef]

55. Schreiner, E.; Nair, N.N.; Wittekindt, C.; Marx, D. Peptide Synthesis in Aqueous Environments: The Role of Extreme Conditions and Pyrite Mineral Surfaces on Formation and Hydrolysis of Peptides. *J. Am. Chem. Soc.* **2011**. [CrossRef] [PubMed]

56. Muñoz-Santiburcio, D.; Marx, D. Chemistry in Nanoconfined Water. *Chem. Sci.* **2017**, *8*, 3444–3452. [CrossRef] [PubMed]

57. Car, R.; Parrinello, M. Unified Approach for Molecular Dynamics and Density-Functional Theory. *Phys. Rev. Lett.* **1985**, *55*, 2471–2474. [CrossRef] [PubMed]

58. Stirling, A.; Rozgonyi, T.; Krack, M.; Bernasconi, M. Prebiotic NH3 Formation: Insights from Simulations. *Inorg. Chem.* **2016**, *55*, 1934–1939. [CrossRef]

59. Pietrucci, F.; Saitta, A.M. Formamide Reaction Network in Gas Phase and Solution via a Unified Theoretical Approach: Toward a Reconciliation of Different Prebiotic Scenarios. *Proc. Natl. Acad. Sci. USA* **2015**, *112*, 15030–15035. [CrossRef]

60. Pérez-Villa, A.; Saitta, A.M.; Georgelin, T.; Lambert, J.-F.; Guyot, F.; Maurel, M.-C.; Pietrucci, F. Synthesis of RNA Nucleotides in Plausible Prebiotic Conditions from Ab Initio Computer Simulations. *J. Phys. Chem. Lett.* **2018**, *9*, 4981–4987. [CrossRef]

61. Sena, M.M.; Morrow, C.P.; Kirkpatrick, R.J.; Krishnan, M. Supercritical Carbon Dioxide at Smectite Mineral–Water Interfaces: Molecular Dynamics and Adaptive Biasing Force Investigation of CO_2/H_2O Mixtures Nanoconfined in Na-Montmorillonite. *Chem. Mater.* **2015**, *27*, 6946–6959. [CrossRef]

62. Martínez, L.; Andrade, R.; Birgin, E.G.; Martínez, J.M.; Martinez, L.; Andrade, R.; Birgin, E.G.; Martínez, J.M. PACKMOL: A Package for Building Initial Configurations for Molecular Dynamics Simulations. *J. Comput. Chem.* **2009**, *30*, 2157–2164. [CrossRef] [PubMed]

63. Evans, D.J.; Holian, B.L. The Nose–Hoover Thermostat. *J. Chem. Phys.* **1985**, *83*, 4069–4074. [CrossRef]

64. Giannozzi, P.; Baroni, S.; Bonini, N.; Calandra, M.; Car, R.; Cavazzoni, C.; Ceresoli, D.; Chiarotti, G.L.; Cococcioni, M.; Dabo, I.; et al. QUANTUM ESPRESSO: A Modular and Open-Source Software Project for Quantum Simulations of Materials. *J. Phys. Condens. Matter* **2009**, *21*, 395502. [CrossRef] [PubMed]

65. Perdew, J.P.; Burke, K.; Ernzerhof, M. Generalized Gradient Approximation Made Simple. *Phys. Rev. Lett.* **1996**, *77*, 3865–3868. [CrossRef]

66. Ahmed Adllan, A.; Dal Corso, A. Ultrasoft Pseudopotentials and Projector Augmented-Wave Data Sets: Application to Diatomic Molecules. *J. Phys. Condens. Matter* **2011**, *23*, 425501. [CrossRef] [PubMed]

67. Verlet, L. Computer "Experiments" on Classical Fluids. I. Thermodynamical Properties of Lennard-Jones Molecules. *Phys. Rev.* **1967**, *159*, 98–103. [CrossRef]

68. Tribello, G.A.; Bonomi, M.; Branduardi, D.; Camilloni, C.; Bussi, G. PLUMED 2: New Feathers for an Old Bird. *Comput. Phys. Commun.* **2014**, *185*, 604–613. [CrossRef]

69. Humphrey, W.; Dalke, A.; Schulten, K. VMD: Visual Molecular Dynamics. *J. Mol. Graph.* **1996**, *14*, 33–38. [CrossRef]

70. Laporte, S.; Finocchi, F.; Paulatto, L.; Blanchard, M.; Balan, E.; Guyot, F.; Saitta, A.M. Strong Electric Fields at a Prototypical Oxide/Water Interface Probed by Ab Initio Molecular Dynamics: MgO(001). *Phys. Chem. Chem. Phys.* **2015**, *17*, 20382–20390. [CrossRef]

71. Ranjan, S.; Todd, Z.R.; Sutherland, J.D.; Sasselov, D.D. Sulfidic Anion Concentrations on Early Earth for Surficial Origins-of-Life Chemistry. *Astrobiology* **2018**, *18*, 1023–1040. [CrossRef] [PubMed]

72. Rydzewski, J.; Nowak, W. Ligand Diffusion in Proteins via Enhanced Sampling in Molecular Dynamics. *Phys. Life Rev.* **2017**, *22–23*, 58–74. [CrossRef] [PubMed]

73. Islam, S.; Powner, M.W. Prebiotic Systems Chemistry: Complexity Overcoming Clutter. *Chem* **2017**, *2*, 470–501. [CrossRef]

life

MDPI

Article

Origin of the Genetic Code Is Found at the Transition between a Thioester World of Peptides and the Phosphoester World of Polynucleotides

Hyman Hartman [1],* and Temple F. Smith [2]

[1]   Earth, Atmosphere, and Planetary Science Department, Massachusetts Institute of Technology, Cambridge, MA 02139, USA
[2]   BioMedical Engineering, Boston University, Boston, MA 02215, USA
*   Correspondence: hhartman@mit.edu

Received: 1 July 2019; Accepted: 14 August 2019; Published: 22 August 2019

Abstract: The early metabolism arising in a Thioester world gave rise to amino acids and their simple peptides. The catalytic activity of these early simple peptides became instrumental in the transition from Thioester World to a Phosphate World. This transition involved the appearances of sugar phosphates, nucleotides, and polynucleotides. The coupling of the amino acids and peptides to nucleotides and polynucleotides is the origin for the genetic code. Many of the key steps in this transition are seen in the catalytic cores of the nucleotidyltransferases, the class II tRNA synthetases (aaRSs) and the CCA adding enzyme. These catalytic cores are dominated by simple beta hairpin structures formed in the Thioester World. The code evolved from a proto-tRNA, a tetramer XCCA interacting with a proto-aminoacyl-tRNA synthetase (aaRS) activating Glycine and Proline. The initial expanded code is found in the acceptor arm of the tRNA, the operational code. It is the coevolution of the tRNA with the aaRSs that is at the heart of the origin and evolution of the genetic code. There is also a close relationship between the accretion models of the evolving tRNA and that of the ribosome.

Keywords: genetic code origin; tRNA accretion model; tRNA-synthetase; metabolism; thioester; nucleotidyltransferases; early peptides; ribosome

1. Introduction

There are two competing theories for the origin of life that are based on Darwinian selection; the RNA World and the Clay World. They both assume a replicating and a mutating entity that has catalytic capacity. The selection in both cases is on the evolution of the catalytic capacities whose products increase the replication and survivability of in the first case, of the RNA world, the replicating RNA and its catalytic capability and in the second case, of the Clay world, the replicating Clay and its catalytic capability [1].

The genetic code and the translation system had to have arisen in an environment containing, at a minimum, the molecular building blocks (e.g., amino acids and nucleotides). "A general scheme is proposed, involving the fixation of CO_2 and N_2, that led to the evolution of intermediary metabolism. The result is the evolution of a complex system from a simple one. Following the logic that core metabolism recapitulates biopoiesis, we begin by describing, an autotrophic origin of metabolism based on self-replicating iron-rich clays, transition-state metals, disulfide and dithiols, and UV radiation" [2].

A scenario for the origin of life consistent with the origin and evolution of metabolism started at the surface of an early outgassing Earth. In the surface hot springs, the water was rich in ferrous ions, magnesium ions, and silicate ions that formed iron-rich clays. The waters of the hot spring also included transition-state metal ions (i.e., Mn, Cu, Zn), adding to the catalytic capabilities of the iron-rich clays. These self-replicating clays would photochemically fix CO_2 into organic acids and gradually evolve into the sulfide-rich region acquiring N_2 fixation in the process.

A Thioester World had come into being. The entry of phosphate into the evolving catalytic network expanded the metabolic network. The Phosphate World had come into being [3,4].

The onion heuristic view claims that a complex system like a metabolic network started simple and grew more complex by layers being added, as for example, the Thioester World evolving into the Phosphate World by adding a new layer dominated by phosphate. Looking at intermediary metabolism, in particular anabolism, one is immediately struck by the central role played by the (reverse) citric acid cycle. If we apply the method of the onion heuristic view, it can be conjectured that the citric acid cycle came first and was followed by the amino acids, lipids, nucleotides, and carbohydrates.

2. Metabolism

The Evolution of Metabolism Can Be Considered to Have Undergone Four Distinct Phases

The first phase began with Iron-rich clays with inputs (CO_2, H_2O). The formation of dicarboxylic acids such as oxalic acid and glyoxylic acid are synthesized from CO_2 and H_2O by iron-rich clays and light which is the source of energy. These dicarboxylic acids are catalysts in the formation of iron-rich clays themselves. The iron-rich clays in the light are the center from which metabolism started [3].

The second phase brought in sulfur; Iron-rich clays with inputs (CO_2 H_2O H_2S, Fe_2S_2, and Fe_4S_4). The entry of sulfur brings in the thioesters and the Fe_2S_2 and Fe_4S_4. This results in the reverse citric acid cycle and the polymerization of acetyl-thioesters, resulting in the biosynthesis of the fatty acids. This step is continuous with the next step, which is the entry of nitrogen.

The third phase brought the addition of nitrogen and created the "Thioester World": Iron-rich clays with inputs (CO_2, H_2O, H_2S, Fe_2S_2, Fe_4S_4, and N_2). Fixation of nitrogen occurred here on the surface of an iron sulfide [4,5]. The amino acids were synthesized from the reverse citric acid cycle giving rise to the metabolic metric —the number of catalytic steps in the biosynthesis of 16 amino acids from that cycle. The four amino acids Phenylalanine, Tyrosine, Tryptophan, and Histidine are not derived from the reverse citric acid. The aromatic amino acids may have been synthesized by Fe-serpentines and Iron saponites [6]. The synthesis of di, tri and tetrapeptides had been formed from their amino acid thioesters [3,4]. These initial peptides and their combinations had been able to provide additional catalytic activities to those of the iron-rich clay surfaces.

The fourth phase brings in Phosphate: Iron clays with inputs (CO_2, H_2O, H_2S, Fe_2S_2 Fe_4S_4, N_2 and N_2PO_4) and the biosynthesis of nucleotides. The formation of pyro-phosphate is followed by the biosynthesis of the sugar phosphates such as 5-phosphoribosyl-1-pyrophosphate (PRPP) the phosphate sugar that initiates the biosynthesis of purine nucleotides.

The biosynthesis of purine bases (Figure 1) begins with phosphoribosyl-1-pyrophosphate (PRPP) the phosphate sugar that initiates the biosynthesis of Purine nucleotides. The biosynthesis of the Purine base involves the glycine molecule (atoms 4, 5, 7), the amino nitrogen of aspartate (atom 1), amide nitrogen of glutamine (atoms 3, 9), components of the folate-one-carbon pool (atoms 2, 8), and carbon dioxide. In the biosynthesis, IMP ((Inosine monophosphate) is the first nucleotide formed. It is then converted to either AMP (adenosine monophosphate) or GMP (guanosine monophosphate). The biosynthesis of Histidine is closely related to Purine biosynthesis.

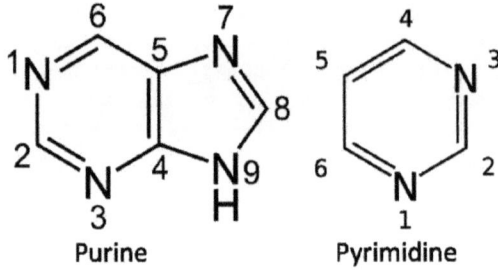

Figure 1. The basic Purine and Pyrimidine bases. The numbering is used in the text to identify the sources of their atoms in biosynthesis.

The pyrimidine biosynthesis begins with carbamoyl phosphate (atoms 2, 3) that condenses with aspartate (atoms 4, 5, 6, 1), resulting in dihydro-orotate which is then added to PRPP to form the nucleotide. Further steps are needed to form Uridylic acid and Cytidylic acid. The Nucleotides Guanylic acid, Cytidylic acid, Adenylic acid, and Uridylic acid are then polymerized by the proto-CCA nucleotidylsynthetases formed in the Thioester World.

3. Genetic Code

The search for the origin of the genetic code began by examining the extant genetic code, which is a mapping of the sixty-four codons (triplet nucleotide sequences), to the twenty amino acids and three termination codons, (Table 1). The genetic code uses a set of four nucleotides normally displayed in a single letter code: A for Adenylic acid, G for Guanylic acid, C for Cytidylic acid, and U for Uridylic acid. There are twenty amino acids encoded by 64 codons and presented in a four-by-four table containing the three letter abbreviations for each of the twenty common amino acids, as shown in Table 1.

Table 1. The standard genetic code table, re-ordered to place the Guanine–Cytosine coding block in the upper left corner, and using the standard abbreviations for the twenty amino acids.

	G	C	A	U	
G	Gly	Ala	Glu Asp	Val	G A ------ C U
C	Arg	Pro	Gln His	Leu	G A ------ C U
A	Arg Ser	Thr	Lys Asn	Met ILe	G A ------ C U
U	Trp Term Cys	Ser	Term Tyr	Leu Phe	G A ------ C U

A structural relationship between the general classes of amino acids and the middle nucleotide of the codon has been well recognized. For example, the middle nucleotide U codes for hydrophobic amino acids, and the middle nucleotide A codes for hydrophilic and the larger neutral amino acids. The middle nucleotide C codes for small neutral amino acids. On the other hand, the middle nucleotide G points to a unique case of Arg as a large and an unlikely amino acid sharing a codon with serine. This uniqueness is eliminated if a smaller positive amino acid is substituted as a biochemical precursor for Arginine.

3.1. The Evolution of the Genetic Code

This can be seen by sequentially omitting the columns headed by U and A, and the rows headed by U and A, one can peel back the genetic code to its earlier form, see Figure 2. For example, if we omit the U column and row from the genetic code: The hydrophobic amino acids Val, Leu, Ile, Met, and Phe are omitted as are Tyr, Trp, and Cys, as well as the three termination codons UAA, UAG, UGG, leaving a GCA code. Next by omitting the A column which then omits Glu/Asp, Gln/Hist, and Lys/Asn. This removes the two negative amino acids Glu and Asp and their neutral versions Gln and Asn and two positive amino acids Lys and Hist.

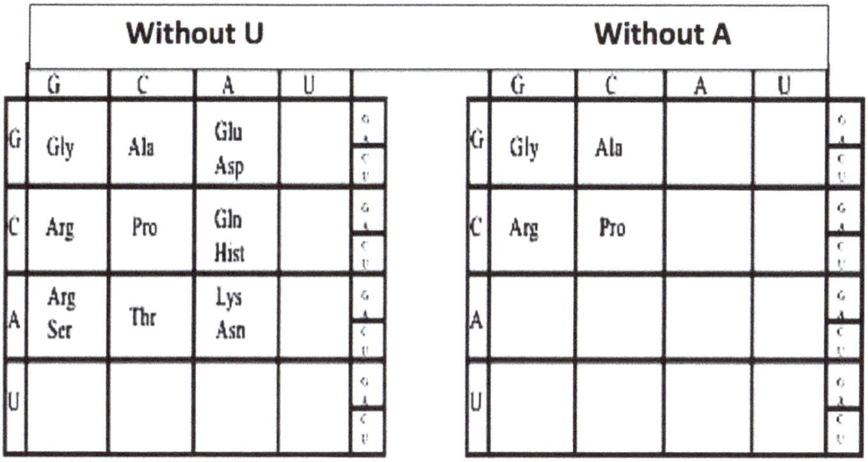

	Without U						Without A				
	G	C	A	U			G	C	A	U	
G	Gly	Ala	Glu Asp		G C U	G	Gly	Ala			G C U
C	Arg	Pro	Gln Hist		G A C U	C	Arg	Pro			G A C U
A	Arg Ser	Thr	Lys Asn		G A C U	A					G A C U
U					G A C U	U					G A C U

Figure 2. The reduced nucleotide genetic coding tables.

The codons in the A column are split due to the wobble in the third base. Omitting the A row: removes Thr, and the split codons for Arg/Ser. This leaves a simple GC code for Gly, Pro, Ala, and Arg. This GC form of the first code was proposed in 1975 [7]. Reversing the sequence implies that the genetic code evolved in an onion-like fashion, beginning as simple GC code and growing by adding layers (e.g., by adding A and forming the GCA code and finally adding U and forming the GCAU code).

The role of Arginine in the evolution of the genetic code appears to need additional discussion, as it is the only one left in the GC code which is activated by a different tRNA synthetase class and requires multiple biosynthetic steps. If Arg is then not considered to be among the early available amino acids, then what might the CGX codons have coded for early on? First, note Arg is related biochemically to a simpler amino acid ornithine as its precursor in its biosynthesis, see Figure 3 [8].

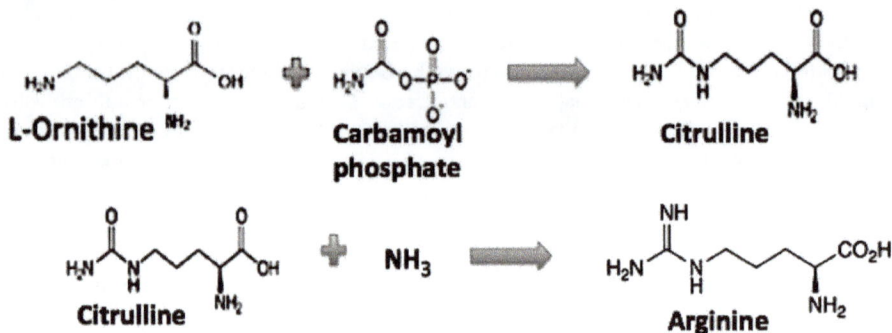

Figure 3. The Ornithine to Arginine pathway.

Why then was Ornithine chosen rather than Lysine as a precursor to Arginine? The answer is that the difference in chain length between Ornithine and Lysine is related to the formation of alpha helices. It was shown experimentally that the helix propensities for the basic amino acids increase with the length of the side-chain in the rank order 2,3-Diamino-L-propionic acids < 2,4-Diamino-L-butyric acid < 2,5-Diaminopentanoic acid (ornithine)< 2,6-Diaminohexanoic acid (Lysine) [9]. The number of side-chain atoms beyond the carboxyl group of the amino acid ending in the NH2 group is the length of the carbon chain in this family of amino acids that determines whether a polypeptide alpha helix will form.

In a GC code coding for Glycine, Proline, Alanine, and 2,3-Diamino-L-propionic acid (Diapr), the earliest coded peptides were not alpha helical, rather mainly simple random coils and beta hairpins as proposed by Soding and Lupas [10]. If one substitutes 2,3 Diamino-L-proprionic acid (Diapr) in place of Arginine (or Ornithine), then there would be a simple chemical relationship among three of the G column amino acids, Diapr, Serine, and Cystine. Diapr has an NH2 group while Cystine has an SH in place of the OH of Serine. This gives us a similarity in their structure and perhaps in their early biosynthesis. As the genetic code evolved to the GCA code (AAA, AAG coding for Lysine) and to the GCAU (hydrophobic amino acids), then the later peptides now incorporating Lysine rather than Diapr were able to form alpha helices. Arginine resulted when an Ornithine reacted with carbamoyl phosphate that resulted in a guanidino group.

We now have a conjecture for the origin and evolution of the genetic code (GC code –> GCA code —> GCAU code) The evidence for this conjecture is to be found in the translational apparatus of the cell in which this code is expressed. The obvious place to start is in the set of aminoacyl-tRNA synthetases.

3.2. Aminoacyl-Transfer RNA Synthetases Contain the Genetic Code

It is in the aminoacyl-tRNA synthetases (aaRS) where each particular amino acid is activated and attached to its cognate tRNA. The majority of synthetases are composed of three domains: The catalytic domain where the amino acids are activated and passed on to tRNA or a thiol; the anticodon domain, which recognizes the anticodon stem loop of the extant tRNA; and the editing domain, which removes a mischarged amino acid from its bound tRNA. The catalytic domain [11] of the aminoacyl-tRNA synthetases (aaRS) is considered to be the original encoding module of the aaRS while the anticodon domain and the editing domains are considered to be later evolutionary additions to the catalytic domain. The reactions carried out by the catalytic domain are:

1. Amino Acid + ATP → Aminoacyl-AMP + PPi
2. Aminoacyl-AMP + tRNA → Aminoacyl-tRNA + AMP

There are two classes of these synthetases, which are characterized by their different catalytic domain structures. Class I's catalytic domain has a Rossman fold and activates Val, Leu, Ile, Met, Cys,

Tyr, Trp, Glu, Gln, and Arg, while class II's catalytic domain is unique and has an antiparallel beta sheet backed by an alpha helix. Class II synthetases activate Gly, Pro, Ala, Thr, Ser, Hist, Asp, Asn, Lys, and Phe. The Class II synthetases are considered to be older than the Class I synthetases. This is based on the fact that the majority of the hydrophobic amino acids are activated by class I synthetases and were unlikely to have been available for early polymerization, while Glycine, Alanine, Proline, and Aspartic acid were available. The evidence for this is based on the metabolic metric, which orders the entry of the amino acids into the genetic code by counting the number of catalytic steps involved in the biosynthesis of the amino acid from the reverse citric acid cycle [12]. It is in the catalytic domain of the class II synthetases and its interaction with the XCCA of the proto-tRNA that evidence for the first GC code arises.

3.3. The Catalytic Domain of the Class II Aminoacyl-tRNA Synthetases

A detailed look at the catalytic domain of these assumed first tRNA synthetases is warranted. This catalytic core is composed primarily of an antiparallel beta sheet constructed from two or three beta hairpin segments [11]. In the extant structures, two of these beta hairpin substructures are connected by an alpha helix, which is on the underside of the active site pocket formed by only two beta hairpins, one hairpin with a large end loop (see Figure 4B). It is this large end loop that forms a lid over the active site pocket and that is evolved into the defining motif II of the class II aminoacyl tRNA synthetase. This motif II loop and the four beta strands in dark gray in Figure 4B, compose the full binding site for the ATP, the amino acid to be activated, and the CCA-3' acceptor end of the tRNA. This structure was proposed as the proto-class II synthetase [11].

If the origin of the genetic code is in the class II synthetase catalytic core, then assuming that core existed before the fully functioning code, a problem presents itself: Where did the peptides forming these beta hairpins come from? The answer proposed here, is in the Thioester World. There the polymerization of amino acid Thioesters generated short peptides, some of which formed catalytic active beta hairpins [13].

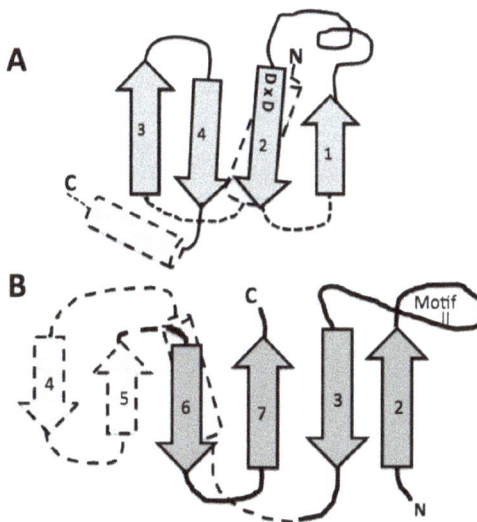

Figure 4. Comparison of the similarity of the double beta hairpin cores of (**A**) the bacterial CCA adding enzyme head domain and (**B**) the class II tRNA synthetase minimum core of the catalytic domain. The minimum catalytic domains are in dark gray. Note the beta hairpin components 3 and 2 in (**B**), which define the motif II loop; and 1 and 2 in (**A**), which contain the two Aspartic acids. The strand numbering for (**B**) is as in reference [11] for the entire traditionally-defined class II catalytic domain.

3.4. The Thioester World and Aminoacyl-tRNA Synthetases

De Duve proposed an early proto-metabolism of amino acids and peptides in a Thioester environment, a world in which there were no Phosphates and thus no nucleotides [14]. He went further and proposed "that clues to the nature of that early proto-metabolism exists within modern metabolism" [15]. In collaboration with the Segre group at Boston University [4] such a proto-network was identified within the modern metabolic network (The Kegg metabolic network). A proto-metabolic network based on Thioesters was obtained when Phosphate was omitted from this Kegg metabolic network. This proto-metabolism had been capable of synthesizing amino acids and fatty acids. The network displays hallmarks of prebiotic chemistry (e.g., iron sulfur cofactors) [4]. The existence of a Thioester environment prior to translation points to the existence of metabolism that preceded the existence of translation. Unlike the phosphate esters which are involved in the activation of amino acids to form peptides, the core metabolism of the Thioester World provides amino acid thioesters allowing for their polymerization to form peptides. In the Thioester World, the metabolic metric can be defined as the number of steps from the reverse citric acid cycle to form an amino acid. The amino acids Glycine, (GG *) Proline (CC *), and Alanine (CG *) are among the earliest available amino acids as based on this metabolic metric. There is a close relationship between core metabolism and the Thioester world, especially in the case of the synthesis of 16 of the 20 amino acids. The exceptions are Histidine, Phenylalanine, Tyrosine, and Tryptophan [12].

Jakubowski has studied the thioester formation by class II Aminoacyl-tRNA synthetases and he summarized these findings as follows, "These and other data support a hypothesis that the present-day Aminoacyl-tRNA synthetases had originated from ancestral forms that were involved in non-coded Thioester-dependent peptide synthesis, functionally similar to the present-day non-ribosomal peptide synthesis by multi-enzyme thiol-template systems" [16]. This finding is also consistent with a recent example of an atypical Seryl-tRNA synthetase found in a methanogenic Archaea, which lacks a tRNA binding site and instead transfers the activated amino acid (Aminoacyl-AMP) to a sulfhydryl group found on the phosphopantetheine (related to Coenzyme A) which is bound to a carrier protein forming a Thioester. The activity of this atypical Seryl-tRNA synthetase is "reminiscent" of the adenylation domains in non-ribosomal peptide synthesis involving a Thioester [16,17]. Thus, before the development of mRNA-coded protein synthesis, ancestral aaRSs facilitated formation of Aminoacyl-Thioesters. Then this early catalytic proto-class II synthetase core, capable of both Thioester and Phosphate ester activation of amino acids, can be considered the link in the transition between a Thioester World and a Phosphate ester World [8].

From such amino acid thioesters, peptide formation is certainly possible. Note that peptide hairpins, such as those forming the catalytic core of the class II synthetases, are considered to have been among the earliest peptide structures [10] and can be formed by the polymerization of amino acid thioesters. The beta hairpins are known to be involved in many other catalytic activities [13]. As noted below such catalytic beta hairpins form a significant part of the core of other key early translational system proteins, such as the CCA nucleotidyltransferases and tRNA ligases. Both of these protein catalytic activities are essential for understanding the coupling of the synthetase's continuing coevolution with that of the tRNA.

4. Origin of the tRNA

The first proto-tRNA is proposed to have been a short three or four polynucleotide, for example, 3'-XCCA-5' [18,19], that interacted with the catalytic core of a proto-class II aminoacyl-tRNA synthetase (aaRS) pictured in Figure 4B. First why CCA? The required base stacking in the extant class II synthetases suggests that the Adenine is essential in base stacking to coordinate the positioning of the incoming ATP and AMP of the activated the amino acid. The other two bases, the Cytosines, thus must not be Adenine.

The likely origin of this proto-tRNA is proposed to involve a CCA-like nucleotidyltransferase beta structure. There are two distinct extant CCA adding enzymes, one in Archaea and a second in

Bacteria. Both CCA enzymes are composed of four domains, which are labeled head, neck, body, and tail. The head or catalytic domain in both of these enzymes are homologous structures, whereas the neck, body, and tail are not. Of interest here, is this common invariant catalytic core in the Head. In the extant CCA enzyme, the catalytic Head behaves like the glycosyltransferases in their mechanism of transferring an activated sugar phosphate in the ligation step. There is a β-DxD-β metal-binding motif in the active center of glycosyltransferases and nucleotidyltransferases [19]. An examination of the CCA catalytic head DxD peptide (Aspartate x Aspartate) structure shows an interesting similarity to that of the class II synthetase catalytic core (Figure 4). Both are constructed from beta hairpins, see Figure 4. Part of this interest here is that both of these enzymes are nucleotidyltransferases, in that they both transfer nucleotides to recipient molecules. In the case of the CCA enzyme the recipient molecule is a 3′ nucleotide of the tRNA, while in the case of the class II aminoacyl-tRNA synthetase, it is to an amino acid forming the Aminoacyl-AMP, which is then transferred to the CCA of the tRNA.

The full catalytic pocket of the modern CCA enzymes is formed by the interaction between the catalytic core, the Head, and an alpha helical domain in the Neck [20]. As these enzymes do not use a polynucleotide template, base specificity is in the neck where amino acids form Watson/Crick-like hydrogen bonds with just the nucleotide base. Now, assuming the earliest proto-transferase was not this complex, the specificity must have come from some other peptides or even a short polynucleotide generated by the similar kinds of enzymes.

The recent determined structure of a tRNA ligase from a T4 phage has in its N-terminal domain a set of beta structures that dominate the catalytic core [21]. This core of an extant tRNA ligase is composed of two beta hairpin substructures backed by a helix or two, almost as if composed of two CCA-like catalytic cores. There is even a functional similarity between the aaRS and the extant tRNA ligase nucleotidyltransferase. Both form an aminoacyl-AMP intermediate: The tRNA ligase forms a Lysyl-AMP intermediate with the AMP that is subsequently added to the 5′ phosphate end of a polynucleotide RNA. This is then followed by a third step when the AMP is removed upon the ligation with the 3′ OH of a second polynucleotide RNA. The independent functioning of the nucleotidyltransferase activity of these ligases has been demonstrated by mutational studies [22]. It is clear that the modern tRNA ligases have undergone extensive evolution, yet the proto-ligase could have been carried out by two primitive proto-CCA enzyme-like entities. Again, the catalytic cores of the tRNA ligases, the nucleotidyltransferases (e.g., CCA enzyme), and the class II aminoacyl-tRNA synthetases are all built from beta hairpins, supporting the idea that their proto forms were formed in the Thioester world [14,15].

4.1. The Accretion Model of the tRNA

The origin and evolution of tRNA is based on the proposed appearance of the proto-nucleotidyltransferase catalytic cores arising in the Thioester that generated nucleotide dimers, trimers and tetramers as well as the trimer CCA. In addition, the ligation of these would have formed single-stranded short RNAs. These coupled with hybridization formed double stranded helices. The modern tRNA structure suggests an assembly from such short helices. For example the extant secondary structure of a typical extant tRNA, shown in Figure 5, can be dissected into five segments or domains: The XCCA attached at the 3′ end of the tRNA; the acceptor arm, which has a stem of seven base pairs; the Anticodon stem loop, which has a stem of five base pairs; and the anticodon loop containing seven unpaired nucleotides. Three of these unpaired nucleotides form the anticodon: the TΨC stem loop, which has a stem of five base pairs, and a loop containing seven unpaired nucleotides in which CGA are isomorphic with the three anticodon bases in the anticodon stem loop; the D stem loop, which has a stem of four or five base pairs and a loop containing seven or eight unpaired nucleotides, with an invariant triplet, GGU, in the loop. Note that the D and T loop names reflect various nucleotide modifications found in the majority of extant tRNAs.

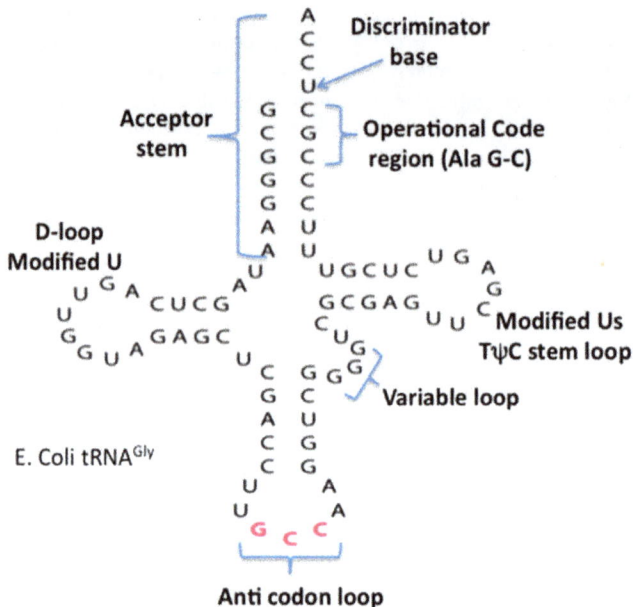

Figure 5. The secondary cloverleaf structure of *E. coli's* tRNA Gly. The discriminator base, the operational code region, and the various stems and loops are labeled in their traditional manner.

4.2. The Operational Code

The initial expansion of the acceptor arm from the XCCA and its relationship to the catalytic domain of the class II synthetases is reflected in the discovery by Hou and Schimmel [23], that a major determinant in the specificity of the class II-tRNA synthetase coding for Alanine was in the G–U pair (3–70) and was found in the acceptor arm. De Duve went on to describe: "A second genetic code still largely un-deciphered is written into the structure of aminoacyl-transfer RNA synthetases" [24].

Schimmel noted: "Because the amino acid-trinucleotide algorithm of the genetic code is established by the specific aminoacylations of tRNAs, the sequences and structure of tRNAs have long been investigated with the idea of learning about the possible origin of the code. The idea is that elements of the early code might [yet] appear in parts of the tRNA structure other than the anticodon. These parts might represent primordial components of the tRNA molecule, which possibly served as structures associated with the aminoacylations of particular amino acids" [25]. He was reporting on a finding by Rodin and Ohno [26] of codons and anticodons in the arms of the tRNA.

Rodin and Ohno, by studying the acceptor arms of numerous tRNAs, identified an expanded view of this code, now called the operational code. They found "In contrast with anticodons, which are built from all four nucleotides, their double-stranded precursors in the 1-2-3 positions of acceptor arms appear as doublets or triplets almost invariably composed of G-C and C-G base pairs. Even in many modern synthetases, including the deep Archaea, specific amino-acylation of the acceptor terminus operates within the G-C variety of first three acceptor base pairs" [26]. It should be pointed out that as early as 1975 it was hypothesized that the first genetic code was a GC code for Gly, Pro, Ala and a positively charged amino acid [7].

Rodin and Ohno proposed [27] a detailed synthesis of the acceptor arm from the trimers 5'-GGC-3' and 5'-GCC-3' and the tetramers 5'-ACCA-3' and 5'-UGGU-3'. Resulting in the proposed initial acceptor arm:

3'-A-C-C-A-C-C-G-U-G-G-U-5'
5'-G-G-C-A-C-C-A-3'.

Here then is a proposed origin of the GC operational code in the tRNA acceptor arm.

In order to access the information in this initial form of the operational code, the catalytic domains of the class II synthetases had to be able to "read" the upper acceptor stem of the primordial tRNA. The minimum catalytic core of the assumed early class II synthetases does not make significant contact with the arm of the tRNA. These minimal enzymes have added relatively short peptides to create N- or C-terminal extensions, often as alpha helices, to their minimal catalytic cores. It is these extensions, which make the explicit contact with the operational code region of the extant t-RNAs [11], (see Figure 6). This is an excellent example of the co-evolution of the proto-tRNA and the proto-class II synthetases in generating the amino acid to RNA coding association of the Archaic GC operational code in the tRNA arm.

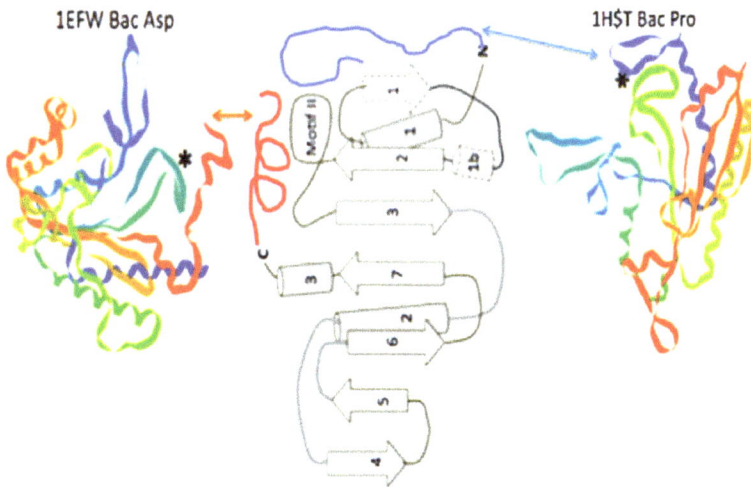

Figure 6. The class II synthetase catalytic core showing two alternate operational code-contacting peptide extensions, one on the N-terminal, as seen in the bacterial Proline aaRS structure 1HST.pdb, and one on the C-terminal, as seen in the bacterial Aspartic acid aaRS structure, 1EFW.pdb. Figure is adapted from [11].

The next step in the evolution of the proto-tRNA is the formation of the stem loops. Schimmel and his group [28] tested the operational code with a variety of truncated arms of the tRNA. These truncated tRNAs were stem helices of various length with the XCCA at the 3' end. These were suggested as possible precursors of the expanding proto-tRNA and as well as the anticodon stem-loop structure of the tRNAs. One of these truncated arms had an overhang of a single-stranded 5-mer at the five prime end of the second strand, yet was still properly charged [28]. This overhanging stem is a reasonable initial model for the synthesis of an anticodon stem loop. The upper triplets of this stem helix contained the archaic GC code (operational code) [28]. In a manner similar to the one found in the upper triplets of the extant tRNA arm, a 5-mer containing a trimer based on GC plus a dimer provided the GC code in the stem helix and also in the overhang at the 5' end. The resulting structure was closed by ligating another dimer to form a seven-nucleotide closed loop with the operational code not only in the stem but also in the related anticodon in the loop. The GC code was in the helix and in the loop. This was the first set of stem loops with the GC code in the loop. The CGA triplet in the loop of the T stem loop and the GGU triplet in the loop of the D stem loop are remnants of a GC code. The later anticodon stem loops incorporated the anticodons for the GCA code and then for the GCAU code [29–31].

4.3. The Maturation of tRNAs

There are two proto-tRNA-type structures available, an arm and stem loop, each containing the CCA end. The ligation of two such sub structures has been referred to as the boomerang, (Figure 7C). A second ligation of two stem-loop substructures formed the so-called dumbbell, (Figure 7D). Such a boomerang functioning as a truncated tRNA has been found in the mitochondria of the worm Enoplea. "There, the mitochondria of some metazoans, contain deviations from the consensus tRNA structure, some tRNAs lack both the D- or T-arm without losing their function. In Enoplea, this miniaturization comes to an extreme, and functional mitochondrial tRNAs can lack both arms, leading to a considerable size reduction, and a flexible link between the anti-codon arm and the CCA acceptor arm. Hence, it is believed that the single connector elements in these transcripts have an increased flexibility in order to allow the formation of an extended 3D tRNA shape, bringing the anticodon loop and 3′-terminus into a conserved distance that is required for acceptance by the ribosome and, consequently, participation in translation. This shape deviates from the classical L form and is more boomerang-like [32].

The ligation between the stem loop (Figure 7B) and the anticodon stem loop (Figure 7B) containing the anticodons generated a second structure referred to as the dumbbell, as shown in Figure 7D. Finally, the ligation of these two substructures, the boomerang and the dumbbell, resulted in the secondary structure of the tRNA, the cloverleaf, as shown in Figure 5.

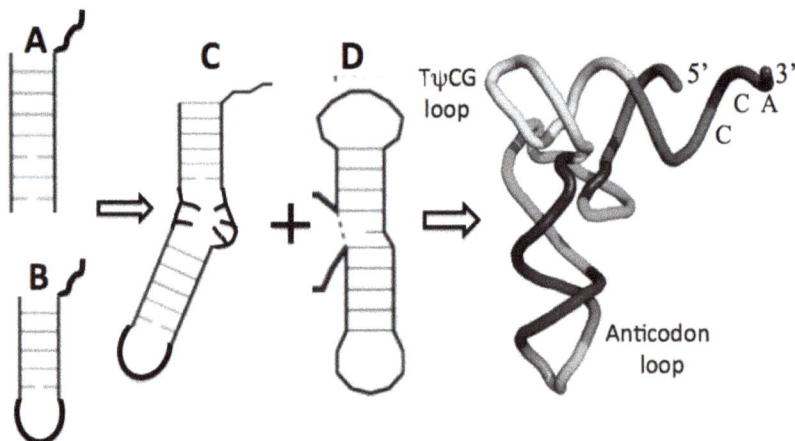

Figure 7. Possible maturation of the full tRNA: (**A**) arm with CCA attachment, fused with (**B**) stem loop with CCA attachment resulting in (**C**) the boomerang that then fused with (**D**) the dumbell resulting in the tRNA cloverleaf secondary structure in Figure 5 and the full 3D, folded tRNA structure pictured here.

The 2D cloverleaf representation of the tRNA in Figure 5, when compared to the 3D structure in Figure 7, shows a complex change in its configuration due to the coevolution of the proto-tRNA with the aminoacyl-tRNA synthetases and the large subunit of the ribosome (**LSU**) and the small subunit (SSU) of the ribosome. The accretion model of the tRNA was implicit in forming the seed for the accretion models of the LSU and the SSU of the ribosome [33].

4.4. The Accretion Models of the tRNA and the Seeds for the Ribosome

The proto-tRNA and its interactions with the proto-ribosome points us to the accretion model for the ribosome as proposed by Petrov et al. [33]: "In phase 1, ancestral RNAs form stem loops and minihelices. In phase 2, the LSU catalyzes the condensation of nonspecific oligomers. The SSU (small subunit of the ribosome) may have a single-stranded RNA-binding function. In phase 3, the subunits associate, mediated by the expansion of tRNA from a minihelix to the modern L shape. LSU and SSU evolution is independent and uncorrelated during Phases 1–3".

In phase 1, the start of the accretion model for the LSU and the SSU began with a set of ancestral RNAs. The source for this set of RNAs was not identified. It is here proposed that the source for their accretion model [33] was the same as that for the accretion of the tRNAs, which began with di-, tri-, and tetra-nucleotides ligated to form various additional stem-loop structures that supplied the seeds for the accretion model of the LSU and SSU of the Ribosome. An example was the minihelix formed by ligation of a seven-member helix (arm), with the GC code in the arm, and five-member stem, with a seven-member loop containing a CGA triplet. The difference between the boomerang and the minihelix is the ligation took place without the XCCA on the arm and the TΨCG stem loop.

The seed for the large subunit of the Ribosome was the minihelix, similar to those forming the proto-tRNAs, Figure 8. A number of suggestions have been made for the formation of the PTC of the LSU. However, the one by Tamura is most relevant as it involves the tRNA. Tamura [34] proposed that the minihelix was an ancestral precursor of both the modern tRNA and of the large ribosomal subunit PTC (peptidyl transferase center). He claimed that "...considering the RNA minihelix as the molecule of origin can definitely lead us to attaining our goal of elucidating the emergence of the modern peptide synthesis machinery of the ribosome". This idea was examined by Farias et al. [35] by reconstructing probable ancestral tRNA sequences from extant tRNAs and comparing them with the large subunit PTC RNA from a range of different organisms. The result showed a remarkable sequence identity.

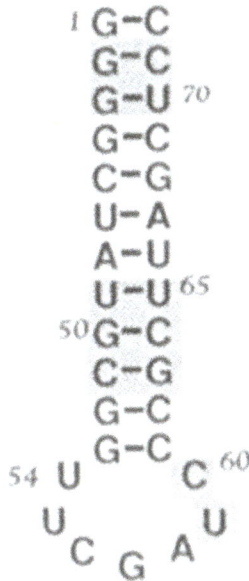

Figure 8. Minihelix example, proposed by Tamura [34] as a seed for the ribosomal subunits.

Note, a recent model derives the tRNA from three minihelices similar to those proposed for the accretion models of both the ribosomal subunits. "The conserved archaeal tRNA core (75-nt) is posited to have evolved from ligation of three proto-tRNA minihelices (31-nt) and two-symmetrical nine-nt deletions within joined acceptor stems (93 − 18 = 75-nt)" [36].

4.5. Early Function of the LSU Proteins

What was the initial function of the proto-LSU? It was examined by Schimmel et al. [31] in testing the operational code with a variety of truncated tRNAs with XCCA at the 3′ ends. The truncated tRNAs were able to interact with the catalytic domain of the class II aminoacyl-tRNA synthetase and add the amino acid to the CCA. However, if the arm was long enough it bound to the PTC of the LSU and transferred the amino acid to a growing peptide. Thus, the most likely early function of the proto-LSU, composed of little more than the PTC, was to have bound two minihelices; and by bringing the activated amino acids attached to the XCCA into contact allowed the formation of dipeptides. It generated longer peptides that would depend on the further formation of the tunnel. There was a GC code (operational code) in the arm of the proto-tRNA that interacted with class II synthetases catalytic modules and perhaps the GC code in the arm stabilized the formation of the dipeptide with the PTC of the proto-LSU.

4.6. The Origin of the SSU

Gulen [37] proposed, that the Domain A of the extant SSU plays a crucial role in the SSU by forming a scaffold linking the other SSU domains, Figure 9. As such, it was proposed as the starting center, or seed, for an accretion model of the SSU. There is an interaction in this substructure, which mimics the elbow of the tRNA. The structure of this proposed SSU seed also has interesting similarities to the boomerang structure in Figure 7C, of the proposed intermediate form of the earliest Proto-tRNAs above.

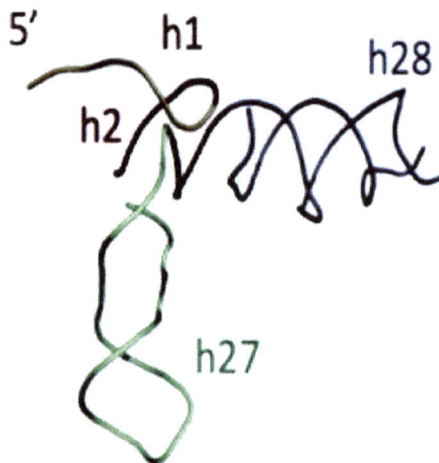

Figure 9. Proposed SSU seed, domain A Gulen et al. [37]. The SSU domain A was proposed to include SSU helices H27 and H28, resulting in a structure similar to the structure of the "boomerang" tRNA intermediate proposed structure in Figure 7C. Figure 9 has been provided by Anton Petrov.

This proposed seed supports the link between the tRNA accretion model and that of the ribosome subunits [38]. The merger of the arm with the stem loops, completes the search for the origin of the GC code. It is in coevolution of the catalytic domain of the class II Aminoacyl-tRNA synthetases and the XCCA and the operational code in the arm of the tRNA.

4.7. What Was the Function of the Proto-SSU?

The extant SSU's functions involve the binding of a messenger RNA and a tRNA whose anticodon stem loop reads the codon on the messenger RNA. This is an evolved form of what happened with the proto-messenger RNA and with the stem loops containing the GC anticodons for the triplets in a proto-messenger RNA. The proto-message was due to the ligation of triplets and provided an alternate method of forming coded short peptides of Glycine, Proline, and Alanine. Then, as proposed above, the anticodon stem loops also contained a CCA-3′ end with an activated amino acid.

It is here proposed that the proto-SSU was where a coded form of peptide synthesis took place before interacting with the LSU. The stem loops which were synthesized from doublets and triplets resulted in stem loops with a GC code in the five base paired helix of the stem and with a GC code in the seven-membered nucleotide loop. This allowed the XCCA on the stem to form an ester with an amino acid catalyzed by a catalytic module of a proto-class II aminoacyl-tRNA synthetase. The amino acid ester on the CCA of a stem loop of a proto-tRNA was coded for by a codon on the proto-messenger RNA. Moreover, the adjacent amino acid ester on the CCA of stem loop proto-tRNA formed a coded dipeptide and, if continued formed a longer coded peptide. The proto-ribosomal subunits (LSU and SSU) began with their key functions independently before merging to form the ribosome. This later interdependence is supported by the full ribosomal accretion models of the Georgia Tech group [33,38].

5. tRNA and Replication

The above accretion model for the tRNA is also a model for the replication of the proto message as well as for the earliest form of the ribosomal subunits. "The earliest catalyst RNA ligase spliced two RNA molecules by forming 3'–5' phosphodiester bond between the 3' hydroxyl of one RNA molecule and the 5' phosphate of another RNA molecule. It is, therefore, relatively easy to use a polymerized set of primitive single-stranded triplets as a template for the polymerization of a complementary set of triplets, which are polymerized by a primitive RNA ligase. The same complementarity between the triplets or doublets in the loops between the stem anticodon loops are used in translation and it would be no accident that replication and translation were similar processes" [39]. Thus, the origin of the GC code was a tale of three peptide catalysts: The nucleotidyltransferases CCA enzyme, the class II catalytic domains, and the RNA ligase. We can continue to trace the evolution of the genetic code with the ribosomal proteins.

Ribosomal Proteins and the Evolution of the Genetic Code

The ribosome is a complex molecular machine found within all living cells, that serves as the site of biological protein synthesis (translation). Ribosomes link amino acids together in the order specified by messenger RNA (mRNA) molecules. Ribosomes consist of two major components: The small ribosomal subunits, which read the messenger RNA; and the large subunits, which join amino acids to form a polypeptide chain. Each subunit comprises one or more ribosomal RNA (rRNA) molecules and a variety of ribosomal proteins (r-protein). The ribosomes and associated molecules (e.g., the aminoacyl-tRNA synthetases) are also known as the translational apparatus.

There has been a coevolution between the tRNA and the Ribosome. This has been observed at two different levels. First there has been an extensive accretion model developed for the evolution of the ribosome. In its early phases, it parallels that of the above tRNA accretion model for the tRNA [33]. Second, an analysis of the ribosomal proteins has provided insight as to a sequence in the emergence of different structural types of peptides and the resulting block structure of the ribosomal proteins. Most important here are the large subunit protein extensions to the PTC (peptidyl transferase center). These are largely without secondary structure and are rich in Alanine, Proline, Glycine, and positive amino acids, implying that they were early-encoded peptides based on a GC code. The other key characteristic of the ribosomal proteins is that they appear to have been assembled from multiple shorter peptides. This is reflected in their taxonomic division block structure [40]. Finally, the structural complexity of the ribosomal proteins increase as each one goes out from the random coils at the PTC to the tunnel, and finally to the surface of the ribosome, changing from simpler random coils, through mostly beta hairpins, to alpha beta domains and finally to alpha domains.

6. Conclusions

(1) The first replicators were self-replicating iron-rich clays which fixed carbon dioxide into oxalic and other dicarboxylic acids.

(2) " … this system of replicating clays and their metabolic phenotype, then evolved into the sulfide-rich region of the hot-spring acquiring the ability to fix nitrogen. The feedback between the clay surfaces and the early metabolic products, oxalic acid, citric acid amino acids, and short peptides result in the synthesis of the clays themselves. It is this feedback, which is involved in the selective forces, that expands the metabolic network. Finally, phosphate was incorporated into the evolving system, which allowed the synthesis of nucleotides and phospholipids. If biosynthesis recapitulates biopoiesis [41], then the synthesis of amino acids preceded the synthesis of the purine and pyrimidine bases. Furthermore, the polymerization of the amino acid thioesters into polypeptides preceded the directed polymerization of amino acid esters by polynucleotides. "Thus, the origin and evolution of the genetic code is a late development and records the takeover of the clay by RNA" [42].

(3) The proteins of the translational apparatus have a memory of their evolution and thus we can read that record.

(4) The arrival of RNA in the biosphere is best understood by the origin and evolution of the tRNA. The remaining problem is the evolution of the GC code to the GCA code and to the GCAU code, as exemplified in the proteins of the translation apparatus.

(5) The origin of the genetic code began with a GC code. The evolution of the genetic code from the GC code to the GCA code is based on the further coevolution of the aminoacyl-tRNA synthetases and the proto-tRNA. For example, the entry of Glutamic acid into the coding system means that the catalytic domain of the class I aaRS, the Rossmann fold, must be introduced and its evolution worked out. The appearance of the wobble is visible in the A column of the genetic code: Glu/Asp, Gln/Hist, Lys/Asn (e.g., Glu (GA pur)/Asp (GA pyr) share the GA* codon). The wobble is related to the introduction of the anticodon domains of the class I and II synthetases. The coevolution of the proto-tRNA evolution with the evolution of the anticodon domain likely explains the origin of the wobble. The further evolution of the 3D structure of the tRNA, especially the elbow, involves the continuing evolution of the LSU and SSU of the ribosome and the interaction of those two subunits. These will be dealt with in our next paper.

Author Contributions: H.H. and T.F.S. contributed equally to the design and writing of this paper.

Funding: This research received no external funding.

Conflicts of Interest: The authors declare no conflicts of interest.

References

1. Dyson, F. ; *Origins of Life*; Cambridge Press: Cambridge, UK, 1999.
2. Hartman, H. Speculations on the origin and evolution of metabolism. *J. Mol. Evol.* **1975**, *4*, 359–370. [CrossRef]
3. Hartman, H. Conjectures and reveries. *Photosynth. Res.* **1992**, *33*, 171–176. [CrossRef]
4. Goldford, J.E.; Hartman, H.; Smith, T.F.; Segre, D. Remnants of an ancient metabolism without phosphate. *Cell* **2017**, *168*, 1126–1134. [CrossRef]
5. Dorr, M.; Kassbohrer, J.; Grunert, R.; Kreisel, G.; Brand, W.A.; Werner, R.A.; Geilmann, H.; Apfel, C.; Robl, C.; Weigand, W. A Possible prebiotic formation of ammonia from dinitrogen on iron sulfide surfaces. *Angew. Chem. Int. Ed.* **2003**, *42*, 1540–1543. [CrossRef]
6. Menez, B.; Pisapia, M.; Andreani, M.; Jamme, Q.P.; Vanbellingen, A.; Brunelle, L.; Richard, P.D.; Refregiers, M. Abiotic synthesis of amino acids in the recesses of the oceanic lithosphere. *Nature* **2018**, *64*, 59–63. [CrossRef]
7. Hartman, H. Speculations on the evolution of the genetic code. *Orig. Life* **1975**, *6*, 423–427. [CrossRef] [PubMed]
8. Hartman, H.; Smith, T.F. The evolution of the ribosome and the genetic code. *Life* **2014**, *4*, 227–249. [CrossRef]
9. Padmanabhan, S.; York, E.J.; Stewart, J.M.; Baldwin, R.L. Helix Propensities of basic amino acids increase with the length of the Side-chain. *J. Mol. Biol.* **1996**, *257*, 726–734. [CrossRef] [PubMed]
10. Soding, J.; Lupas, A.N. More than the sum of their parts: On the evolution of proteins from peptides. *BioEssays* **2003**, *25*, 837–846. [CrossRef] [PubMed]

11. Smith, T.F.; Hartman, H. The evolution of class II Aminoacyl-tRNA synthetases and the first code. *FEBS Lett.* **2015**, *589*, 3499–3507. [CrossRef] [PubMed]
12. Klipcan, L.; Safro, M. Amino acid biogenesis, evolution of the genetic code and Aminoacyl-tRNA synthetases. *J. Theor. Biol.* **2004**, *228*, 389–396. [CrossRef] [PubMed]
13. Metrano, A.J.; Abascal, N.C.; Mercado, B.Q.; Paulson, E.K.; Hurtley, A.E.; Miller, S. Diversity of secondary structure in catalytic peptides with-turn-biased Sequences. *J. Am. Chem. Soc.* **2012**, *139*, 492–516. [CrossRef] [PubMed]
14. De Duve, C. Clues from Present-Day Biology: The Thioester World. In *The Molecular Origins of Life: Assembling Pieces of the Puzzle*; Brack, A., Ed.; Cambridge University Press: Cambridge, UK, 1988; pp. 219–236.
15. De Duve, C. The beginning of life on Earth. *Am. Sci.* **1995**, *83*, 428–430.
16. Jakubowski, H. Aminoacyl-tRNA synthetases and the evolution of coded peptide synthesis in the Thioester World. *FEBS Lett.* **2016**, *590*, 469–481. [CrossRef] [PubMed]
17. Mocibob, M.; Ivic, N.; Bilokapic, S.; Maier, T.; Luic, M.; Ban, N.; Weygand-Durasevic, I. Homologs of aminoacyl-tRNA synthetases acylate carrier proteins and provide a link between ribosomal and nonribosomal peptide synthesis. *Proc. Natl. Acad. Sci. USA* **2010**, *107*, 14585–14590. [CrossRef] [PubMed]
18. Aravind, L.; Koonin, K.V. DNA polymerase beta-like nucleotidyltransferase superfamily: Identification of three new families, classification and evolutionary history. *Nucleic Acids Res.* **1999**, *27*, 1609–1618. [CrossRef] [PubMed]
19. Liu, J.; Mushegian, A. Three monophyletic super families account for the majority of the known glycosyltransferases. *Protein Sci.* **2003**, *12*, 1418–1431. [CrossRef]
20. Vortler, S.; Morl, M. Trna-nucleotidyltransferases: Highly unusual RNA polymerases with vital functions. *FEBS Lett.* **2010**, *584*, 297–302. [CrossRef]
21. Ho, K.; Wang, L.K.; Lima, C.D.; Shuman, S. Structure and mechanism of RNA ligase. *Structure* **2004**, *12*, 327–339. [CrossRef]
22. Zhelkovsky, A.M.; McReynolds, L.A. Structure-function analysis of Methanobacterium thermoautotrophicum RNA ligase–engineering a thermostable ATP independent enzyme. *BMC Mol. Biol.* **2012**, *13*, 24. [CrossRef]
23. Hou, Y.M.; Schimmel, P. A simple structural feature is a major determinant of the identity of a transfer RNA. *Nature* **1988**, *333*, 140–145. [CrossRef]
24. De Duve, C. Transfer RNAs: The second genetic code. *Nature* **1988**, *333*, 117–118. [CrossRef] [PubMed]
25. Schimmel, P. Origin of genetic code: A needle in the haystack of tRNA sequences. *Proc. Natl. Acad. Sci. USA* **1996**, *93*, 4521–4522. [CrossRef]
26. Rodin, S.N.; Ohno, S. Four primordial modes of tRNA-synthetase recognition, determined by the (GC) operational code. *Proc. Natl. Acad. Sci. USA* **1997**, *94*, 5183–5188. [CrossRef]
27. Rodin, A.S.; Szathmary, E.; Rodin, S.N. On origin of genetic code and tRNA before translation. *Biol. Direct* **2000**, *6*, 1–24. [CrossRef] [PubMed]
28. Schimmel, P.; Giege, R.; Moras, D.; Yokoyamat, S. An operational RNA code for amino acids and possible relationship to genetic code. *Proc. Natl. Acad. Sci. USA* **1993**, *90*, 8763–8768. [CrossRef] [PubMed]
29. Komatsu, R.; Sawada, R.; Umehara, T.; Tamura, K. Proline might have been the first amino acid in the primitive genetic code. *J. Mol. Evol.* **2014**, *78*, 310–312. [CrossRef]
30. Tamura, K. Origin and early evolution of the early tRNA. *Life* **2015**, *5*, 1687–1690. [CrossRef] [PubMed]
31. Tamura, K. Beyond the frozen accident: Glycine assignment in the genetic code. *J. Mol. Evol.* **2015**, *81*, 69–71. [CrossRef]
32. Juhling, T.; Duchardt-Ferner, E.; Bonin, S.; Wohnert, J.; Putz, J.; Florentz, C.; Betat, H.; Sauter, C.; Morl, M. Small but large enough: Structural properties of armless mitochondrial tRNAs from the nematode *Romanomermis culicivorax*. *Nucleic Acids Res.* **2018**, *46*, 9170–9180. [CrossRef]
33. Petrov, A.S.; Gulen, B.; Norris, A.M.; Kovacs, N.A.; Bernier, C.R.; Lanier, K.A.; Fox, G.E.; Harvey, S.C.; Wartell, R.M.; Hud, N.V.; et al. History of the ribosome and the origin of translation. *Proc. Natl. Acad. Sci. USA* **2015**, *112*, 15396–15401. [CrossRef]
34. Tamura, K. Ribosome evolution: Emergence of peptide synthesis machinery. *J. Biosci.* **2011**, *36*, 921–928. [CrossRef] [PubMed]
35. Farias, S.T.; Rego, T.G.; Jose, M. Origin and evolution of the Peptidyl Transferase Center from proto-tRNAs. *FEBS Open Bio* **2014**, *4*, 175–178. [CrossRef]

36. Root-Bernstein, R.; Kim, Y.; Sanjay, A.; Burton, Z.F. tRNA evolution from the proto-tRNA minihelix world. *Transcription* **2016**, *7*, 153–163. [CrossRef] [PubMed]
37. Gulen, B.; Petrov, A.S.; Kafor, C.D.; Wood, D.V.; O'Neill, E.B.; Hud, N.V.; Williams, L.D. Ribosomal small subunit domains radiate from a central core. *Sci. Rep.* **2016**, *6*, 20885. [CrossRef]
38. Hsiao, C.S.; Mohan, B.K.; William, L.D. peeling the onion: Ribosomes are ancient molecular fossils. *Mol. Biol. Evol.* **2009**, *26*, 2415–2425. [CrossRef] [PubMed]
39. Hartman, H. Speculations on the evolution of the of the genetic code III: The evolution of the tRNA. *Orig. Life* **1984**, *14*, 643–648. [CrossRef] [PubMed]
40. Vishwanath, P.; Favaretto, P.; Hartman, H.; Mohr, S.C.; Smith, T.F. Ribosomal protein-sequence block structure suggests complex prokaryotic evolution with implications for the origin of eukaryotes. *Mol. Phylogenet. Evol.* **2004**, *33*, 615–625. [CrossRef]
41. Granick, S. Speculations on the origins and evolution of photosynthesis. *Ann. N. Y. Acad. Sci.* **1957**, *69*, 292–308. [CrossRef] [PubMed]
42. Hartman, H. Photosynthesis and the origin of Life. Origins of life and evolution of biospheres. *Orig. Life Evol. Biosph.* **1998**, *28*, 515–521. [CrossRef]

Article

Low-Digit and High-Digit Polymers in the Origin of Life

Peter Strazewski

Institut de Chimie et Biochimie Moléculaires et Supramoléculaires (Unité Mixte de Recherche 5246), Université de Lyon, Claude Bernard Lyon 1, 43 bvd du 11 Novembre 1918, 69622 Villeurbanne CEDEX, France; strazewski@univ-lyon1.fr; Tel.: +33-472-448-234

Received: 4 January 2019; Accepted: 26 January 2019; Published: 2 February 2019

Abstract: Extant life uses two kinds of linear biopolymers that mutually control their own production, as well as the cellular metabolism and the production and homeostatic maintenance of other biopolymers. Nucleic acids are linear polymers composed of a relatively low structural variety of monomeric residues, and thus a low diversity per accessed volume. Proteins are more compact linear polymers that dispose of a huge compositional diversity even at the monomeric level, and thus bear a much higher catalytic potential. The fine-grained diversity of proteins makes an unambiguous information transfer from protein templates too error-prone, so they need to be resynthesized in every generation. But proteins can catalyse both their own reproduction as well as the efficient and faithful replication of nucleic acids, which resolves in a most straightforward way an issue termed "Eigen's paradox". Here the importance of the existence of both kinds of linear biopolymers is discussed in the context of the emergence of cellular life, be it for the historic orgin of life on Earth, on some other habitable planet, or in the test tube. An immediate consequence of this analysis is the necessity for translation to appear early during the evolution of life.

Keywords: digit multiplicity; information; transmission; encoding; translation; diversity; function

---

1. Introduction

The analogy between the role of linear biological polymers in cellular life, and that of strings composed of digits in the elaboration and transmission of discrete information, is as old as the foundations of information theory [1–3] and the Central Dogma of molecular biology [4–6]. The definition and understanding of "information" in the former theory finds its analogy in the latter through a shift on a superordinate level, in a metaphor [7]. Classical information theory presupposes that the transmission of information is based on a purely one-way "Laplacian" deterministic transfer mechanism whereby the instruction encoded in a one-dimensional ("linear") digital information determines fully the outcome; it implies causality and, in principle, "bottom-up" predictability. This of course is not the case in living organisms. Even in the most simple organism there are, despite the Central Dogma stating a strict irreversibility of information transmission, feedback mechanisms that "impose" changes in the "programme" of the descendants of the very organism, through the selection of stochastic errors, and thus permit its "evolution" in a Darwinian sense, hence, a change in frequency and abundance of a heritable trait of a population, through adaptation and in competition ("fitness") with others, as opposed to through random drift, migration or molecular changes per se. Therefore, the metaphorical use of information theoretical or mathematical terms like "programme", "code", "signal", "noise", "random", "algorithm" for the description of physical, chemical and biological phenomena and processes need to be taken with uttermost care and a full awareness of the pros and cons of metaphors across these research domains.

Having said that, the advent of an upcoming new non-deterministic information theory, owing to expected developments in the artificial intelligence (AI) field, will probably induce a general overhaul of classical information theoretical terms, since life-like extrinsic feedback mechanisms are not only inevitably emerging but also at the very heart of AI [8]. Likewise, in a number of expected processes, that are thought to have taken place shortly before and during the (or any other) orgin of life, the differences between genotype and phenotype are expected to be less pronounced and the strength of intrinsic and extrinsic feedback mechanisms weaker than they are in extant biota [9,10]. Therefore, it is worthwhile to keep analysing the effects of linear biological polymers in the light of information theory [11]. Here the consequences of different multiplicities of digits in linear polymers (strings) are examined with respect to the information transmission from one to another biopolymer and their respective functional competences. It will become apparent that both low-digit and high-digit linear polymers are likely to be prerequisite for any life form to emerge from complex prebiotic chemical systems.

2. Information Transmission and Capacity of Digital Strings

According to Claude Shannon [3], the information content of a string of M digits to be transmitted, termed here *Shannon Information (SI)*, is inversely related to its expectancy of appearing by chance from a random alignment of its digital components. *SI* is thus a measure of "unpredictability" and "randomness", since the more unlikely a particular sequence of M digits is to self-assemble by pure chance the more its occurrence is "informative" [7] and "patterened" [11]. The amount of this information stored in a string of binary digits is proportional to the logarithm of N possible states of that system, denoted $log_2 N$ (Equation (1)). Changing the base of the logarithm to a different number b has the effect of multiplying the value of the logarithm by a fixed constant $log_2 b$. The choice of the base b determines the unit used to measure information, for which different unit names are used: *bit/shannon* (binary digit) for $b = 2$, *nit/nat/nepit* (natural or Neperian digit) for $b = e \approx 2.718$, *trit* (trinary or ternary digit) for $b = 3$, *quit* (quaternary digit) for $b = 4$, *dit/ban/hartley* (decimal digit) for $b = 10$, and so forth. One *nit* ≈ 1.443 *bits*, 1 *trit* ≈ 1.585 *bits*, 1 *quit* = 2 *bits*, 1 *dit* ≈ 3.322 *bits*, etc. M denotes the number of digits in a string (Equation (2)). The longer the string the proportionally higher the *SI* (Equation (3)).

$$SI(N) = log_2 N = log_b N \cdot log_2 b \qquad (1)$$

$$N = b^M \qquad (2)$$

$$SI(M) = M \cdot log_2 b \qquad (3)$$

The digit multiplicity b, that is, the number of different digits in a string, can in principle vary without limitation. The higher the multiplicity b the shorter the string with identical information storage capacity and *SI* (Table 1). For example, 40-meric *bit* strings (*SI* = 40 *bits*) can realise about a (long-scale) trillion different variants. Roughly the same information storage capacity is realised by their "compression" to 26-meric *trit* strings, 20-meric *quit* strings, 18-meric *quint* strings, and so forth. The higher multiplicities chosen in Table 1 ($b > 5$) refer to the possible generation of "secondary" high-digit strings from the "primary" low-digit strings through the usage of a code that enhances the multiplicities b to higher b' values by integer exponents {2, 3, 4}: $2^3 = 8$, $3^2 = 9$, $2^4 = 4^2 = 16$, $3^3 = 27$, $4^3 = 64$, $3^4 = 81$, $4^4 = 256$ (see also Figure 1). Another multiplicity denoted b'_{eff} refers to a partly redundant use of a higher multiplicity b', *vide infra*.

The usage of information theoretical *SI* for the calculation of the "unpredictability" in biopolymers of extant biota, such as polynucleotides and polypeptides, is in that sense questionable as all evolved life forms did not emerge from random self-assemblies of its monomeric residues; on the contrary, they evolved from deletions, insertions and rearrangments of whole DNA segments from "horizontal gene transfer", and from "random walks through sequence-space" being carried over in minute mutation steps from antecedent organisms to their progeny that happened to be "fit" enough to give viable offspring again. However, in the context of an origin of first cellular life from complex chemical systems, "random" self-assemblies—with all constraints imposed by the real atomistic chemistry that differentiates this kind of randomness from true randomness in an abstract mathematical sense—could be reasonably plausible events happening in a prebiotic environment, given recurrent chemical potential gradients and the prebiotic availability of sufficiently large amounts of linear polymers of some realistic distribution of limited lengths *M*. Therefore, it makes sense to analyse processes that could have occurred on the early Earth, or that might occur under other comparable circumstances, in the light of information transmission according to the formalism pioneered by Shannon, as well as purely combinatorial information storage capacity assuming equal probabilities for all string sequences.

Table 1. String lengths *M*, binary string compression factors [‡] and approximate *SI* values for strings of different integer-digit multiplicities [#] bearing approximately the same storage capacity $N \approx 10^{12}$ [§].

Digit Multiplicity [#]			String Length *M*	String Compression [‡] $M_{b=2}/M_{b \geq 2}$	Shannon Information (Equation (3), $b'_{eff}: \to b': \to b$)	Storage Capacity N [§] (Equation (2), $b'_{eff}: \to b': \to b$)
b	*b'*	*b'_{eff}*				
2			40	1.00	40 *bits*	$2^{40} \approx 10^{12}$
3			26	1.54	≈ 41.2 *bits* = 26 *trits*	$3^{26} \approx 10^{12}$
4			20	2.00	40 *bits* = 20 *quits*	$4^{20} \approx 10^{12}$
5			17-18	2.35–2.22	≈ 39.5–41.8 *bits*	$5^{17-18} \approx 10^{11-12}$
	8		13-14	3.08–2.85	≈ 39.0–42.0 *bits*	$8^{13-14} \approx 10^{11-12}$
	9		12-13	3.33–3.08	≈ 38.0–41.2 *bits*	$9^{12-13} \approx 10^{11-12}$
	16		10	4.00	40 *bits*	$16^{10} \approx 10^{12}$
		20	9-10	4.44–4.00	≈ 38.9–43.2 *bits*	$20^{9-10} \approx 10^{11-13}$
	27		9	4.44	≈ 42.8 *bits*	$27^{9} \approx 10^{12}$
	64		6-7	6.67–5.71	≈ 36.0–42.0 *bits*	$64^{6-7} \approx 10^{11-12}$
	81		6	6.67	≈ 38.0 *bits*	$81^{6} \approx 10^{11-12}$
	256		5	8.00	40 *bits*	$256^{5} \approx 10^{12}$

$b' = b^2$, b^3, b^4 (three different codon lengths); b'_{eff} = reduced through redundancy from higher *b*, cf. Section 3.
‡ cf. complexity Ψ = *bits* per monomer (e.g. per nucleotide, codon, amino acid) [9]. § precision within ± 1 order of magnitude.

3. Low-Digit Memory Polymers

The bricks that biotic nature—as we know it on Earth—uses to maintain a systemic memory throughout many generations of reproduction of individual system units, that themselves individually almost fully degrade and ultimately vanish, are composed of nucleic acids. Nucleic acids can harbor and transmit an astonishingly large number of information through more or less faithfully copying long strings termed "linear polymers". In biotic nature these strings are very soluble and solvent-accessible polyanionic linear polymers composed of 4 different (but similar) "letters". In information theoretical terms these are *quits* (not Qbits) realised by the nucleotides A, G, C and U or T. Natural *quits* are pairwise complementary to one another through the Watson–Crick rules (G–C, A–U or A–T), which gives the grounds for faithful template-directed copying as during cellular replication (double-copying of complementary single-stranded DNA), or complement-copying, as for the transcription or reverse-transcription of strings of nucleic acids of virtually deliberate length (from DNA to RNA or vice versa). Upon translation, in contrast, specific "coding fractions" of these strings of *quits*, rather than being recognized one by one as during complement-copying, can be read out by anticodons—parts of transfer RNA bound to ribosomes—as a series of consecutive *3-letter* "words" termed "base triplets", that is, information theoretical unitary blocks of 3-*quit* "quytes" (3Q), that are chained up in heterogeneously and almost deliberately long "sentences" termed "reading frames" (genes). The grammar, syntax and dialects (gene regulation, message editing, epigenetics) used in these sentences are then a matter of *system unit type* and *network organization*, for instance, cell (germ line or somatic), organism, species, interaction with other species, ecological traits and niches, and so forth.

Biotic nucleic acids such as RNA and DNA, whether translated or not, are used in known animate systems as *quit* carriers that are relatively easy to copy through molecular templating, irrespective of whether this copying is assisted by enzymes or not. The information theoretical difference between enzyme-catalyzed or enzyme-free ("spontaneous") template copying is merely the fidelity resulting in a more or less complete carry over of the information from template polymer to product polymer. Complementary or self-complementary read-outs, that is, the copying and encoding rules as we know them from biotic genome replication, transcription and the "universal genetic code", are reducible to the hydrogen bond donor-acceptor patterns that are being exposed from the so-called Watson–Crick face of the natural nucleobases of each nucleotide [12,13]. These patterns are not limited to the natural nucleobases. Other N-heterocycles may furnish different patterns and thus distinct pairing preferences [14]. Therefore, from a purely chemical point of view, molecular template variants like binary- or ternary-digit memory polymers composed of strings of subsequent *bits* or, respectively, *trits* are well imaginable (Figure 1). These *bits* or *trits* could be complementary through different pairing modes. In principle, each *bit* or *trit* could be strictly self-complementary, bearing an exclusively self-recognising pairing property: 0 pairs only with 0, 1 pairs only with 1, 2 pairs only with 2. Chemically much more likely, alternative *bit* genomes could be composed of only, for instance, G and C or only A and U, where one digit (0 or 1) is complementary to the other (Figure 1A). Alternative *trit* genomes could bear two digits that are complementary to one another (e.g. 0 pairs with 2) and a third strictly self-complementary digit (1 pairs only with 1), thus being composed of, say, G, C and X, the latter being an exclusively self-recognising nucleotide (Figure 1B).

In addition, the coding fractions of such low-digit memory polymers could be read out, for example, as 4-*bit bytes* (4B) or 3-*trit trytes* (3T). Biotic 3-*quit quytes* (3Q: natural base triplets) comprise $b' = b^3 = 4^3 = 64$ different values, thus offer 64 different triplet "codons" (large frame in Figure 1C). So do 6-*bit bytes* of binary memory polymers (6B: $b' = 2^6 = 64$, not shown) but such long codons would necessitate hexaplet anticodons for translational read-out. Shorter 5-*bit bytes* (5B) generate $b' = 32$ different pentaplet codons. Chemically more realistic are 4-*bit bytes* (4B) giving rise to $b' = 16$ different quadruplet codons and 3-*bit bytes* (3B) giving merely $b' = 8$ different triplet codons (Figure 1A). In ternary-digit memory polymers, blocks of 3-*trit trytes* (3T) produce $b' = 3^3 = 27$ different triplet codons, whereas 4-*trit trytes* (4T) generate $b' = 3^4 = 81$ different quadruplet codons. The latter set of codons would suffice for an even larger than natural (biotic) diversity of translated digits b', compare the 4T code in Figure 1B with the 3Q code in Figure 1C. Of note, the information storage capacity is invariant irrespective of the type of code used to compact the low-digit into a high-digit string, cf. identical left and right values b^M and *SI* before and, respectively, after translation, e.g. $b^M = 2^{28} = 4^{14} = 16^7; 3^{28} = 9^{14} = 81^7; 4^{28} = 16^{14} = 256^7$ (see Figure 1A–C for *SI* values).

These are simply numerical-combinatorial guidelines that exempt "degenerate" (redundant) and "stop" codons. The modern-day ribosomal translation mechanism has established a universal genetic code based on 64 different 3-*quit quytes*, i.e., triplet codons that are currently occupied by merely 20 "proteinogenic" amino acids and usually 3 stop codons, unless a biocompatible "expanded alphabet" for triplet codons has been artificially introduced at selected positions using synthetic nucleotides that offer a distinct "orthogonal" pairing selectivity that may differ from the natural Watson–Crick rules [12,15–20]. Most of the twenty proteinogenic amino acids are encoded by a set of faster and slower, thus, more or less erroneously translated, redundant codons being read by more abundant and, respectively, rarer "isoaccepting" anticodon triplets all carrying the same amino acid. Hence, the multiplicity b' in the secondary "condensed" high-digit polymer is reduced to an effective high-digit value $b'_{eff} = 20$. Already the fact that the effectively used multiplicity in extant biota is less than a third of the theoretically possible (20 amino acids + 1 stop/64 codons) hints at a limit that organic molecules encounter. It is the recognition selectivity, the uniqueness and reliability of a specific molecular recognition that becomes increasingly ambiguous and error-prone with growing diversity of the digits [9,11]. This is the information theoretical ground for the "central dogma" of molecular biology to be a correct assumption [4–6]. Nature can reliably transmit information, being imprinted into molecular atom arrangements under liquid water-conditions only from low-digit to low-digit, or from low-digit to high-digit polymers, never from high-digit to low-digit polymers. These low-digit read-outs from high-digit polymers would immediately loose their informational identity. Molecular recognition of high-digit polymers, thus from highly diverse molecular variants, is too ambiguous.

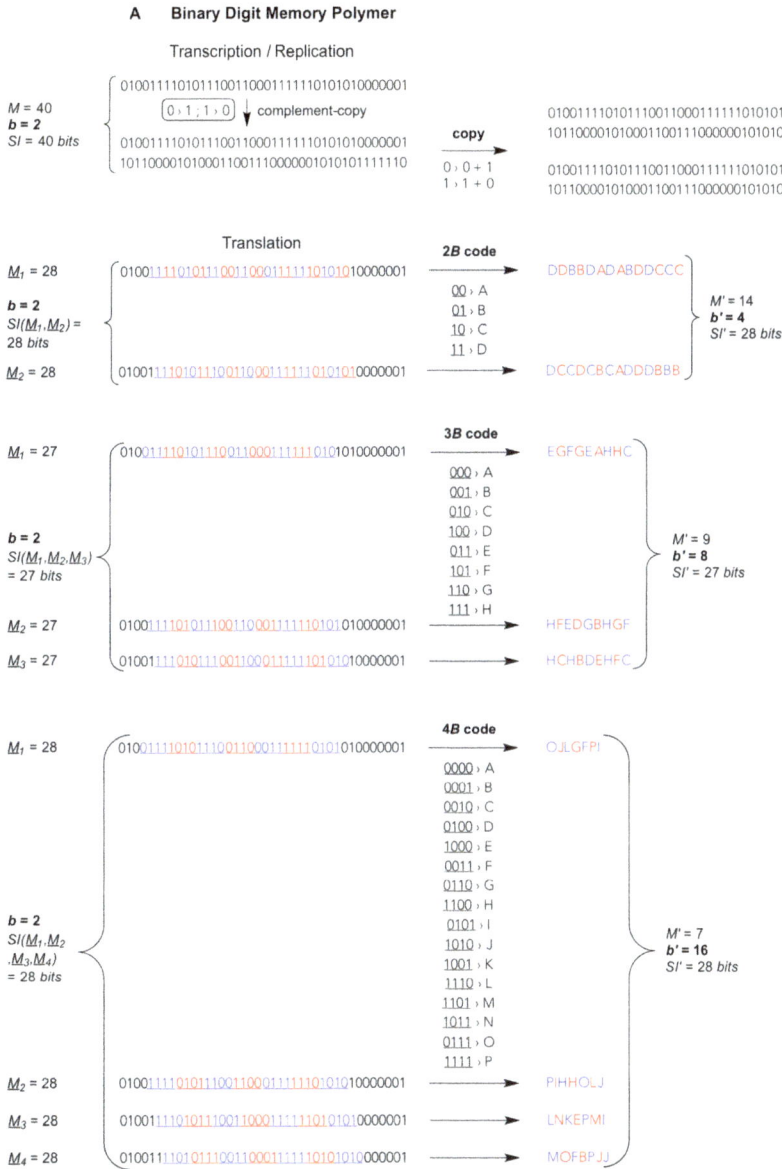

Figure 1. *Cont.*

B   **Ternary Digit Memory Polymer**

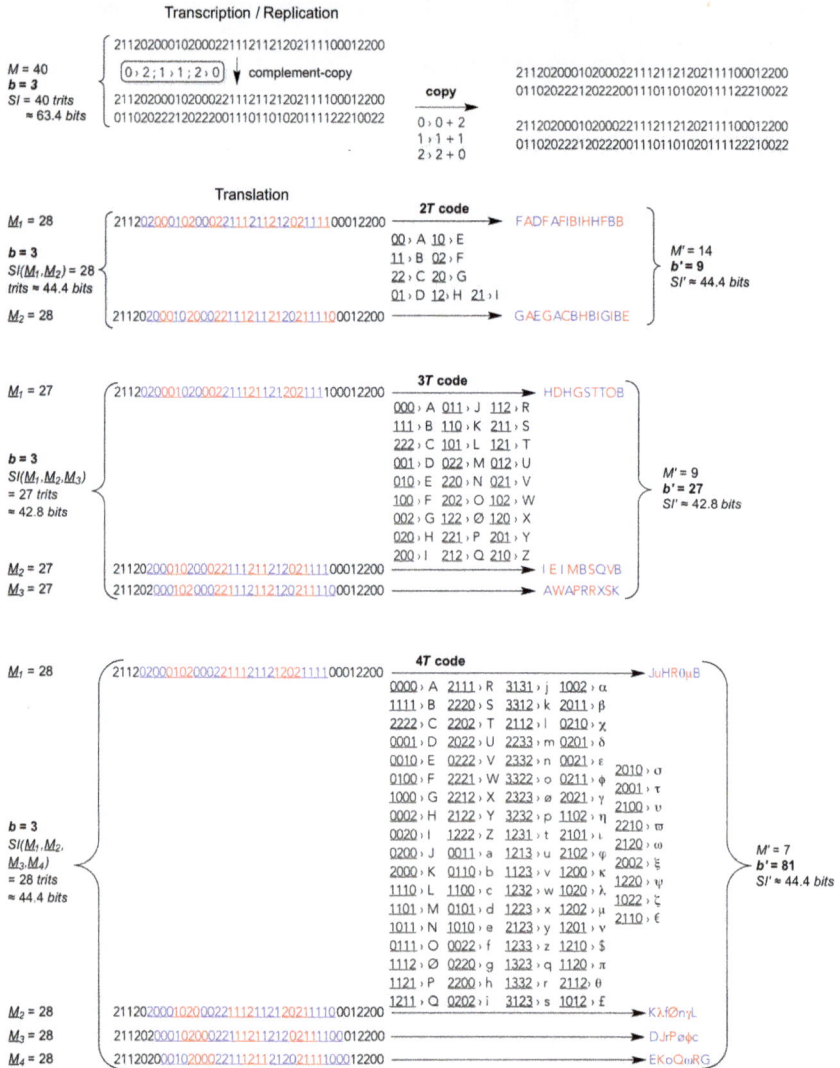

Transcription / Replication

M = 40
b = 3
SI = 40 *trits*
≈ 63.4 *bits*

2112020001020002211121121202111100012200

[0 › 2 ; 1 › 1 ; 2 › 0] ↓ complement-copy

2112020001020002211121121202111100012200
0110202221202220011101101020111122210022

copy →

0 › 0 + 2
1 › 1 + 1
2 › 2 + 0

2112020001020002211121121202111100012200
0110202221202220011101101020111122210022

2112020001020002211121121202111100012200
0110202221202220011101101020111122210022

Translation

2T code

M_1 = 28

2112020001020002211121121202111100012200 ——→ FADFAFIBIHHFBB

00 › A   10 › E
11 › B   02 › F
22 › C   20 › G
01 › D   12 › H   21 › I

b = 3
SI(M_1,M_2) = 28
trits ≈ 44.4 *bits*

M_2 = 28

2112020001020002211121121202111100012200 ——→ GAEGACBHBIGIBE

M' = 14
b' = 9
SI' ≈ 44.4 *bits*

3T code

M_1 = 27

2112020001020002211121121202111100012200 ——→ HDHGSTTOB

000 › A   011 › J   112 › R
111 › B   110 › K   211 › S
222 › C   101 › L   121 › T
001 › D   022 › M   012 › U
010 › E   220 › N   021 › V
100 › F   202 › O   102 › W
002 › G   122 › Ø   120 › X
020 › H   221 › P   201 › Y
200 › I   212 › Q   210 › Z

b = 3
SI(M_1,M_2,M_3)
= 27 *trits*
≈ 42.8 *bits*

M_2 = 27

2112020001020002211121121202111100012200 ——→ I E I MBSQVB

M_3 = 27

2112020001020002211121121202111100012200 ——→ AWAPRRXSK

M' = 9
b' = 27
SI' ≈ 42.8 *bits*

4T code

M_1 = 28

2112020001020002211121121202111100012200 ——→ JuHR0μB

0000 › A   2111 › R   3131 › j   1002 › α
1111 › B   2220 › S   3312 › k   2011 › β
2222 › C   2202 › T   2112 › l   0210 › χ
0001 › D   2022 › U   2233 › m   0201 › δ
0010 › E   0222 › V   2332 › n   0021 › ε
0100 › F   2221 › W   3322 › o   0211 › φ
1000 › G   2212 › X   2323 › ø   2021 › γ
0002 › H   2122 › Y   3232 › p   1102 › η
0020 › I   1222 › Z   1231 › t   2101 › ι
0200 › J   0011 › a   1213 › u   2102 › φ
2000 › K   0110 › b   1123 › v   1200 › κ
1110 › L   1100 › c   1232 › w   1020 › λ
1101 › M   0101 › d   1223 › x   1202 › μ
1011 › N   1010 › e   2123 › y   1201 › ν
0111 › O   0022 › f   1233 › z   1210 › $
1112 › Ø   0220 › g   1323 › q   1120 › π
1121 › P   2200 › h   1332 › r   2112 › θ
1211 › Q   0202 › i   3123 › s   1012 › £

2010 › σ
2001 › τ
2100 › υ
2210 › ϖ
2120 › ω
2002 › ξ
1220 › ψ
1022 › ζ
2110 › €

b = 3
SI(M_1,M_2,
M_3,M_4)
= 28 *trits*
≈ 44.4 *bits*

M_2 = 28

2112020001020002211121121202111100012200 ——→ KλfØnγL

M_3 = 28

2112020001020002211121121202111100012200 ——→ DJrPøφc

M_4 = 28

2112020001020002211121121202111100012200 ——→ EKoQωRG

M' = 7
b' = 81
SI' ≈ 44.4 *bits*

Figure 1. *Cont.*

C   **Quaternary Digit Memory Polymer**

Transcription / Replication

M = 40
b = 4
SI = 40 *quits*
= 80 bits

3010222130001211203330001211021312303121

| 0 › 2 ; 1 › 3 ; 2 › 0 ; 3 › 1 | ↓ complement-copy

3010222130001211203330001211021312303121
1232000312223033021112223033203130121303

copy →

0 › 0 + 2
1 › 1 + 3
2 › 2 + 0
3 › 3 + 1

3010222130001211203330001211021312303121
1232000312223033021112223033203130121303

3010222130001211203330001211021312303121
1232000312223033021112223033203130121303

Translation

2Q code

M_1 = 28   3010222130001211203330001211021312303121 → CLNAIBKDNAIBFJ

b = 4
$SI(\underline{M_1},M_2)$ = 28
quits = 56 bits

00 › A 01 › E 12 › I 23 › M
11 › B 02 › F 13 › J 30 › N
22 › C 03 › G 20 › K 31 › O
33 › D 10 › H 21 › L 32 › P

M' = 14
b' = 16
SI' = 56 bits

M_2 = 28   3010222130001211203330001211021312303121 → CJAELIGDAELHLO

3Q code

M_1 = 27   3010222130001211203330001211021312303121 → qcB!j x BIJ

000 › A 100 › Q 200 › g 300 › w
001 › B 101 › R 201 › h 301 › x
002 › C 102 › S 202 › i 302 › y
003 › D 103 › T 203 › j 303 › z
010 › E 110 › U 210 › k 310 › €
011 › F 111 › V 211 › l 311 › £
012 › G 112 › W 212 › m 312 › ¥
013 › H 113 › X 213 › n 313 › $
020 › I 120 › Y 220 › o 320 › α
021 › J 121 › Z 221 › p 321 › β
022 › K 122 › a 222 › q 322 › γ
023 › L 123 › b 223 › r 323 › δ
030 › M 130 › c 230 › s 330 › ε
031 › N 131 › d 231 › t 331 › ρ
032 › O 132 › e 232 › u 332 › φ
033 › P 133 › f 233 › v 333 › π

b = 4
$SI(\underline{M_1},M_2,M_3)$
= 27 *quits*
= 54 bits

M' = 9
b' = 64
SI' = 54 bits

M_2 = 27   3010222130001211203330001211021312303121 → pwGWPwGUn
M_3 = 27   3010222130001211203330001211021312303121 → nAZYπAZSd

b = 4
$SI(\underline{M_1},M_2,M_3,M_4)$ = 28 *quits* = 56 bits

M_1 = 28   3010222130001211203330001211021312303121
M_2 = 28   3010222130001211203330001211021312303121
M_3 = 28   3010222130001211203330001211021312303121
M_4 = 28   3010222130001211203330001211021312303121

4Q code M' = 7 b' = 256 SI' = 56 bits

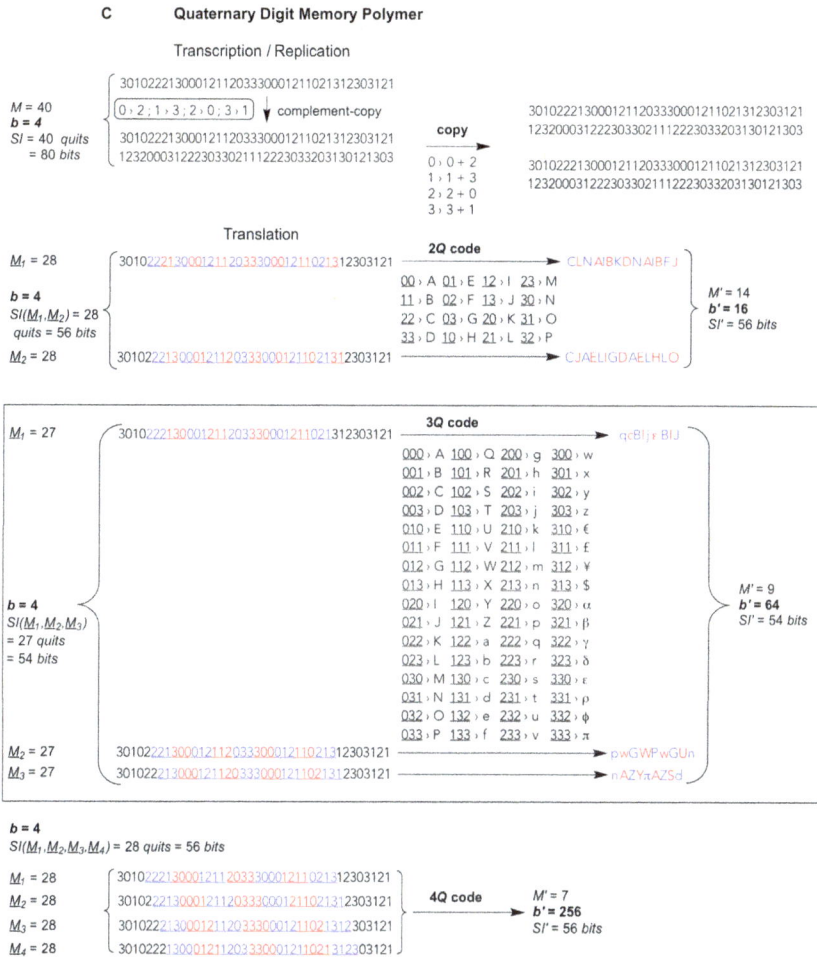

Figure 1. **Template-copying and translating an exemplary string of total M = 40 residues** composed of b different digits transmitting, according to Equation (3), a given amount of Shannon Information $SI(M)$ from one parental string to others. **Upper part**: Transcription/Replication. A single complement-copying event (vertical arrow) is realised during, both, transcription (or reverse transcription) and the first step of replication; the second (horizontal) step applies to replication only. Complement-copying rules (small round-edged shaded frames) by virtue of a minimal requirement for self-complementary digits, i.e., *bits* {0,1}, *trits* {0,1,2} and *quits* {0,1,2,3}. **Lower part**: Translation of consecutive blocks (*B* = *bytes*, *T* = *trytes*, *Q* = *quytes*) of 2–4 digits (**A** *bits*, **B** *trits*, **C** *quits*). The reading frame (underligned coloured digits) is translated into products (strings of letters) of a condensed residue number M', higher digit multiplicity (diversity) $b' = b^{2-4}$, and unchanged SI'. Frameshifts $\underline{M}_{1-2}$ for 2-digit blocks, $\underline{M}_{1-3}$ for 3-digit blocks and $\underline{M}_{1-4}$ for 4-digit blocks generate alternative translation products of the same length M', diversity b' and SI' but radically different sequences. (**A**) Binary digit (*bit*) strings, replicated and translated from 2–*bit bytes* (2B), 3–*bit bytes* (3B) and 4–*bit bytes* (4B). (**B**) Ternary

digit (*trit*) strings, replicated and translated from 2–*trit trytes* (2*T*), 3–*trit trytes* (3*T*) and 4–*trit trytes* (4*T*). (C) Quaternary digit (*quit*) strings, replicated and translated from 2–*quit quytes* (2*Q*), 3–*quit quytes* (3*Q*) and 4–*trit quytes* (4*Q*, code and translation products not shown). Shadowed large frame: the current natural (biotic) memory system are *quit* strings being translated from reading frames of consecutive 3*Q* utilizing $b' = 64$ codons that are reduced, mainly for fidelity reasons, to $b'_{eff} = 20$ effectively translated digits, viz. the "universal genetic code" for 20 different amino acids and a stop signal (i.e., lack of amino acid). Reproduced and modified from The Handbook of Astrobiology; published by CRC Press, 2019 © Taylor & Francis [21].

4. High-Digit Functional Polymers

The copying and encoding principles shown in Figure 1 insinuate that unbranched molecular strings (1D polymers) composed of a limited number of different monomeric complementary residues (monomers), that is, strings that bear a relatively low multiplicity of digits (low *b*), are likely to be a general feature of memory keepers in any animate system. The lower the digit multiplicity the simpler the composition of the template and less ambiguous it is to copy and replicate the string on a molecular level [22]. This generates fewer errors in template-copied memory polymers, thus, a higher replication error threshold for a given spreading "quasi-species" (similar genome population), and eventually imposes a weaker selection pressure on the maximal genome length of any evolved organism [23–25].

The opposite is true when it comes to functional translation products, in which the higher their digit multiplicity is (high *b'*) the stronger the compression, the shorter the resulting string lengths (lower *M'*). A comparison between primary low-digit and secondary high-digit polymers of the same string length (when $M = M'$) reveals a much higher structural diversity of the latter. This high diversity is further multiplied by the number of reading frame shifts (M_b) that give rise to an encoded set of a completely different choice of translated string sequences (Figure 1A–C, below each code). This generates translated string polymers that are inherently difficult to copy through direct templating, since the complement rules—analogous to the Watson–Crick base pairing rules—needed to be as manifold and exclusive as the digits are diverse. On the other hand, the longer the translation blocks (codons, translated words) in the messenger nucleic acids the more compact is the generated diversity of the secondary polymer, which allows for more diverse "molecular functions" at a given secondary string length *M'*. A higher compositional diversity means a wider, more versatile and fine-grained (higher dimensional) sequence space, thus lending such polymers easier access to their folding and assembly into structurally more defined, more rigid, catalytically more competent functional objects [9].

5. Discussion: What to Expect from Linear Biopolymers of Unknown Biota

At unchanged *SI* the compression $M_{b=2}/M_{b \geq 2}$ of a string of digits upon enhancement of the digit multiplicity *b* follows a binary-logarithmic dependence (Equation (4), numerical examples in Table 1).

$$M_{b=2}/M_{b \geq 2} = log_2 b \tag{4}$$

All digital devices are based on *bit* string information storage and transmission systems. The lowest possible digit multiplicity works best despite the resulting longest possible string length *M*. Not only are uncompressed *bit* strings highly unpredictable in Shannon's sense. Historically, electronic devices work most reliably when the digits are encoded by a "weak current" of whatever strength {1} and "no current" {0}. The storage and transmission of this kind of string is least error-prone, since the difference between "zero" and "more than zero" is the largest possible, so a binary digital read-out delivers the highest signal-to-noise ratio [11]. If the digits were "tension", one could feed computers with *trit* string instructions based on sequences of "no tension" {0}, "positive tension" {1} and "negative tension" {2} of whatever strength. The resulting strings would be $log_2 3$-fold shorter, the information more compact, but also more error-prone to transmit. Dangerously unreliable would be the usage of *quit* strings in digital devices. The digits would have to be realised from a "highest current", "high current", "low current" and "no current" code. Replace "current" and "tension" with "amplitude" and, respectively, "frequency (wavelength)" or "phase transition" (liquid-solid, amorphous-crystalline, absorbing-reflective, and so forth), and the same applies to optical data storage devices. Human minds, at the other extreme, can easily distinguish all *dits* from one another (and more). It is all a matter of the distinguishability and of the similarity of the digits.

Biotic nature has hitherto evolved *quit* strings as macromolecular memory carriers, why $b = 4$? With respect to *bit* strings this means a two-fold compression. As long as we cannot precisely measure the similarity of molecular digits (but see [11]), a general quantified answer remains elusive. These primary *quit* strings code in parts for *vigintit* strings, $b'_{eff} = 20$, which means that the translated parts are furthermore two-fold compressed isoinformational secondary polymers (four-fold with respect to *bit* strings). To obtain a quantitative answer, why *quit*-to-*vigintit* and not any other low digit-to-high digit translation, is impossible by analytical means owing to the complexity of extrinsic and intrinsic feedback networks, as mentioned in the introduction. However, as for the population dynamics of replicators [26], the dynamics of the stochastic generation of translation products is best approached by simulation methods from differential equations, particularly of the kind, where the translation fidelity comes out "impedance-matched" to that of the replication of the whole genome [9,10].

Generally, the reason why *quit* strings have evolved to reach a stable dynamic optimum in extant biota has a strong bearing with "Eigen's paradox", which states: there is no accurate replicase without a large genome and there could be no large genome without an accurate replicase. Thus, the information that can be reliably replicated is less than the information necessary to code for the replicating machinery being composed of strings of the same digit multiplicity. Various ways of resolving this paradox have been proposed and are being worked at. One of the current difficulties in modeling evolutionary population dynamics is to properly outline the scope of "selectability" of replicators, that is, the emergence of Darwinian selection through the extinction of competing sub-populations of coexisting replicators while maintaining the survivors stable, for example, stable against parasites, yet still evolvable over space and time in the sense that the survivors may integrate more different replicating (memory) polymers without making the whole system collapse. It turns out from the research of the past decade that spatially explicit systems of cooperating replicators, that are irrevocably coupled to ("fed by") metabolic reaction networks, are incomparingly more robust than replicators devoid of metabolism. The intrinsic coupling of translation and replication in reflexive genetic information systems, thus comprising genes whose expression by rules can, in turn, execute those expression rules, are particularly effective and fast in dynamically stabilising the robustness of evolving replicator systems.

One of the most remaining problems is the intrinsic molecular trait of macromolecular low-digit polymers originating from the obligatory mutual affinity between template molecule and product molecule. Macromolecules usually replicate in the parabolic growth regime in which every generation of replicators produces on the average fewer complementary products per template than the previous generation (per template) [27], a general phenomenon termed "strand inhibition". The very attribute of low-digit polymers, that makes them relatively easy to replicate through template copying, renders them too slow growing in numbers required to open the gates for truly competitive population dynamics, thus for Darwinian selection to apply. What transpires most out of this dilemma is the need for polymers of high catalytic potential, much higher than that of low-digit memory polymers (see Supplement to ref. [9]). Not only are low-digit polymers too inefficient in catalysing the attachment of codon-cognate high-digit monomers to low-digit polymers needed for an operational genetic code (specific aminoacylation of transfer RNA), this requires a high degree of selectivity with respect to the recognition of, both, a high variety of high-digit monomers and low-digit polymers (amino acids and transfer RNA), which can only be accomplished by high-digit polymers (proteic aminoacyl RNA synthetases). In addition and most importantly, only high-digit polymers, by virtue of providing efficient replication machineries (proficiently selective catalysts), can bring down the residual intrinsic error-proneness of low-digit polymers to levels that resolve Eigen's paradox and heave the units that harbour them (cells) into the exponential growth regime, where different units can compete with one another and thrive through mutation and selection. In an origin of life context, this is the most fundamental reason for translation to occur at an early stage of evolution. In the wording of a biochemist: nucleic acid helicases and polymerases (protein enzymes), that open up nucleic acid double-strands and, respectively, insert highly selectively the complementary mononucleotides to each template-bound primer strand within the same generation—resulting in exponential growth of dynamically stable populations—are needed very early on in an evolutionary timescale. For this to happen, the production of highly diverse gene products is extremely advantageous (Figure 2), if not mandatory [9,10].

On the other hand, the extant translation machinery itself is mainly composed of few long low-digit polymers (ribosomal RNA) and an optimal number of uniformly small high-digit polymers (ribosomal proteins). It turns out that the highest possible efficiency of production of the translation machinery, that is, the need to sequester as little autocatalytic enzymatic time as possible to synthesise this machinery, in order to have as much available time as possible to produce other catalysts than itself, is the guiding concept for the fact that ribosomes are mainly composed of RNA with a high rRNA/r-protein ratio [28]. The time ribosomes invest in r-protein synthesis can be up to two orders of magnitude longer than for an equivalent mass of rRNA, especially in fast growing organisms.

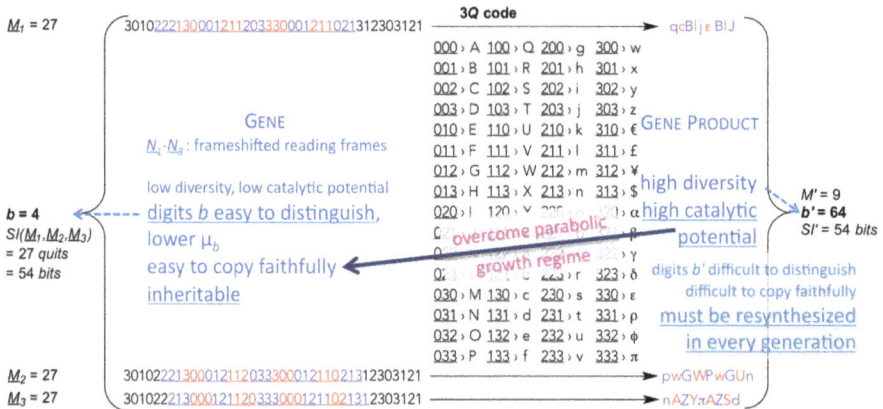

Figure 2. Translation of parts of low-digit memory polymers into high-digit functional polymers as a means to achieve inheritable exponential population growth. The expectedly much higher catalytic potential of high-digit polymers (gene products) allows for more efficient use of the templating ability of the low-digit memory polymers (containing genes), both in terms of copying fidelity $Q = (1-\mu_b)^M$, where μ_b denotes mutation probability of every digit, and population growth order p, where $0 < p < 1$ defines the parabolic growth regime and $p = 1$ the exponential growth regime, in populations of replicators x_i that grow in time t, as in $dx_i/dt = k_i\, x_i{}^p$, where each replicator population i replicates with an apparent replication rate konstant k_i.

Alternative nucleic acids composed of fewer letters, *bits* or *trits* rather than *quits*, could be considered in extra-terrestrial biota and/or during early periods of the origin of life on Earth. They might encode a smaller or larger choice of proteinogenic amino acids—or some other molecular equivalent of a functionally more diverse polymer than nucleic acids—by translating from shorter or, respectively, longer *bytes* or *trytes* as mentioned above and shown in Figure 1A,B. In principle, alternative nucleic acids could also form triple complements through triple-strand formation or even higher-order supramolecular string associations, which would change the stoichiometry of transcription and replication. The chemical reality, as expressed in pairing/tripling/quadrupling/ ... properties of such alternative nucleic acids, would be expected to impose grave consequences on their copying and translation fidelity, and thus on the number of genes and maximal genome length [23–25]. In principle, memory strings could also be extended to higher than quaternary digit multiplicities (not to be confused with a locally "expanded alphabet" of triplet codons) and translated using longer than 4-digit blocks (pentaplet, hexaplet etc. codons). In the reality of macromolecules offered by nature, however, more diverse higher-digit memory polymers are likely to be copied more erroneously, since the monomers would necessarily be more similar to one another, again limiting the replication error threshold, maximal number of genes and total genome length. In addition, longer codons than quadruplets are at higher risk of being misread due to spontaneous frameshifting and mispairing, which would produce more erroneously assembled proteins (secondary polymers) and necessitate a more elaborate and costly error correction effort by the system.

Yet alien biota that would provide linear memory polymers that were markedly more rigid than "natural" RNA or DNA, thus perhaps less prone to frameshifting and mispairing, should not be ruled out a priori, not for chemical reasons. The overall energetic cost at the available energy influx needed to generate such polymers, to keep their replication error threshold high, also to keep the erroneously produced secondary polymers under a liveable limit, are probably much more preventive factors than the huge choice of bricks that chemistry can in principle offer.

6. Conclusions

The chemistry on our planet apparently produced prebiotic bricks that could condense under prebiotic reaction conditions into 1D polymers (nucleic acids) that could form double-strands, at least locally in certain string zones, through the spontaneous association (hydrization) of pairwise complementary digits, as shown for the complement-copying in Figure 1C. The digit multiplicity of the first replicating nucleic acids (*bit, trit, quit,* etc.) is unknown, although there is a consensus on *bit* polymers having preceded modern natural (biotic) nucleic acids that are generally *quit* polymers. These prebiotic bricks are purine and pyrimidine ribonucleoside pairs that, under appropriate prebiotic reaction conditions being present on this planet some 3.6 Gya, could condense with phosphate and polymerize into RNA and similar RNA-like linear polymers [21]. At least a part of the early nucleic acid single strands could synthesize 3Q-translated secondary polymers (Figure 2), viz. polypeptides and proteins very early on, or else we would hardly expect the genetic code to be universal [29]. Apparently, on Earth, RNA proved to be the most successful "primary" memory polymer. Not only can its monomer sequence be easily copied and faithfully reproduced. More faithful and streamlined information storage carriers can be derived from RNA by its deoxygenation to DNA. Most importantly, RNA not DNA can direct and catalyse the linking of amino acids into defined strings of polypeptides, that is, take an essential part in catalysing the controlled dehydration of amino acids to produce amide bonds, a process termed peptidyl transfer (PT). Strong evidence suggests that uncoded PT preceded coded PT, thus, that RNA could grow polypeptide chains from amino acids before a recognition system eventually emerged—from RNA, too—that allowed RNA-directed PT to profit from specific codon–anticodon interactions, and thus to translate genetic information [30–32].

The arguments presented in this work insinuate that in other prebiotic environments perhaps different kinds of linear polymers could become dominant and evolve in reproducing entities, and this should not be excluded a priori from a chemical-molecular perspective. But we should expect alien and very early biota to evolve right from the start string polymers of both kinds, low-digit and high-digit variants, where the more diverse latter is encoded by the simpler former.

Funding: This research was co-funded by the *Volkswagen* Foundation, grant number Az 92850, title "Molecular Life".

Acknowledgments: I thank Laurent Boiteau (Univ. Montpellier) for pointing out the work of Giuseppe Longo et al. (ref. [7]), and Jerzy Górecki (Univ. Warsaw) for discussions on information theory, which made me prevent an error propagation on Shannon information and the units of information from refs. [21,30] in this work.

Conflicts of Interest: The author declares no conflict of interest.

References

1. Turing, A. On computable numbers with an application to the entscheidungs problem. *Proc. London Math. Soc. Ser.* **1936**, *2*, 230–265. [CrossRef]
2. Schrödinger, E. *What is Life? The Physical Aspect of the Living Cell*; The University Press: Cambridge, UK, 1945.
3. Shannon, C. A mathematical theory of communication. *Bell Syst. Tech. J.* **1948**, *27*, 279–423. [CrossRef]
4. Crick, F.H.C. On protein synthesis. *Symp. Soc. Exp. Biol.* **1958**, *12*, 138–163. [PubMed]
5. Crick, F.H.C. Central dogma of molecular biology. *Nature* **1970**, *227*, 561–563. [CrossRef] [PubMed]
6. Cobb, M. 60 years ago, Francis Crick changed the logic of biology. *PLoS Biol.* **2017**, *15*, e2003243. [CrossRef] [PubMed]
7. Longo, G.; Miquel, P.-A.; Sonnenschein, C.; Soto, A.M. Is information a proper observable for biological organization? *Prog. Biophys. Mol. Biol.* **2012**, *109*, 108–114. [CrossRef] [PubMed]
8. Pearl, J.; Mackenzie, D. *The Book of Why: The New Science of Cause and Effect*; Basic Books: New York, NY, USA, 2018.
9. Carter, C.W., Jr.; Wills, P.R. Interdependence, reflexivity, fidelity, impedance matching, and the evolution of genetic coding. *Mol. Biol. Evol.* **2018**, *35*, 269–286. [CrossRef] [PubMed]
10. Wills, P.R.; Carter, C.W., Jr. Insuperable problems of the genetic code initially emerging in an RNA world. *BioSystems* **2018**, *164*, 155–166. [CrossRef] [PubMed]
11. Schneider, T.D. A brief review of molecular information theory. *Nano Commun. Networks* **2010**, *1*, 173–180. [CrossRef]
12. Strazewski, P.; Tamm, C. Replication experiments with nucleotide base analogues. *Angew. Chem. Int. Ed. Engl.* **1990**, *29*, 36–57. [CrossRef]
13. Strazewski, P. The biological equilibrium of base pairs. *J. Molec. Evol.* **1990**, *30*, 116–124. [CrossRef] [PubMed]
14. Krishnamurthy, R. Role of pK_a of nucleobases in the origins of chemical evolution. *Acc. Chem. Res.* **2012**, *45*, 2035–2044. [CrossRef] [PubMed]
15. Picirilli, J.A.; Benner, S.A.; Krauch, T.; Moroney, S.E. Enzymatic incorporation of a new base pair into DNA and RNA extends the genetic alphabet. *Nature* **1990**, *343*, 33–37. [CrossRef] [PubMed]
16. Cornish, V.W.; Mendel, D.; Schultz, P.G. Probing protein structure and function with an expanded genetic code. *Angew. Chem. Int. Ed. Engl.* **1995**, *34*, 621–633. [CrossRef]
17. Liu, H.; Gao, S.; Lynch, R.; Saito, Y.D.; Maynard, L.; Kool, E.T. A four-base paired genetic helix with expanded size. *Science* **2003**, *302*, 868–871. [CrossRef]
18. Malyshev, D.A.; Dhami, K.; Lavergne, T.; Chen, T.; Dai, N.; Foster, J.M.; Corrêa, I.R., Jr.; Romesberg, F.E. A semi-synthetic organism with an expanded genetic alphabet. *Nature* **2014**, *509*, 385–388. [CrossRef]
19. Georgiadis, M.M.; Singh, I.; Kellett, W.F.; Hoshika, S.; Benner, S.A.; Richards, N.G.J. Structural basis for a six nucleotide genetic alphabet. *J. Am. Chem. Soc.* **2015**, *137*, 6947–6955. [CrossRef]
20. Winiger, C.B.; Kim, M.-J.; Hoshika, S.; Shaw, R.W.; Moses, J.D.; Matsuura, M.F.; Gerloff, D.L.; Benner, S.A. Polymerase interactions with wobble mismatches in synthetic genetic systems and their evolutionary implications. *Biochemistry* **2016**, *55*, 3847–3850. [CrossRef]
21. Strazewski, P. Prebiotic chemical pathways to RNA and the importance of its compartmentation. In *The Handbook of Astrobiology*; Kolb, V., Ed.; CRC Press: Boca Raton, FL, USA, 2019; pp. 235–263.
22. Szathmáry, E. What is the optimum size for the genetic alphabet? *Proc. Natl. Acad. Sci. USA* **1992**, *98*, 2614–2618. [CrossRef]
23. Eigen, M. Self-organization of matter and the evolution of biological macromolecules. *Naturwissenschaften* **1971**, *58*, 465–523. [CrossRef]
24. Biebricher, M.; Eigen, C.K. The error threshold. *Virus Res.* **2005**, *107*, 117–127. [CrossRef] [PubMed]
25. Kun, Á.; Santos, M.; Szathmáry, E. Real ribozymes suggest a relaxed error threshold. *Nature Genet.* **2005**, *37*, 1008–1011. [CrossRef] [PubMed]
26. Szilágyi, A.; Zachar, I.; Scheuring, I.; Kun, Á.; Könnyű, B.; Czárán, T. Ecology and evolution in the RNA world dynamics and stability of prebiotic replicator systems. *Life* **2017**, *7*, 48. [CrossRef] [PubMed]
27. Strazewski, P. The Beginning of Systems Chemistry. *Life* **2019**, *9*, 11. [CrossRef] [PubMed]
28. Reuveni, S.; Ehrenberg, M.; Paulsson, J. Ribosomes are optimized for autocatalytic production. *Nature* **2017**, *547*, 293–297. [CrossRef] [PubMed]

29. Theobald, D.L. A formal test of the theory of universal common ancestry. *Nature* **2010**, *465*, 219–222. [CrossRef]
30. Strazewski, P. Omne vivum ex vivo ... omne? How to feed an inanimate evolvable chemical system so as to let it self-evolve into increased complexity and life-like behaviour. *Isr. J. Chem.* **2015**, *55*, 851–864. [CrossRef]
31. van der Gulik, P.T.S.; Speijer, D. How amino acids and peptides shaped the RNA world. *Life* **2015**, *5*, 230–246. [CrossRef]
32. Carter, C.W., Jr. What RNA world? Why peptide/RNA partnership merits renewed experimental attention. *Life* **2015**, *5*, 294–320. [CrossRef]

life

MDPI

Review

Exploring the Emergence of RNA Nucleosides and Nucleotides on the Early Earth

Annabelle Biscans

RNA Therapeutics Institute, University of Massachusetts Medical School, Worcester, 01605 MA, USA;
annabelle.biscans@umassmed.edu

Received: 2 October 2018; Accepted: 3 November 2018; Published: 6 November 2018

Abstract: Understanding how life began is one of the most fascinating problems to solve. By approaching this enigma from a chemistry perspective, the goal is to define what series of chemical reactions could lead to the synthesis of nucleotides, amino acids, lipids, and other cellular components from simple feedstocks under prebiotically plausible conditions. It is well established that evolution of life involved RNA which plays central roles in both inheritance and catalysis. In this review, we present historically important and recently published articles aimed at understanding the emergence of RNA nucleosides and nucleotides on the early Earth.

Keywords: origin of life; prebiotic chemistry; nucleotide and nucleoside synthesis

1. Introduction

How did life begin? Where did the essential components to life—nucleic acids, proteins, and lipids—come from? To answer these fundamental questions, efforts have been made to understand the prebiotic synthesis of these biomolecules through chemical processes [1]. Nucleic acids, proteins, and lipids share a similar atomic composition, which includes hydrogen, carbon, oxygen, nitrogen, phosphorous, and sulfur. Therefore, we can assume that they have generated from common natural constituents present on Earth. A large number of astrophysicists, physicists, and mathematicians succeeded in identifying the plausible chemical composition of the early Earth using radio telescopes and spacecraft, such as the Atacama large millimeter/submillimeter array (ALMA) implanted in Chile's Atacama Desert [2] and the Rosetta space probe [3,4]. They reported on morphological, thermal, mechanical, and electrical properties and composition of the surface of satellites, planets, and comets. Apart from water, carbon monoxide, and carbon dioxide, mixtures of fifteen compounds from the chemical groups of alcohols, amines, carbonyls, nitriles, amides, and isocyanates were detected. They showed that the favored geochemical conditions for life to arise involve volcanic activities and/or the impact of meteorites, with complex organic chemistry; several sources of energy; and dynamic light–dark, cold–hot, and wet–dry cycles [5]. Thus, an important amount of chemistry is potentially possible to favor synthesis of biomolecules or their precursors from simple feedstock molecules.

Among the biomolecules, it is well established that RNA may have played a central role in the early evolution of life. Indeed, RNA can not only act as an enzyme and perform catalytic reactions, but it can also store and transfer genetic information [6–10]. Using knowledge on the availability of starting materials on primitive Earth and the geological conditions when life began, the prebiotic synthesis of RNA building blocks has been explored. Miller–Urey experiments [11], which mark the beginning of prebiotic chemistry, inspired Oro et al. to analyze the products formed when ammonium cyanide was refluxed in aqueous solution, leading to the discovery that adenine can be formed by cyanide polymerization [12]. Since this discovery, the prebiotic synthesis of RNA has been intensely investigated. This review gives an overview of the plausible origin of RNA. Different possible routes for the formation of nucleosides and nucleotides, RNA building blocks, under prebiotic conditions will be discussed.

2. Prebiotic Synthesis of Nucleotides from the Assembly of a Nucleobase, a Ribose, and a Phosphate

First efforts to understand the prebiotic synthesis of ribonucleotides, the building blocks of RNA, have been based on the hypothesis that they should be formed from three distinct entities: a nucleobase (uracil, cytosine, adenine, or guanine), a ribose sugar, and a phosphate, which have been formed separately and combined (Figure 1A) [13].

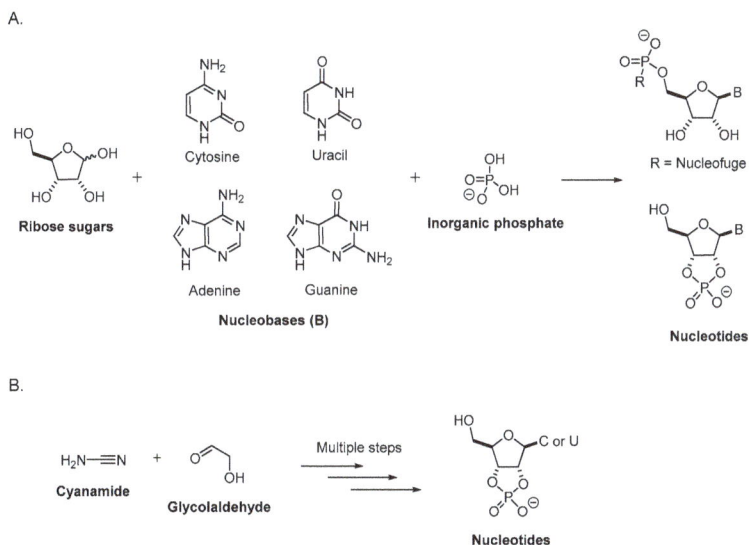

Figure 1. Possible routes for the synthesis of nucleotides. (**A**) The traditional RNA disconnection route, which is based on the hypothesis that nucleotides are formed from a ribose sugar, nucleobases, and inorganic phosphate, each prepared separately and assembled. (**B**) Alternative approach to synthesize nucleotides where sugars and nucleobases are formed during the same process.

2.1. Sugar Synthesis

Prebiotic chemists suggested that sugar formation relied on the synthesis of formose, discovered by Butlerow in 1861 [14]. This reaction consists of the polymerization of formaldehyde in the presence of calcium hydroxide. For a long time, the mechanism of how this reaction initiates assumed that a homocoupling of formaldehyde occurred to produce glycolaldehyde, later converted in glyceraldehyde. However, such a direct dimerization has been considered chemically unfavored [15]. Recently, Schreiner et al. demonstrated that glycolaldehyde could have been formed from formaldehyde reacting with its isomer hydroxymethylene in the absence of base and solvent at cryogenic temperature [16]. Hydroxymethylene could have been generated from the pyrolysis of glyoxylic acid in the gas phase or on surfaces. During sugar formation, a variety of reactions can occur (myriad aldol; Cannizzaro; or Lobry de Bruyn–Alberda van Ekenstein reactions, which involves transforming an aldose into the ketone isomer or vice versa) leading to a complex mixture of linear and branched aldo- and ketosugars [17]. This uncontrolled reactivity forms ribose with less than 1% yield [18].

Significant efforts have been made to search efficient plausible prebiotic routes to favor sugar synthesis (Figure 2). Indeed, the formose reaction is a catalyzed reaction [15], and thus many groups focused on the identification of a prebiotic catalyst, which could have explained ribose formation. Zubay et al. showed that more than 30% of the formaldehyde can be converted to a mixture of aldopentoses using a lead catalyzed formose reaction [13,19]. The presence of lead in the incubation mixture also accelerated a number of other reactions including the interconversion of the aldopentoses into ribose. Also, it has been shown that hydroxyapatite, which consists of phosphate and calcium ions, increased ribose formation from formaldehyde and glycolaldehyde in hot water (80 °C) [20]. Hydroxyapatite enhanced cross-aldol reactions and Lobry de Bruyn–Alberda van Ekenstein transformations, utilizing the effective positioning of calcium ions on the surface of hydroxyapatite.

Moreover, ribose is unstable under the alkaline conditions required for the formose reaction [21]. Borate addition has been investigated as borate could have been available on early Earth and as ribose can form borate complexes, stabilizing the molecule under the harsh prebiotic formation while other sugars would degrade [22–24]. However, even though ribose–borate complexes are more stable than other pentoses, the stabilization is modest and an excess of glycolaldehyde over formaldehyde is required to inhibit borate and prevent sugar isomers with no selectivity for ribose. Eschenmoser et al. showed that the variety of products induced by the formose reaction and the destructive effects that the reaction conditions have on the ribose can be significantly suppressed by phosphorylation of glycolaldehyde [25]. Indeed, under alkaline conditions, glycolaldehyde phosphate leads to a simple mixture of tetrose-2,4-diphosphates and hexose-2,4,6-triposphates. The presence of phosphate groups prevents the Lobry de Bruyn–Alberda van Ekenstein reaction and stabilizes the sugars, providing a plausible prebiotic route to the synthesis of ribose if the ribose-2,4-diphospate could later be converted to a 5-phosphate or a 1,5-diphosphate.

Alternative routes leading to the formation of ribose are possible. Indeed, sugar formation could be coming from hydrogen cyanide irradiated by ultraviolet light in the presence of copper cyanide complexes [26]. Moreover, bisulfite salts could have played a role in sugar formation as certain sulfidic anions could have been available on the early Earth [27]. Sutherland et al. found that bisulfite could be used to form a glycolaldehyde–bisulfite adduct from glycolnitrile. This bisulfite adduct formation, allowing the stabilization of the aldehyde, could have led to the formation of ribose [28,29].

In addition, aldoses including ribose pentose sugars have been found in interstellar ice analogs (composed of water, methanol, and ammonia) after their irradiation by ultraviolet light [30]. These results suggest that the formation of numerous sugars, including the ribose, may be possible from photochemical and thermal treatment of cosmic ices in the late stages of the solar nebula.

Even if some progress has been made to understand the ribose formation under prebiotic conditions, each suggested route presents obstacles, limiting ribose yield and purity necessary to form nucleotides. A selective pathway has yet to be elucidated.

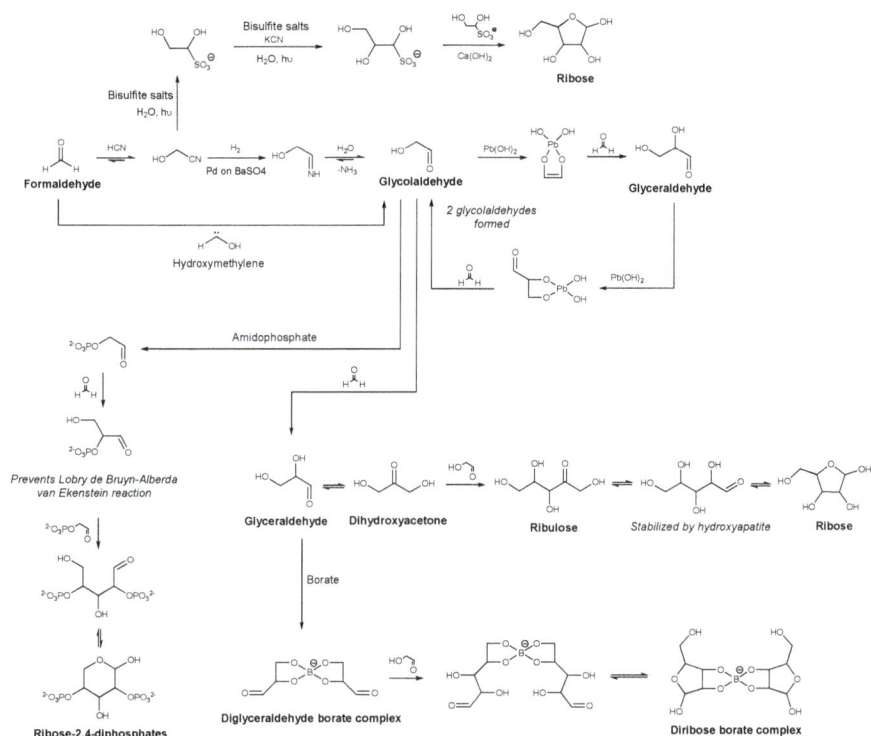

Figure 2. Possible routes favoring ribose formation.

2.2. Nucleobase Synthesis

Starting with the prebiotic purine formation (adenine and guanine), it has been shown that the nucleobase adenine could be formed by mixing hydrogen cyanide and ammonia in solution [31,32]. Indeed, formamidine can be produced by addition of ammonia to hydrogen cyanide. It can then react with hydrogen cyanide tetramer or diaminomaleodinitrile, resulting from hydrogen cyanide polymerization in aqueous solution, to form 4-amino-5-cyano-imidazole. The latter product could then react with a second formamidine molecule to lead to adenine (Figure 3) [33–35]. As the hydrolysis of 4-amino-5-cyano-imidazole to form 4-amino-imidazole-5-carboxamide can occur, these results suggest that a high concentration of hydrogen cyanide and ammonia should have been present on Earth. Even though hydrogen cyanide could be found in high concentration in frozen environments [31,36], a significant amount of ammonia on Earth is questionable. Ferris and Orgel suggest an alternative route where the production of 4-amino-5-cyano-imidazole could have been possible by photochemical isomerization of hydrogen cyanide tetramer, a mechanism that does not involve ammonia [32].

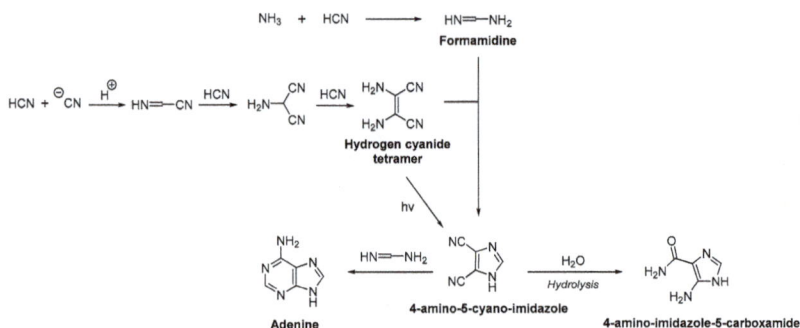

Figure 3. Possible prebiotic synthesis of adenine from hydrogen cyanide.

Other hypotheses on adenine and purine accumulation on Earth have been discussed. Miyakama et al. suggest that purines have been formed in the atmosphere in the absence of hydrogen cyanide [37]. They reported that guanine could have been generated from a gas mixture (nitrogen, carbon monoxide, and water) after cometary impacts. Also, it has been proposed that adenine was formed in the solar system (outside of Earth) and brought to Earth by meteorites, given the fact that adenine was found in significant quantity in carbonaceous chondrites [38].

Most of the work on the prebiotic synthesis of pyrimidines (cytosine and uracil) suggest reactions between cyanoacetylene or its hydrolysis product, cyanoacetaldehyde, and cyanate ions, cyanogen, or urea [39–42]. Indeed, in concentrated urea solution, which could have been found in an evaporating lagoon on the early Earth, cyanoacetaldehyde could have reacted to form cytosine with a yield of 30–50%, from which uracil could be generated by hydrolysis [41]. As the cyanoacetylene can be formed when an electrical discharge is passed through a mixture of nitrogen and methane [43], and as its hydrolysis into cyanoacetaldehyde occurs spontaneously [39], these molecules can be considered prebiotic [44]. For example, pyrimidine formation has been observed when an energy source was applied on a urea solution in the presence of methane and nitrogen at low temperatures [45].

2.3. Nucleoside Synthesis from Sugars and Nucleobases

The synthesis of nucleosides from sugars and nucleobases in prebiotic conditions is one of the major difficulties encountered, when attempting to resolve the early formation of nucleosides. The reaction between the ribose and nucleobase is thermodynamically unfavorable, leading to poor yields and little selectivity [46]. Only a few examples showing successful synthesis have been reported in the literature.

The formation of adenosine (4% yield) has been observed from the condensation of adenine with ribose in the presence of inorganic salts, providing complex mixtures of purine ribosides [47]. However, regioselectivity problems were encountered owing to the reactivity of all N atoms of the purine skeleton. To overcome this regiospecificity limitation, Becker et al. showed that N-formamidopyrimidines could be used to generate purine nucleosides with absolute nucleobase regioselectivity [48]. Recently, they reported a plausible prebiotic synthesis of formamidopyrimidines, which can be generated from 5-nitroso-pyrimidine in the presence of formic acid and elementary metals (Ni or Fe). When combined with ribose, formamidopyrimidines can react and lead to efficient production of canonical and non-canonical purine bases in parallel [49].

In addition, adenine nucleoside phosphate has been formed from the direct coupling reaction of cyclic carbohydrate phosphate with the free nucleobase. The reaction is stereoselective and regioselective, giving the N-9 nucleotide as a major product [50]. It is unknown if pyrimidine nucleosides that could be formed with the same strategy as pyrimidines are more resistant to ribosylation [46]. Moreover, a feasible prebiotic pathway to synthesize both purine and pyrimidine simultaneously under the same conditions in an aqueous microdroplet containing ribose, phosphoric acid, nucleobases, and small amounts of magnesium ion has been reported [51]. Indeed, microdroplets allow organization of the molecules at the air–water interface of their surfaces, which possess a strong electrical field [52], diminishing the thermodynamic barrier for chemical reactions [53].

Even though important efforts are made to determine the possible prebiotic conditions for the nucleoside formation from sugars and nucleobases, this strategy leads to significant problems and an alternative approach has been suggested to form nucleosides [54].

3. The Revisited Approach for the Synthesis of Nucleosides

The alternative approach for the synthesis of nucleosides is based on the hypothesis that nucleobases and sugars emerged from a common precursor. This would mean that both the sugar and the nucleobase are formed during the same process (Figure 1B).

3.1. Pyrimidine Nucleoside Synthesis

Many years ago, Orgel et al. reported the synthesis of α-cytidine from ribose, cyanamide, and cyanoacetylene in aqueous solution. They showed that the replacement of the ribose by a ribose-5-phosphate allows the obtention of α-cytidine-5′-phosphate, which can be photoanomerized into β-cytidine-5′-phosphate required for RNA synthesis [55]. However, during this process, the yields were low (5%) and too many side products were formed to consider this route as prebiotic. Moreover, the photoanomerization destroyed most of the nucleosides [56,57]. Therefore, the formation of pyrimidines under plausible prebiotic conditions had to be further investigated.

Remarkably, one nucleotide, β-cytidine-2′-3′-cyclic phosphate, showed great stability under irradiation; only partial conversion to the corresponding uridine was observed [58]. Irradiation could thus provide a mechanism to destroy undesired products and partially convert cytidine into uridine. Therefore, the synthesis of this nucleotide raised a lot of interest. However, the determination of the pathway to obtain this cyclic phosphate under prebiotic conditions was a challenge.

Inspired by previously reported work of Navigary et al., who demonstrated that β-cytidine-2′-3′-cyclic phosphate can be obtained from the 3′-phosphate of the anhydronucleoside arabinose (which was prepared by conventional synthesis) [59], Sutherland et al. showed that it is possible to prebiotically obtain the desired 3′-phosphate of the anhydronucleoside arabinose from the corresponding arabinose in the presence of urea melts and formamide (Figure 4) [58].

The latter intermediate can be formed from the aminooxazoline arabinose and cyanoacetylene [55]. The presence of a phosphate during the reaction is essential to induce the right reactivity and to allow pH buffering, making the conversion of the aminooxazoline arabinose into the anhydronucleoside arabinose clean and with good yield (>90%).

Aminooxazoline arabinose could be formed by the reaction of 2-aminooxazole and glyceraldehyde in excellent yields [60]. This reaction supports construction of a five-carbon pentose backbone with complete furanosyl selectivity and regiospecific glycosylation in one step. The differential solubilities of pentose aminooxazolines facilitated a direct crystallization of pure compounds from the reaction mixture [60,61].

Synthesis of 2-aminooxazole by reaction of glycolaldehyde with cyanamide was achieved [62]. Once again, the presence of a phosphate during this step was crucial to obtain good yields (~90%). The phosphate mediated a neutral pH and first catalyzed the production of 2-aminooxazole. It also permitted the hydration of the excess cyanamide to urea, which in turn catalyzed the formation of 2-aminooxazole [63]. This synthetic route can be truly considered as prebiotic as the pyrimidines were formed from glycolaldehyde and cyanamide.

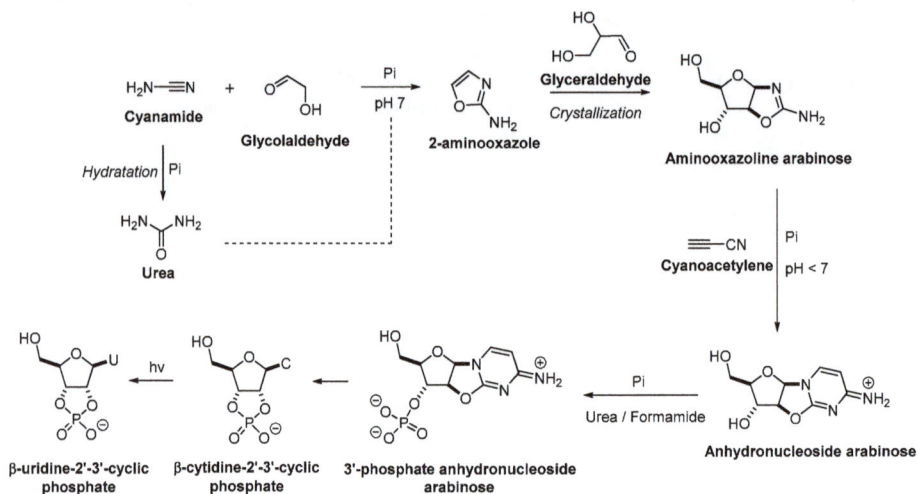

Figure 4. Possible prebiotic route for the synthesis of pyrimidine nucleotides. U = uridine; C = cytidine; Pi = inorganic phosphate.

Another route to obtain β-cytidine-2'-3'-cyclic phosphate by photoanomerization from ribose aminooxazoline instead of arabinose aminooxazoline was investigated (Figure 5). In fact, ribose aminooxazoline has a greater propensity to crystallize than its stereoisomers, making it extremely attractive in prebiotic chemistry. As mentioned previously, Orgel et al. reported that the photoanomerization of α-cytidine into β-cytidine results in a low yield (about 5%) [55]. Indeed, the low yield can be partially explained by a combination of nucleobase loss and oxazolidinone formation [56]. To improve this process, it has been suggested to incorporate a 2'-phosphate, leading to a 10-fold improvement of the photoanomerization [64]. However, a prebiotic synthesis of α-cytidine-2'-phosphate has not yet been demonstrated. Moreover, acetylation of α-cytidine-5'-phosphate was investigated to block oxazolidinone formation and a four-fold improvement of photoanomerization to produce β-cytidine-5'-phosphate has been observed [65]. Another solution to the inefficient photoisomerization has been found, and consisted of reacting anhydronucleoside ribose with hydrosulfide to form α-2-thiocytidine. Irradiation of the latter compounds led to the β stereochemistry. In addition, phosphorylation in the presence of urea produced the 2'-3'-cyclic phosphate nucleotide and converted the nucleobase thiocarbonyl to a carbonyl in one step to form β-cytidine-2'-3'-cyclic phosphate [66].

These two routes, exploring both ribose and arabinose variants, are key working examples for understanding pyrimidine formation under prebiotic conditions.

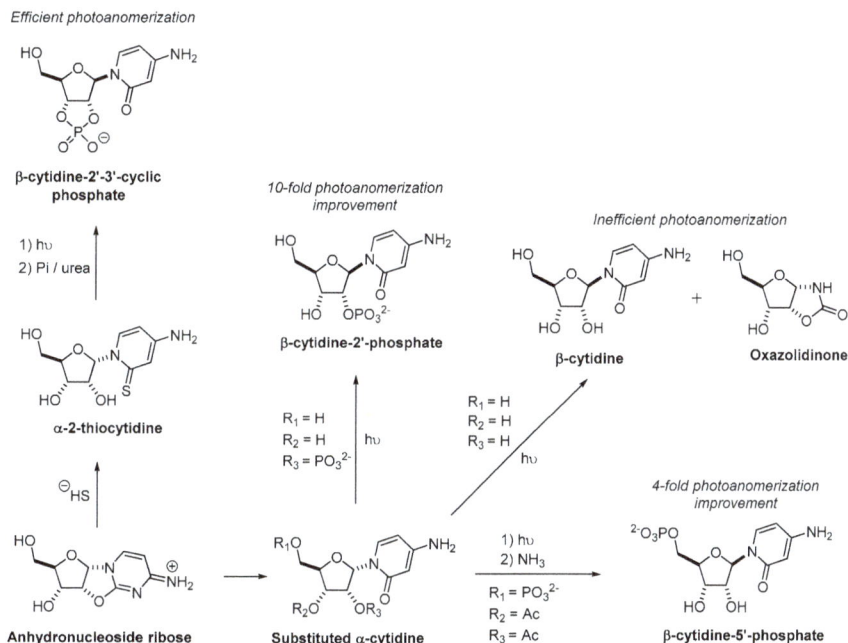

Figure 5. Alternatives to improve photoanomerization α→β for the formation of pyrimidine nucleotides. Pi = inorganic phosphate; Ac = acetyl.

3.2. Purine Nucleoside Synthesis

Possible routes for purine formation still have to be investigated. However, preliminary studies provide potential leads. Indeed, a purine precursor has been successfully assembled from 4-amino-5-cyanoimidazole or 5-aminoimidazole-4-carboxamide, 2-aminooxazole, and glyceraldehyde (Figure 6A). This Mannich-type reactivity results in N^9-glycolysation with absolute regiospecificity [67]. This route is particularly interesting because it provides the opportunity to produce both purine and pyrimidine precursors from the same environment. Moreover, the preference between purine or pyrimidine precursor formation can be controlled by pH. At pH 7, pentoses aminooxazolines (pyrimidine precursors) are predominant, whereas at pH 4–5, purine precursors are dominant. At a pH between 5 and 7, a mixture of both precursors are observed.

Another prebiotically plausible reaction for the synthesis of both pyrimidine and purine nucleotides from an oxazoline scaffold has been reported (Figure 6B). An oxazolidinone thione provided the chemical differentiation required for divergent pyrimidine and 8-oxo-purine nucleotide synthesis from one common precursor, the 2-thiooxazole [68]. Even though the transformation of the oxo-purine 2′,3′-cyclic phosphate nucleotides to the canonical nucleotides remains to be determined, it is possible that oxo-purine nucleotides were tolerated during template-directed RNA synthesis [69].

While promising, these proposed synthetic pathways demand further investigation.

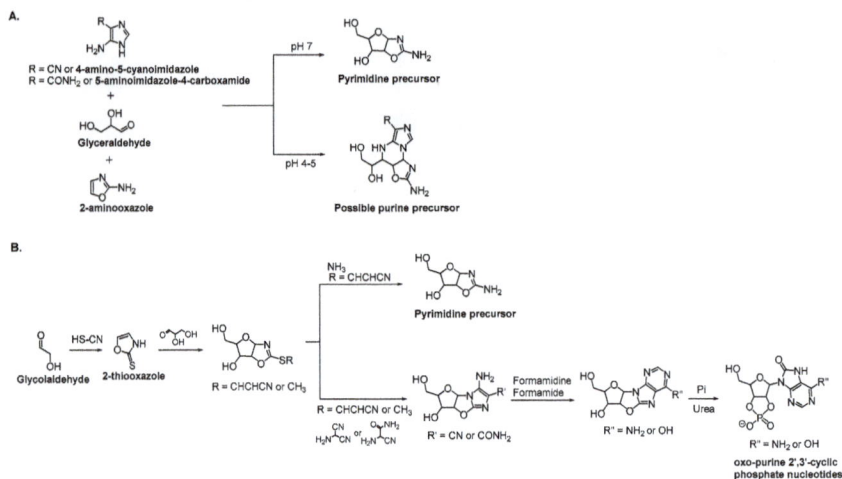

Figure 6. Possible prebiotic formation of both pyrimidine and purine precursors from a common environment. (**A**) Possible purine precursor from 4-amino-5-cyanoimidazole or 5-aminoimidazole-4-carboxamide, 2-aminooxazole, and glyceraldehyde. (**B**) Possible pyrimidine and oxo-purine from 2-thiooxazole. Pi = inorganic phosphate.

4. Conclusions and Outlook

The role of RNA in the origin of life is well established, and understanding how RNA emerged on the early Earth is one of the first steps in understanding the origins of life. Despite great efforts and impressive advancements in the study of nucleoside and nucleotide abiogenesis, further investigation is necessary to explain the gaps in our understanding of the origin of RNA.

The comprehension of nucleotide formation under prebiotic conditions is only one of the steps to understand the complex production of RNA as nucleotides must be oligomerized to generate RNA. Assuming that RNA must be 5′-3′-linked, regioselectivity issues have to be overcome. Polymerization of activated nucleotides has been studied intensely as a model for non-enzymatic oligomerization of RNA and has been considered a plausible scenario for the emergence of RNA during the origin of life [70–75]. Furthermore, for its genetic role to be realized, RNA must be able to evolve and replicate [6,76]. Unfortunately, the chemical processes that sustain RNA oligomerization and replication remain unclear [77].

Other than RNA, cells require various chemical subsystems, including peptides for functional support and lipids for compartmentalization. The assumption that one subsystem came first and then generated the others is debated [78–80]. Consequently, a search for a chemistry that can concurrently deliver nucleotides, peptides, and lipids or for chemistries that can be compatible with each other within the same geochemical environment could provide the most compelling explanation for the origins of life.

Funding: This research received no external funding.

Acknowledgments: I would like to thank Michele Fiore for the invitation to write this review. Also, I would like to thank Peter Strazewski for his advice about this review.

Conflicts of Interest: The author declares no conflict of interest.

References

1. Eschenmoser, A. Etiology of potentially primordial biomolecular structures: From vitamin B12 to the nucleic acids and an inquiry into the chemistry of life's origin: A retrospective. *Angew. Chem. Int. Ed.* **2011**, *50*, 12412–12472. [CrossRef] [PubMed]

2. Palmer, M.Y.; Cordiner, M.A.; Nixon, C.A.; Charnley, S.B.; Teanby, N.A.; Kisiel, Z.; Irwin, P.G.J.; Mumma, M.J. Alma detection and astrobiological potential of vinyl cyanide on titan. *Sci. Adv.* **2017**, *3*, e1700022. [CrossRef] [PubMed]

3. Boehnhardt, H.; Bibring, J.-P.; Apathy, I.; Auster, H.U.; Finzi, A.E.; Goesmann, F.; Klingelhöfer, G.; Knapmeyer, M.; Kofman, W.; Krüger, H.; et al. The philae lander mission and science overview. *Philos. Trans. A Math. Phys. Eng. Sci.* **2017**, *375*, 20160248. [CrossRef] [PubMed]

4. Altwegg, K.; Balsiger, H.; Bar-Nun, A.; Berthelier, J.-J.; Bieler, A.; Bochsler, P.; Briois, C.; Calmonte, U.; Combi, M.R.; Cottin, H.; et al. Prebiotic chemicals—Amino acid and phosphorus—In the coma of comet 67p/churyumov-gerasimenko. *Sci. Adv.* **2016**, *2*, e1600285. [CrossRef] [PubMed]

5. Powner, M.W.; Sutherland, J.D. Prebiotic chemistry: A new modus operandi. *Philos. Trans. R. Soc. Lond. B Biol. Sci.* **2011**, *366*, 2870–2877. [CrossRef] [PubMed]

6. Szostak, J.W.; Bartel, D.P.; Luisi, P.L. Synthesizing life. *Nature* **2001**, *409*, 387–390. [CrossRef] [PubMed]

7. Pace, N.R. The universal nature of biochemistry. *Proc. Natl. Acad. Sci. USA* **2001**, *98*, 805–808. [CrossRef] [PubMed]

8. Robertson, M.P.; Joyce, G.F. The origins of the RNA world. *Cold Spring Harb. Perspect. Biol.* **2012**, *4*, a003608. [CrossRef] [PubMed]

9. Kruger, K.; Grabowski, P.J.; Zaug, A.J.; Sands, J.; Gottschling, D.E.; Cech, T.R. Self-splicing RNA: Autoexcision and autocyclization of the ribosomal RNA intervening sequence of tetrahymena. *Cell* **1982**, *31*, 147–157. [CrossRef]

10. Guerrier-Takada, C.; Gardiner, K.; Marsh, T.; Pace, N.; Altman, S. The RNA moiety of ribonuclease p is the catalytic subunit of the enzyme. *Cell* **1983**, *35*, 849–857. [CrossRef]

11. Miller, S.L. A production of amino acids under possible primitive earth conditions. *Science* **1953**, *117*, 528–529. [CrossRef] [PubMed]

12. Oro, J.; Kimball, A. Synthesis of adenine from ammonium cyanide. *Biochem. Biophys. Res. Commun.* **1960**, *2*, 407–412. [CrossRef]

13. Zubay, G.; Mui, T. Prebiotic synthesis of nucleotides. *Orig. Life Evol. Biosph.* **2001**, *31*, 87–102. [CrossRef] [PubMed]

14. Butlerow, A. Bildung einer zuckerartigen substanz durch synthese. *Liebigs Ann. Chem.* **1861**, *120*, 295–298. [CrossRef]

15. Breslow, R. On the mechanism of the formose reaction. *Tetrahedron Lett.* **1959**, *1*, 22–26. [CrossRef]

16. Eckhardt, A.K.; Linden, M.M.; Wende, R.C.; Bernhardt, B.; Schreiner, P.R. Gas-phase sugar formation using hydroxymethylene as the reactive formaldehyde isomer. *Nat. Chem.* **2018**, *10*, 1141–1147. [CrossRef] [PubMed]

17. Mizuno, T.; Weiss, A.H. Synthesis and utilization of formose sugars. *Adv. Carbohydr. Chem. Biochem.* **1974**, *29*, 173–227.

18. Decker, P.; Schweer, H.; Pohlamnn, R. Bioids: X. Identification of formose sugars, presumable prebiotic metabolites, using capillary gas chromatography/gas chromatography—MAS spectrometry of n-butoxime trifluoroacetates on ov-225. *J. Chromatogr.* **1982**, *244*, 281–291. [CrossRef]

19. Zubay, G. Studies on the lead-catalyzed synthesis of aldopentoses. *Orig. Life Evol. Biosph.* **1998**, *28*, 13–26. [CrossRef] [PubMed]

20. Usami, K.; Okamoto, A. Hydroxyapatite: Catalyst for a one-pot pentose formation. *Org. Biomol. Chem.* **2017**, *15*, 8888–8893. [CrossRef] [PubMed]

21. Larralde, R.; Robertson, M.P.; Miller, S.L. Rates of decomposition of ribose and other sugars: Implications for chemical evolution. *Proc. Natl. Acad. Sci. USA* **1995**, *92*, 8158–8160. [CrossRef] [PubMed]

22. Ricardo, A.; Carrigan, M.A.; Olcott, A.N.; Benner, S.A. Borate minerals stabilize ribose. *Science* **2004**, *303*, 196. [CrossRef] [PubMed]

23. Furukawa, Y.; Horiuchi, M.; Kakegawa, T. Selective stabilization of ribose by borate. *Orig. Life Evol. Biosph.* **2013**, *43*, 353–361. [CrossRef] [PubMed]

24. Kim, H.-J.; Ricardo, A.; Illangkoon, H.I.; Kim, M.J.; Carrigan, M.A.; Frye, F.; Benner, S.A. Synthesis of carbohydrates in mineral-guided prebiotic cycles. *J. Am. Chem. Soc.* **2011**, *133*, 9457–9468. [CrossRef] [PubMed]

25. Müller, D.; Pitsch, S.; Kittaka, A.; Wagner, E.; Wintner, C.E.; Eschenmoser, A.; Ohlofjgewidmet, G. Chemistry of α-aminonitriles. Aldomerisation of glycolaldehyde phosphate to RAC-hexose 2,4,6-triphosphates and (in presence of formaldehyde) RAC-pentose 2,4-diphosphates: RAC-allose 2,4,6-triphosphate and RAC-ribose 2,4-diphosphate are the main reaction products. *Helv. Chim. Acta* **1990**, *73*, 1410–1468.

26. Ritson, D.; Sutherland, J.D. Prebiotic synthesis of simple sugars by photoredox systems chemistry. *Nat. Chem.* **2012**, *4*, 895–899. [CrossRef] [PubMed]

27. Ranjan, S.; Todd, Z.R.; Sutherland, J.D.; Sasselov, D.D. Sulfidic anion concentrations on early earth for surficial origins-of-life chemistry. *Astrobiology* **2018**, *18*, 1023–1040. [CrossRef] [PubMed]

28. Ritson, D.; Battilocchio, C.; Ley, S.V.; Sutherland, J.D. Mimicking the surface and prebiotic chemistry of early earth using flow chemistry. *Nat. Commun.* **2018**, *9*, 1–24. [CrossRef] [PubMed]

29. Xu, J.; Ritson, D.J.; Ranjan, S.; Todd, Z.R.; Sasselov, D.D.; Sutherland, J.D. Photochemical reductive homologation of hydrogen cyanide using sulfite and ferrocyanide. *ChemComm* **2018**, *54*, 5566–5569. [CrossRef] [PubMed]

30. Meiner, C.; Myrgorodska, I.; de Marcellus, P.; Buhse, T.; Nahon, L.; Hoffmann, S.V.; d'Hendecourt, L.S.; Meierhenrich, U.J. Ribose and related sugars from ultraviolet irradiation of interstellar ice analogs. *Science* **2016**, *352*, 208–212. [CrossRef] [PubMed]

31. Schwartz, A.W.; Joosten, H.; Voet, A.B. Prebiotic adenine synthesis via HCN oligomerization in ice. *Biosystems* **1982**, *15*, 191–193. [CrossRef]

32. Ferris, J.P.; Orgel, L.E. An unusual photochemical rearrangement in the synthesis of adenine from hydrogen cyanide. *J. Am. Chem. Soc.* **1966**, *88*, 1074. [CrossRef]

33. Ferris, J.P.; Orgel, L.E. Aminomalononitrile and 4-amino-5-cyanoimidazole in hydrogen cyanide polymerization and adenine synthesis. *J. Am. Chem. Soc.* **1965**, *87*, 4976–4977. [CrossRef] [PubMed]

34. Ferris, J.P.; Orgel, L.E. Studies in prebiotic synthesis. I. Aminomalononitrile and 4-amino-5-cyanoimidazole. *J. Am. Chem. Soc.* **1966**, *88*, 3829–3831. [CrossRef] [PubMed]

35. Sanchez, R.A.; Ferris, J.P.; Orgel, L.E. Studies in prebiotic synthesis. Iv. Conversion of 4-aminoimidazole-5-carbonitrile derivatives to purines. *J. Mol. Biol.* **1968**, *38*, 121–128. [CrossRef]

36. Sanchez, R.; Ferris, J.; Orgel, L.E. Conditions for purine synthesis: Did prebiotic synthesis occur at low temperatures? *Science* **1966**, *153*, 72–73. [CrossRef] [PubMed]

37. Miyakawa, S.; Murasawa, K.-I.; Kobayashi, K.; Sawaoka, A.B. Abiotic synthesis of guanine with high-temperature plasma. *Orig. Life Evol. Biosph.* **2000**, *30*, 557–566. [CrossRef] [PubMed]

38. Chyba, C.; Sagan, C. Endogenous production, exogenous delivery and impact-shock synthesis of organic molecules: An inventory for the origins of life. *Nature* **1992**, *355*, 125–132. [CrossRef] [PubMed]

39. Ferris, J.P.; Sanchez, R.A.; Orgel, L.E. Studies in prebiotic synthesis: Iii. Synthesis of pyrimidines from cyanoacetylene and cyanate. *J. Mol. Biol.* **1968**, *33*, 693–704. [CrossRef]

40. Ferris, J.P.; Zamek, O.S.; Altbuch, A.M.; Freiman, H. Chemical evolution. 18. Synthesis of pyrimidines from guanidine and cyanoacetaldehyde. *J. Mol. Biol.* **1974**, *3*, 301–309.

41. Robertson, M.P.; Miller, S.L. An efficient prebiotic synthesis of cytosine and uracil. *Nature* **1995**, *375*, 772–774. [CrossRef] [PubMed]

42. Nelson, K.E.; Robertson, M.P.; Levy, M.; Miller, S.L. Concentration by evaporation and the prebiotic synthesis of cytosine. *Orig. Life Evol. Biosph.* **2001**, *31*, 221–229. [CrossRef] [PubMed]

43. Sanchez, R.A.; Ferris, J.P.; Orgel, L.E. Cyanoacetylene in prebiotic synthesis. *Science* **1966**, *154*, 784–785. [CrossRef] [PubMed]

44. Orgel, L.E. Is cyanoacetylene prebiotic? *Orig. Life Evol. Biosph.* **2002**, *32*, 279–281. [CrossRef] [PubMed]

45. Menor-Salván, C.; Ruiz-Bermejo, D.M.; Guzmán, M.I.; Osuna-Esteban, S.; Veintemillas-Verdaguer, S. Synthesis of pyrimidines and triazines in ice: Implications for the prebiotic chemistry of nucleobases. *Chemistry* **2009**, *15*, 4411–4418. [CrossRef] [PubMed]

46. Sutherland, J.D. Ribonucleotides. *Cold Spring Harb. Perspect. Biol.* **2010**, *2*, a005439. [CrossRef] [PubMed]

47. Fuller, W.D.; Sanchez, R.A.; Orgel, L.E. Studies in prebiotic synthesis. Vii solid-state synthesis of purine nucleosides. *J. Mol. Evol.* **1972**, *1*, 249–257. [CrossRef] [PubMed]

48. Becker, S.; Thoma, I.; Deutsch, A.; Gehrke, T.; Mayer, P.; Zipse, H.; Carell, T. A high-yielding, strictly regioselective prebiotic purine nucleoside formation pathway. *Science* **2016**, *352*, 833–836. [CrossRef] [PubMed]

49. Becker, S.; Schneider, C.; Okamura, H.; Crisp, A.; Amatov, T.; Dejmek, M.; Carell, T. Wet-dry cycles enable the parallel origin of canonical and non-canonical nucleosides by continuous synthesis. *Nat. Commun.* **2018**, *9*, 1–9. [CrossRef] [PubMed]

50. Kim, H.-J.; Benner, S.A. Prebiotic stereoselective synthesis of purine and noncanonical pyrimidine nucleotide from nucleobases and phosphorylated carbohydrates. *Proc. Natl. Acad. Sci. USA* **2017**, *114*, 11315–11320. [CrossRef] [PubMed]

51. Nam, I.; Nam, H.G.; Zareb, R.N. Abiotic synthesis of purine and pyrimidine ribonucleosides in aqueous microdroplets. *Proc. Natl. Acad. Sci. USA* **2018**, *115*, 36–40. [CrossRef] [PubMed]

52. Donaldson, D.J.; Vaida, V. The influence of organic films at the air−aqueous boundary on atmospheric processes. *Chem. Rev.* **2006**, *106*, 1445–1461. [CrossRef] [PubMed]

53. Kathmann, S.M.; Kuo, I.F.; Mundy, C.J. Electronic effects on the surface potential at the vapor-liquid interface of water. *J. Am. Chem. Soc.* **2008**, *130*, 16556–16561. [CrossRef] [PubMed]

54. Fiore, M.; Strazewski, P. Bringing prebiotic nucleosides and nucleotides down to earth. *Angew. Chem. Int. Ed.* **2016**, *55*, 13930–13933. [CrossRef] [PubMed]

55. Sanchez, R.A.; Orgel, L.E. Studies in prebiotic synthesis. V. Synthesis and photoanomerization of pyrimidine nucleosides. *J. Mol. Biol.* **1970**, *47*, 531–543. [CrossRef]

56. Powner, M.W.; Anastasi, C.; Crowe, M.A.; Parkes, A.L.; Raftery, J.; Sutherland, J.D. On the prebiotic synthesis of ribonucleotides: Photoanomerisation of cytosine nucleosides and nucleotides revisited. *ChemBioChem* **2007**, *8*, 1170–1179. [CrossRef] [PubMed]

57. Liu, F.-T.; Yang, N.C. Photochemistry of cytosine derivatives. 2. Photohydration of cytosine derivatives. Proton magnetic resonance study on the chemical structure and property of photohydrates. *Biochemistry* **1978**, *17*, 4877–4885. [CrossRef] [PubMed]

58. Powner, M.W.; Gerland, B.; Sutherland, J.D. Synthesis of activated pyrimidine ribonucleotides in prebiotically plausible conditions. *Nature* **2009**, *459*, 239–242. [CrossRef] [PubMed]

59. Tapiero, C.M.; Nagyvary, J. Prebiotic formation of cytidine nucleotides. *Nature* **1971**, *231*, 42–43. [CrossRef] [PubMed]

60. Anastasi, C.; Crowe, M.A.; Powner, M.W.; Sutherland, J.D. Direct assembly of nucleoside precursors from two- and three-carbon units. *Angew. Chem. Int. Ed.* **2006**, *45*, 6176–6179. [CrossRef] [PubMed]

61. Springsteen, G.; Joyce, G.F. Selective derivatization and sequestration of ribose from a prebiotic mix. *J. Am. Chem. Soc.* **2004**, *126*, 9578–9583. [CrossRef] [PubMed]

62. Cockerill, A.F.; Deacon, A.; Harrison, R.G.; Osborne, D.J.; Prime, D.M.; Ross, W.J.; Todd, A.; Verge, J.P. An improved synthesis of 2-amino-1,3-oxazoles under basic catalysis. *Synthesis* **1976**, *1976*, 591–593. [CrossRef]

63. Lohrmann, R.; Orgel, L.E. Prebiotic synthesis: Phosphorylation in aqueous solution. *Science* **1968**, *161*, 64–66. [CrossRef] [PubMed]

64. Powner, M.W.; Sutherland, J.D. Potentially prebiotic synthesis of pyrimidine β-D-ribonucleotides by photoanomerization/hydrolysis of α-D-cytidine-2′-phosphate. *ChemBioChem* **2008**, *9*, 2386–2387. [CrossRef] [PubMed]

65. Fernández-García, C.; Grefenstette, N.M.; Powner, M.W. Selective aqueous acetylation controls the photoanomerization of α-cytidine-5′-phosphate. *Chem. Comm.* **2018**, *54*, 4850–4853. [CrossRef] [PubMed]

66. Xu, J.; Tsanakopoulou, M.; Magnani, C.J.; Szabla, R.; Šponer, J.E.; Šponer, J.; Góra, R.W.; Sutherland, J.D. A prebiotically plausible synthesis of pyrimidine β-ribonucleosides and their phosphate derivatives involving photoanomerization. *Nat. Chem.* **2017**, *9*, 303–309. [CrossRef] [PubMed]

67. Powner, M.W.; Sutherland, J.D.; Szostak, J.W. Chemoselective multicomponent one-pot assembly of purine precursors in water. *J. Am. Chem. Soc.* **2010**, *132*, 16677–16688. [CrossRef] [PubMed]

68. Stairs, S.; Nikmal, A.; Bučar, D.-K.; Zheng, S.-L.; Szostak, J.W.; Powner, M.W. Divergent prebiotic synthesis of pyrimidine and 8-oxo-purine ribonucleotides. *Nat. Commun.* **2017**, *8*, 1–12. [CrossRef] [PubMed]

69. Kamiya, H.; Miura, H.; Murata-Kamiya, N.; Ishikawa, H.; Sakaguchi, T.; Inoue, H.; Sasaki, T.; Masutanl, C.; Hanaoka, F.; Nishimura, S.; et al. 8-hydroxyadenine (7, 8-dihydro-8-oxoadenine) induces misincorporation in in vitro DNA synthesis and mutations in NIH 3t3 cells. *Nucleic Acids Res.* **1995**, *23*, 2893–2899. [CrossRef] [PubMed]

70. Orgel, L. Prebiotic chemistry and the origin of the RNA world. *Crit. Rev. Biochem. Mol. Boil.* **2004**, *39*, 99–123.
71. Li, L.; Prywes, N.; Tam, C.P.; O'Flaherty, D.K.; Lelyveld, V.S.; Izgu, E.C.; Pal, A.; Szostak, J.W. Enhanced nonenzymatic RNA copying with 2-aminoimidazole activated nucleotides. *J. Am. Chem. Soc.* **2017**, *139*, 1810–1813. [CrossRef] [PubMed]
72. Zhang, W.; Tam, C.P.; Zhou, L.; Soo Oh, S.; Wang, J.; Szostak, J.W. A structural rationale for the enhanced catalysis of nonenzymatic RNA primer extension by a downstream oligonucleotide. *J. Am. Chem. Soc.* **2018**, *140*, 2829–2840. [CrossRef] [PubMed]
73. Jheeta, S.; Joshi, P.C. Prebiotic RNA synthesis by montmorillonite catalysis. *Life* **2014**, *4*, 318–330. [CrossRef] [PubMed]
74. Burcar, B.T.; Jawed, M.; Shah, H.; McGown, L.B. In situ imidazole activation of ribonucleotides for abiotic RNA oligomerization reactions. *Orig. Life Evol. Biosph.* **2014**, *45*, 31–40. [CrossRef] [PubMed]
75. Mariani, A.; Russell, D.A.; Javelle, T.; Sutherland, J.D. A light-releasable potentially prebiotic nucleotide activating agent. *J. Am. Chem. Soc.* **2018**, *140*, 8657–8661. [CrossRef] [PubMed]
76. Joyce, G.F.; Szostak, J.W. Protocells and RNA self-replication. *Cold Spring Harb. Perspect. Biol.* **2018**. [CrossRef] [PubMed]
77. Szostak, J.W. The eightfold path to non-enzymatic RNA replication. *J. Syst. Chem.* **2012**, *3*, 2. [CrossRef]
78. Carter, C.W. What RNA world? Why a peptide/RNA partnership merits renewed experimental attention. *Life* **2015**, *5*, 294–320. [CrossRef] [PubMed]
79. Wills, P.R.; Carter, C.W. Insuperable problems of the genetic code initially emerging in an RNA world. *Biosystems* **2018**, *164*, 155–166. [CrossRef] [PubMed]
80. Islam, S.; Powner, M.W. Prebiotic systems chemistry: Complexity overcoming clutter. *Chem* **2017**, *2*, 470–501. [CrossRef]

Article

The Origin of Prebiotic Information System in the Peptide/RNA World: A Simulation Model of the Evolution of Translation and the Genetic Code

Sankar Chatterjee [1],* and Surya Yadav [2]

[1] Department of Geosciences, Museum of Texas Tech University, Box 43191, 3301 4th Street, Lubbock, TX 79409, USA

[2] Rawls College of Business, Texas Tech University, Box 42101, 703 Flint Avenue, Lubbock, TX 79409, USA; surya.yadav@ttu.edu

* Correspondence: sankar.chatterjee@ttu.edu; Tel: +1-806-787-4332

Received: 13 December 2018; Accepted: 25 February 2019; Published: 1 March 2019

Abstract: Information is the currency of life, but the origin of prebiotic information remains a mystery. We propose transitional pathways from the cosmic building blocks of life to the complex prebiotic organic chemistry that led to the origin of information systems. The prebiotic information system, specifically the genetic code, is segregated, linear, and digital, and it appeared before the emergence of DNA. In the peptide/RNA world, lipid membranes randomly encapsulated amino acids, RNA, and peptide molecules, which are drawn from the prebiotic soup, to initiate a molecular symbiosis inside the protocells. This endosymbiosis led to the hierarchical emergence of several requisite components of the translation machine: transfer RNAs (tRNAs), aminoacyl-tRNA synthetase (aaRS), messenger RNAs (mRNAs), ribosomes, and various enzymes. When assembled in the right order, the translation machine created proteins, a process that transferred information from mRNAs to assemble amino acids into polypeptide chains. This was the beginning of the prebiotic *information* age. The origin of the genetic code is enigmatic; herein, we propose an evolutionary explanation: the demand for a wide range of protein enzymes over peptides in the prebiotic reactions was the main selective pressure for the origin of information-directed protein synthesis. The molecular basis of the genetic code manifests itself in the interaction of aaRS and their cognate tRNAs. In the beginning, aminoacylated ribozymes used amino acids as a cofactor with the help of bridge peptides as a process for selection between amino acids and their cognate codons/anticodons. This process selects amino acids and RNA species for the next steps. The ribozymes would give rise to pre-tRNA and the bridge peptides to pre-aaRS. Later, variants would appear and evolution would produce different but specific aaRS-tRNA-amino acid combinations. Pre-tRNA designed and built pre-mRNA for the storage of information regarding its cognate amino acid. Each pre-mRNA strand became the storage device for the genetic information that encoded the amino acid sequences in triplet nucleotides. As information appeared in the digital languages of the codon within pre-mRNA and mRNA, and the genetic code for protein synthesis evolved, the prebiotic chemistry then became more organized and directional with the emergence of the translation and genetic code. The genetic code developed in three stages that are coincident with the refinement of the translation machines: the GNC code that was developed by the pre-tRNA/pre-aaRS/pre-mRNA machine, SNS code by the tRNA/aaRS/mRNA machine, and finally the universal genetic code by the tRNA/aaRS/mRNA/ribosome machine. We suggest the coevolution of translation machines and the genetic code. The emergence of the translation machines was the beginning of the Darwinian evolution, an interplay between information and its supporting structure. Our hypothesis provides the logical and incremental steps for the origin of the programmed protein synthesis. In order to better understand the prebiotic information system, we converted letter codons into numerical codons in the Universal Genetic Code Table. We have developed a software, called CATI (**C**odon-**A**mino Acid-**T**ranslator-**I**mitator), to translate randomly chosen numerical codons into corresponding amino acids and vice versa. This conversion has granted us insight into how the genetic code might have evolved in the peptide/RNA world. There is great potential in the application of numerical codons to bioinformatics, such as barcoding, DNA mining, or DNA fingerprinting. We constructed the likely biochemical pathways for the origin of translation and the genetic code using the Model-View-Controller (MVC) software framework, and the translation machinery step-by-step. While using AnyLogic software, we were able to simulate and visualize the entire evolution of the translation machines, amino acids, and the genetic code.

Keywords: peptide/RNA world; prebiotic information system; translation and the genetic code; bridge peptide and aaRS; ribozyme and tRNA; tRNA and mRNA; coevolution of translation machine and the genetic code; MVC architecture pattern and biological information; numerical codons; AnyLogic software for computer simulation of translation machine

1. Introduction

The origin of life on early Earth remains one of the deepest mysteries in modern science. Recent evidence suggests that life may have emerged about four billion years ago through the spontaneous interaction of biomolecules in steaming hydrothermal environments, but the actual pathways of biogenesis are still shrouded in mystery [1]. Life's first building blocks had their origin in the tiny ice granules of interstellar space and they can be found on carbonaceous chondrites, comets, and the Murchison meteorite [2–4]. Asteroids were continuously battering the Hadean Earth [5]. As a result, thousands of craters probably pocked the surface of the Eoarchean crust, like the surface of the Moon and Mercury. Unlike our planetary neighbors, however, the crater basins of Eoarchean Earth filled with water and biomolecules, and developed a complex network of hydrothermal systems [6]. Carbonaceous chondrites delivered both water and the building blocks of life to the planetary surface, thus creating innumerable crater basins [7]. The meteorite collisions that created hydrothermal crater lakes in the Eoarchean crust filled with water, organic molecules, and various hydrothermal fluids, gases, and energy; inadvertently, these became the perfect crucibles for prebiotic chemistry [6–13]. There is now evidence that the Late Heavy Bombardment impact spike (4.1–3.8 Ga) during the Hadean–Eoarchean interval may not have happened; most likely, there was a continuous decrease of the bolide flux during this interval [14]. Minerals, such as zircons, and water-lain sediments in the ancient Hadean/Archean crust indicate that liquid water was prevalent as early as four billion years ago. Earth was no longer an alien inhospitable world, but it was transforming into a life-supporting environment [15].

The early atmosphere of Eoarchean Earth was dominated by CO_2 and N_2, not by CH_4 and NH_3. Moreover, the main source of carbon on the primitive Earth was atmospheric CO_2, which might have contributed to the formation of many organic compounds [16]. In the hydrothermal crater lake, cosmic and terrestrial chemicals were mixed, concentrated, and linked together by convective currents in these sequestered crater lakes, which were powered by hydrothermal, solar, tidal, and chemical energies; here, life began to brew [6–13]. Both the chemicals and the energy that were found in these hydrothermal crater lakes fueled most of the chemical reactions that are necessary for prebiotic synthesis and the resulting emergence of life [13]. Monomers, such as nucleotides and amino acids, were selected from random assemblies of molecular pools and then polymerized on the pores and pockets of the mineral substrate to create RNAs and peptides, heralding the peptide/RNA world [9,17–26]. Most likely, pores and crevices of the mineral substrate of the crater floor acted as receptacles for concentrations of simple RNA and peptide molecules [21,22]. The establishment of a symbiotic relationship between peptides and RNA was a landmark threshold in the evolution of life. These two biopolymers, with distinct structures and functions, became codependent and partner. The prebiotic peptides functioned as stabilizing and catalytic agents in the chemical reactions and they adapted to the high temperature vent environment.

The ability of the lipid membranes to encapsulate various monomers and biopolymers was crucial in terms of efficiency, stability, and molecular symbiosis. Encapsulation further ensured the concentration and protection of life-encouraging ingredients from the vent environment, thus enhancing further biosynthesis [1,9,22]. Molecular symbiosis among membranes, RNAs, amino acids, and peptides was the driving force for the origin of complex cellular components. Lipid membranes were randomly encapsulated RNA and peptide molecules from the mineral substrates of the crater floor to initiate a molecular symbiosis inside the protocells that led to the hierarchical emergence of several cell components and their functionalities: first plasma membranes, then peptides and RNAs, then transfer RNAs, messenger RNAs, and then ribosomes; these cooperative molecules created the prebiotic information system step-by-step for programmed protein synthesis [9].

Three classes of RNA molecules, messenger RNAs (mRNA), transfer RNAs (tRNA), and rRNA were the prime players in the expression of genetic information: mRNA was the initial storage molecule of genetic information, tRNA was the carrier of specific amino acids, and rRNA was the essential constituent of the protein producing ribosomes. The interactions between diverse RNA molecules and the myriad of amino acids and enzymes led to the gradual evolution of translation and the genetic code [27]. The peptide/RNA partnership performed two major functions during the origin of translation: storing information and stabilizing and catalyzing chemical reaction. As these two molecules began to develop in concert, the mRNA specified, in triplet code, the amino acid sequence of proteins. RNA molecules and amino acids began to communicate in different languages via bilingual enzymes that allowed for biomolecules to cooperate with each other, leading to information systems and translation. The key processes of the information flow from mRNA to proteins emerged during this stage. As information was stored in the symbolic languages of nucleotides and amino acids, biosynthesis became less random and more organized and directional. With the advent of DNA, genetic information began to flow from DNA, to mRNA, to protein by a two-step process: transcription and translation [27,28].

Recently, it has been argued that the genetic software provides a singular definition regarding what life is [29]. In this view, life emerged in that instant when information gained control over the biomolecules. The information-directed protein synthesis is a unique signature of life. Biological information separates life from nonlife. Although it is difficult to define what makes life so distinctive and remarkable, there is a general agreement that its informational aspect is a key property, perhaps being the key property [30].

We agree with the view of an algorithmic origin of life that indicates a complex system that is comprised of informational networks [31]. However, we suggest that life is more sophisticated than any man-made computer system where the software/hardware dichotomy is blurred and integrated. We find that this computer analogy too simplistic. Both the informational and functional biopolymers in the translational machinery can be viewed as highly mobile molecular nanobots, which are fully equipped with both the information and the material that are needed to accomplish their tasks. These nanobots 'know' how to put themselves together by self-assembly or by cooperation with other molecules. It is our proposition that these complex molecular characteristics of life actually appeared before first life. These molecular nanobots are complex, self-replicating, and self-managing information systems in themselves, being analogous to the 'Universal Constructor' (UC) conceived by von Neumann [32].

To begin to understand how nature invented highly complex and specialized information systems from the vast array of disparate possibilities, we began this quest by running computer simulations of the major biosynthetic steps that might help to explain the emergence of the system. In this paper, we use the Model-View-Controller (MVC) architecture [33] to reconstruct the molecular translation machinery; we built our model from the components that are known to have existed in the prebiotic environment, such as amino acids, nucleotides, and various peptides. In information systems, the MVC architectural pattern has been used for consolidating information together, processing it into a model, isolating it from its manipulation (controller), and then presenting the component (the view) that determines the output form of the product (the artifact). The major premise of the pattern is the modularity and distribution of processing. MVC separates the three different aspects of information processing: the data (the model), the visual representation of the data (the view), and the interface between the view and the model (the controller). The purpose of this article is to review the latest views on the origin of an information system in the prebiotic world during the emergence of translation and the genetic code.

2. Peptide/RNA World

Among several competing hypotheses regarding how life arose on early Earth, the 'RNA world' model is widely accepted [27,28,34–39]. The RNA world has become the main paradigm in the current origin of life research in which RNA assumed informational and functional roles. RNA molecules, such as ribozymes, can act as catalysts for chemical reactions between other RNA molecules. The discovery of catalytic RNAs and the revelation that the ribosome is, in fact, a ribozyme, together added strong circumstantial evidence for the RNA world theory [39].

Despite the conceptual elegance of the RNA world, this hypothesis faces formidable difficulties, primarily the immense challenge of RNA synthesis under plausible prebiotic conditions [22,40,41]. Various building blocks of RNA molecules, such as sugar, phosphorous, and the purine and pyrimidine nucleobases have been identified in carbonaceous chondrites, comets, and interplanetary dust particles [2,3]. During the polymerization of activated nucleotides on the surface of the clay substrates to from primitive RNA molecules, a steady input of peptides was essential [22]. Conversely, amino acids could be easily polymerized on the mineral surface to form peptide molecules [42]. The RNA molecule is inherently fragile in the natural environment and constantly degrades into smaller fragments through hydrolysis, preventing the faithful reproduction. Peptides provided stability to RNA molecules.

The RNA world might have existed, but the exclusivity of RNA and the neglect of peptide and lipid membrane could have been overstated. Vent environments that could support RNA synthesis no doubt also spawned many other organic compounds. It is irrational to think that vent environments exclusively created a load of nucleotides or RNA. Amino acids are easier to synthesize than RNA, as the Miller-type experiment suggests. The versatility of RNA molecules does not prevent the formation of peptides concurrently in the vent environments, especially when peptides were the likely outcome in prebiotic synthesis [9]. Peptides were easy to synthesize than RNAs in the primordial environment. Moreover, amino acids were probably among the most abundant biogenetic building blocks available on both the prebiotic Earth and meteorites [2,4]. There is a growing consensus that RNAs and peptides simultaneously appeared during prebiotic synthesis. The primordial vent environment that could support RNA synthesis no doubt also created many other organic compounds—for example, peptides—and lipid-like membranes, which are much less chemically challenging to generate [43]. In recent times, the RNA world paradigm is shifting to a peptide/RNA world paradigm [9,17–26].

There is increasing evidence that RNA and peptide molecules interacted very early on in the origin of the genetic code (rather than RNA and RNA worlds giving rise to proteins), even short peptides had significant catalytic capabilities. Recent experiment suggests that ribozyme recruits an assortment of proteins to the RNA world as it evolves [44]. Ribozyme, such as RNase P, recognizes pre-tRNA and processes to generate mature tRNAs in collaboration with an assemblage of proteins, thus favoring peptide/RNA world in the early stage of RNA evolution. RNA and protein, two complementary molecules that exist in the prebiotic environment, mingled and interacted to form a dynamic system; one cannot exist without the other. Given that life depends on a diversity of molecule types in a symbiotic effort, each interacting with the other in complicated ways, it is hard to imagine that it would have started with just a single type of molecule.

The duality of replication and metabolism is the intrinsic property of life and it must have simultaneously appeared before the origin of the DNA [45]. RNAs provide instructions to build proteins with the help of various enzymes. The establishment of symbiotic relationships between peptides and RNAs was a fundamental threshold in the evolution of life. These two biopolymers, with distinct structures and functions, became codependent. RNA and peptides worked in tandem to expand their informational, structural, and functional repertoires. The fundamental property of life, replication, and metabolism is believed to have evolved in the peptide/RNA world, where RNA stores genetic information and peptide enzymes function as catalysts [45]. The direct evolution of inherited genetic information coupled to encoded functional proteins, as is observed in real-world molecular biology, is far more plausible than any scenario in which there was, as initial RNA, a world of ribozymes sophisticated enough to operate a genetic code [19].

3. The Age of Information

The age of information arose as an emergent property in the peptide/RNA world before the origin of DNA. The informational and functional molecules, such as RNAs and peptides, have a high degree of specific complexity. The age of *information* introduces molecular communication and complementarity—the lock-and-key relationship—between RNA and peptides. The base pairing of RNA and its replication played a crucial role in building the information system. Different species of RNA would evolve to specify different functions in the information system.

The genetic information is essentially a digital data, plus meaning [46]. The base sequences of mRNA provide the data and the meaning is the translation of the data into a functional protein. Translation is transferring information from the language of mRNA to the language of proteins [27,28]. During translation, mRNAs serve as a data-storage system, transmitting digital instruction to molecular machines, the ribosomes, which manufacture protein molecules. RNAs are essential in encoding information. Three kinds of RNA molecules play major roles in translation: The messenger RNAs (mRNAs) carry genetic information to the ribosomes where the proteins are synthesized. The transfer RNAs (tRNAs) function as adaptors between amino acids and the codons in mRNA during translation. The tRNAs also carry the specific amino acids to the ribosome during protein synthesis; they are the handler by which the mRNA is pulled through the ribosome via-codon anticodon interactions in the course of translocation. The ribosomal RNAs (rRNAs) are the structural and catalytic components of the ribosomes. Arguably, the ribosomes are the most intricate and sophisticated nanomachines in nature that translate the nucleotide sequences of mRNAs into amino acid-sequences of proteins.

The RNA-based information system mostly depends on enzymic peptides for the replication and translation of the nucleic acids. However, the specificity of the enzyme depends on their amino acid sequences, which are determined by the sequences of nucleotides in RNAs. In the beginning, the amino acids were utilized by ribozymes as cofactors, developing complex interactions between different RNAs and amino acids that led to the origin of translation and genetic code. Several enzymes were essential for protein synthesis, including ribozymes, peptides, peptidyl transferase, and aminoacyl-tRNA synthetase (aaRS).

4. The Use of Information Theory in Biology

The discovery of genetic encoding of the DNA molecule, and its mode of translation into protein structures, secured the modern view of biology as an information science [30–32]. Biological systems have embedded information structure for supporting their functions [47–50]. An information system can be defined as a set of related components that work together for storing and processing data and for providing information [51]. This definition of an information system views it as an open system. Like an open system, an information system has a purpose and it interacts with its environment. It differentiates and elaborates itself in dealing with the changing environmental conditions just like biological systems. Terms, such as 'automata' and 'machine', refer to a form of an information system.

Even though the ideas of artificial automata came from the observation and study of natural automata, such as biological systems, we tend to use man-made machines and other artifacts as a metaphor to better understand the biological systems. Various metaphors, such as nanobot, bio-nanobot, etc. have been used in the literature to refer to different types of information system.

Metaphors are useful because they are efficient: they transfer a complex meaning in a few words. Some of the popular metaphors, including robot and nanomachine, are deeply entrenched in our social life, news, and literature. Figure 1 relates these terms to the biological information system. Biological systems exhibit characteristics that relate to processing and using messages that convey information.

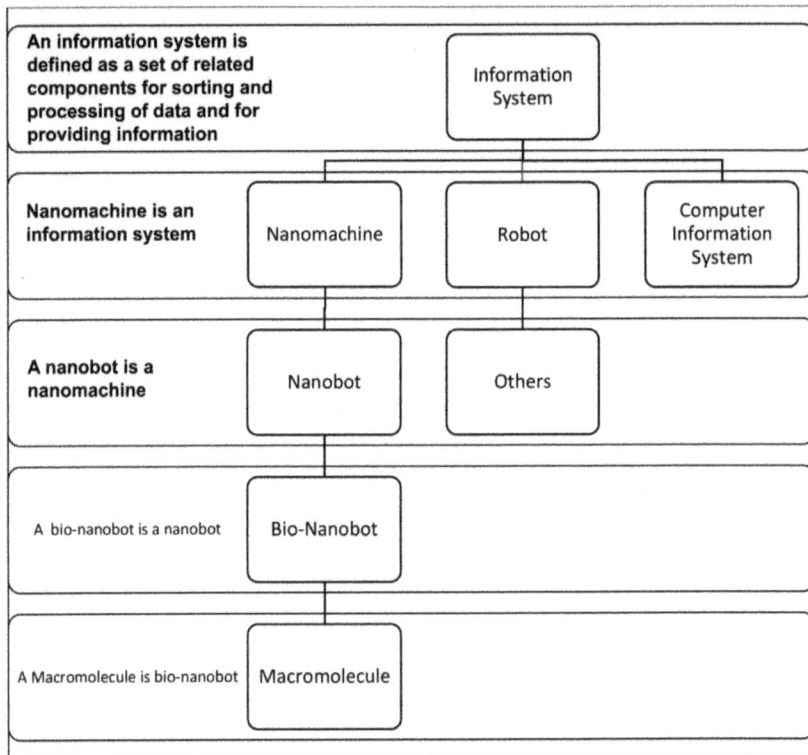

Figure 1. A hierarchy of Information Systems and Nanobots. This diagram shows a unified definition of various terms used for molecular systems. It related the idea of an information system with terms like 'nanomachines', etc.

We use metaphors, such as nanobots and computers, for information systems in cells. This practice may be fine as long as we understand the limitations of these metaphors. It's obvious that a cell is more complex than a computer system. The metaphors and analogies only explain a portion of the activities of the biological systems. We believe that, in most cases, a basic-level metaphor is more useful and discriminatory in explaining difficult concepts by association than a higher-level or system-level metaphor. For example, to say 'a cell is a nanobot' is not very revealing about the complexity of a cell. However, it is more meaningful if we say that a cell is a combination of assembler, transcriptor, translator, adapter, pattern-recognizer, pattern-copier, builder, inventory of materials, etc. Metaphors, like assembler, translator, adapter, etc., can be called as the basic-level metaphors. Taken together, they reveal closely the functions and structure of a cell. Another example, a ribosome can be metaphorically described as an assembler that assembles a protein with the help of charged tRNAs. This also points out the fact that a ribosome is part of a cell. It also illustrates the hierarchical nature of relationships between biological systems and between metaphors. To our knowledge, there is no higher-level metaphor that adequately describes a cell or, for that matter, any other biological system. A basic-level metaphor or a combination of basic-level metaphors can better describe a biological system. It is important to note that any of these basic-level metaphors can be viewed (modeled) as an information system.

Many fundamental biological processes involve the flow of information. The potential for new biological knowledge arises from investigating the complex interaction of many different levels of biological information from DNA to mRNA to protein to cells to organs to individuals. All macromolecules, organelles and cells, no matter how rudimentary, use information and material to conduct their tasks. Information that is used by them is in various forms such as attractiveness, proximity, pattern, match, symmetry, sequence, rule, and feedback, etc. These informational terms have the usual meanings. Attractiveness between modules relate to various chemical bonds that form easily between them. Proximity refers to the closeness between molecules. A pattern is a configuration of things in a certain way. Match involves the similarity and complementarity between molecular elements and surfaces. Symmetry relates to the shape of molecules and organisms. A sequence is a specific order in which related things follow each other. A biological sequence is a molecule that includes smaller molecules, such as nucleotides in RNA or amino acids in proteins. Rules specify conditional information. Feedbacks are information in the form of signals. By the time of DNA-mRNA-Protein synthesis, the cells had developed a very advanced, stable, and streamlined biological information system to help carry out the translation. Our discussion here is limited to the emergence of the information system in the peptide/RNA world, before the appearance of DNA and the first cells.

The characteristics of biological systems were identified and recognized by early pioneers in information systems. They have made tremendous contributions to our understanding of the biological processes by envisioning and proposing the concepts, frameworks, and models that help to imitate the biological processes. Von Neumann [32] proposed the idea of natural and artificial automata and developed detailed models to emulate the behavior and actions of natural automata. Turing [52,53] was instrumental in recognizing organized shapes, patterns, forms, and decision making in biological organisms. Shannon [54] formalized the concepts of information as a message, the transmission of message, and the semantic aspect of communication and information. The Shannon equation is practical for characterizing a signal (or message) and estimating the physical space that it may occupy; most random sequences give the highest possible entropy value (bits). Shannon's information entropy (H) is often confused with the physical entropy (S), because both concepts have a very similar mathematical formulation, but different meanings. Thermodynamic entropy characterizes a statistical ensemble of molecular states, while Shannon's entropy characterizes a statistical ensemble of messages [55,56]. For Shannon, information can be defined through entropy as a discrete set of probabilities to a receiver that reduces uncertainties. In biology, there is another dimensional aspect: information has both a probabilistic and linguistic context over an observable data set. Information in a biological context must exist within 'meaning' [46]. Genetically encoded biological information appears to be somewhat different from Shannon entropy.

Wiener [57] enunciated the concepts of control and feedback in systems. Bertlanffy's general systems theory [58] suggests that all the systems share some common organizing principles. All of these pioneering works in the form of theory, framework, and model have given rise to many advances in technology and biological knowledge. These advances have allowed us to develop better methods to design information systems for the simulation and visualization of biological information systems.

Evolution of Biological Information System

Life may be defined operationally as an information processing system—a structural hierarchy of various functional units—that has acquired through evolution, the ability to store and process the *information* that is necessary for its own accurate reproduction. Here, it is very useful to take a wider meaning of the word 'information' as opposed to just the classical definition of information based upon the information theory [31,52]. There are various definitions of the word information over the years. Historically, the word information has represented three different types of meanings [59]:

(1)　information as the process of being informed,

(2)　information as a state of an object, and

(3)　information as the disposition to inform.

Information as a process includes the ideas of message communication, meaning, and error due to a noisy channel [54]; information as a state of an object covers the idea of knowledge; and, information as the disposition to inform includes the ideas regarding the capacity of an object to inform another object and information as a specific thing [60].

Information can be signals, natural patterns (including shape, space, size, etc.), match, proximity, attractiveness (i.e., hydrophobicity and hydrophilicity), symmetry, sequence, rules, feedback, instructions, algorithms, content, and knowledge, etc. [60–63]. Biological information involves all of the above types of information.

In this paper, we first reconstruct the plausible biochemical pathways in the prebiotic world for the origin of the translation machines and the genetic code. Later, we apply a biological information system to simulate the origin of translation and the genetic code using different stages of translation machines.

5. Temporal Order of Emergence of the Translation Machines

The molecular translation machine consists of various parts and accessories, such as ribozymes, amino acids, tRNAs, aaRS, mRNAs, ribozymes, peptides, and various enzymes. In the modern translation machine, mRNA is decoded in a ribosome to produce a specific amino acid chain or polypeptide. The polypeptide then folds into an active protein and performs its function in the cell. The list of parts of translation machine is not sufficient condition for understanding its biologic function, such as programmed protein synthesis. Understanding how the parts work in unison is also important. However, it is not enough. We have to do reverse engineering to reconstruct how these parts might have evolved and interacted in the prebiotic environment. The origin and evolution of the translation machine may shed new light on how the information system emerged in the peptide/RNA world.

We have now some idea about the molecular milieu in the prebiotic environment in which the genetic code originated in the peptide/RNA world. The prebiotic soup was a rich collection of biomolecules in a highly reactive environment, owing to the constant input of hydrothermal energy. Some of these biomolecules were selected and encapsulated in protocells. The origin of the translation system is the central and the most difficult problem in the study of the origin of life [27,28,64–68]. All of the hypotheses for the origin of the genetic code incorporate peptides as the stabilizing factor in prebiotic reaction [17–21,23,24]. We propose that the transition from peptide to protein in the peptide/RNA world might have given rise to the translation system and the genetic code. This remarkable development could have incrementally occurred by natural selection as the demand for more and more efficient and specific protein enzymes became greater than the supply for protocellular functions. The solution that arose was the translation system—the recipe for making custom-made proteins.

In the peptide/RNA world, peptides played significant roles in accelerating the chemical reactions, by lowering activation energy. Some long peptides are good catalysts and show some enzymic activity. Most likely, ten proteinogenic amino acids were abiotically synthesized [67]. These ancestral amino acids gave rise to a limited variety of random peptides and polypeptides; most were useless, without much specificity, but a few were specifically selected for their catalytic activity. The need for both specific and a wide range of protein enzymes became essential in the peptide/RNA world for biogenesis. Peptides are distinguished from proteins on the basis of size and origin. The peptide is short (only few amino acids long), the protein is long (more than ~40 amino acids), folded, and it forms the catalytic center with a fixed start and end. As a result, the protein enzyme is much more versatile for catalytic reaction than the primitive peptide. Peptides and polypeptides could abiotically form in the prebiotic environment, but the proteins could not arise by chance but are coded, because it is achieved after a long evolutionary process. The evolution of peptides to proteins occurred from a small motif of short peptides to longer folded peptides, and finally to proteins that form complex catalytic centers with almost unlimited possible functions [24]. Darwinian selection provided the driving force for the evolution of specific protein over peptide, so that protein synthesis became essential to the protocell function. These coded proteins were custom-made by translation machines consisting of a repertoire of RNA and protein molecules, and they were highly specified for protocellular functions. The evolution of protein was a long evolutionary process that was driven by incremental advances of the translation machinery, which facilitated the transition from random, simple, peptide produced through an abiotic process, to the eventual production of specific, complex, proteins by RNA-directed protein synthesis.

One of the early manifestations of the transition from peptide to protein is the emergence of the 'bridge peptide' that facilitated the aminoacylation of RNA to specific amino acid. The bridge peptide is a short peptide and stereochemical interactions mediate its property to bind a specific RNA with a specific amino acid. Hybridization-induced proximity of short aminoacylated RNAs led to the emergence of bridge peptides, which were capable of stimulating the interaction between specific RNAs and specific amino acids [24]. Eventually, bridge peptides would give rise to protozymes, urzymes, pre-aaRS, and aaRS. The proposed transition mechanism from peptide to proteins, aided by the translation machines, provided a continuity of functions so that each subsequent step was an improvement.

The evolution of complex information system must consist of plausible, elementary steps, with each conferring a distinct advantage on the evolving ensemble of genetic elements. Here, we map the emergence of potential informational and catalytic oligomers, derived from the assembly of building blocks, and reconstruct the probable steps that lead to the translation machinery and the genetic code. It is well-known that modern protein synthesis proceeds with the participation of 20 amino acids, ribozymes, tRNA, various enzymes, including aminoacyl tRNA synthetase (aaRS), mRNA, ribosomal RNA, ribosomal proteins, ribosome, a considerable number of proteinous factors, ATP, GTP, etc. More than 120 species of RNAs and proteins are involved in the process of protein synthesis [65]. The most important steps include: base pair complementarity, the origin of ribozyme, the origin of tRNA, the origin of aminoacyl-tRNA synthetase, the origin of mRNA, the origin of ribosome, the synthesis of protein, and the origin of the genetic code and translation.

The translation system is ancient and highly conserved, and it must have started with protobiopolymers [41,64–67]. It is most likely that the translation system employed by the cell today has undergone the most extensive and involved evolution; but we do not know this process because the transitional stages have been lost in time. Modern translation requires at least five kinds of macromolecules and amino acids: the set of tRNAs, the set of activating enzymes, the set of amino acids, mRNAs, and ribosomes. Most likely, the evolutionary beginnings of translation could not have involved the interaction of all these components. Thus, the first assumption that we need to make is that the beginning of translation involved, plausibly, a small number of ancestral macromolecules with similar functional capabilities [66]. In our view, the ancestral forms of tRNAs, mRNAs, and aaRS, along with amino acids, were used in the initial stage of translation. Most likely, the ribosome was the last molecular component to appear in the translation machine assembly.

Perhaps, aminoacylated ribozymes, as discussed later (Section 5.3), was the first crude translation machine. Eventually, the ribozyme would give rise to pre-tRNA, and the bridge peptide to pre-aaRS [24]. Later, variants would appear and evolution would produce different by specific aaRS-tRNA-amino acid combinations. Here we start the translation system with two distinct evolutionary precursor macromolecules: pre-tRNA and pre-aaRS that would design and tailor small pre-mRNA molecules for storage information and initiate translation. Eventually, these macromolecules would evolve into tRNA, mRNA, and aaRS. Finally, as the fidelity of translation was refined, ribosomes appeared on the scene, making protein synthesis more efficient.

We identify nine major stages for the origin and evolution of the translation machinery complex and the genetic code leading to the protein synthesis. These possible biochemical pathways are: (1) the selection of amino acids; (2) the origin of RNA; (3) the origin of ribozyme; (4) the origin of transfer

RNA; (5) the origin of metabolism; (6) the origin of aminoacyl-tRNA synthetase; (7) the origin of messenger RNA and translation; (8) the origin of ribosome; and finally, (9) protein synthesis. During the emergence of these biochemical pathways, the genetic code and the translation system coevolved with the translation machine.

5.1. Selection of Amino Acids

Carbonaceous chondrites carry a large number of amino acids and they were probably the major source of naturally occurring amino acids in the prebiotic cradle. A large pool of amino acids was available in the prebiotic environment. 70 amino acids have been identified in Murchison meteorite [1,4]. It is likely that similar numbers of amino acids were present in the prebiotic environment. Similarly, Miller's experiments have produced more than 40 different amino acids [69]. Out of 70 amino acids that were likely present in the prebiotic environment, only four L-amino acids, which were most easily formed in the primordial soup (such as alanine, glycine, aspartic acid, and valine), were selected and recruited through molecular recognition, by pre-tRNA and its corresponding pre-aminoacyl-tRNA synthetase. Later, six more amino acids were recruited from the prebiotic environment (such as glutamic acid, leucine, proline, histidine, arginine, and glutamic acid) for tRNA-mRNA-aaRS interactions. These ten amino acids were precursors for the formation of other ten amino acids along prebiotic pathways [69]. The choice of ten primordial amino acids from prebiotic soup for the synthesis of peptides may have been the first product of molecular selection in the *information* age.

5.2. The Origin of RNA

The basic constituents of RNA molecules, such as D-ribose, phosphate, and the four bases—adenine (A), guanine (G), cytosine (C), and uracil (U), along with unused nucleotides, were delivered to the hydrothermal crater lake by meteorites [2–4]. The polymerization of RNA molecules occurred by mineral catalysis in the prebiotic environment. Nucleotide monomers were linked on the montmorillonite clay substrates of the crater floor in an ATP-rich environment [1,9,22]. The accumulation of phosphates in the vent environment was an important requirement in making the sugar-phosphate backbone of RNA. Nucleotides underwent spontaneous polymerization on the mineral substrate with the loss of water. The resulting product was a mixture of polynucleotides that were random in length and sequence.

Six hypothetical stages for the formation of the RNA molecules in the prebiotic environment is shown in Figure 2. It seems unlikely that the prebiotic soup in the prebiotic environment produced only the four bases that were found in RNA—A, U, G, and C—which formed the polynucleotide chain. Certainly, there were other nucleotides (including hypothetical F and N bases in Figure 2), which were incapable of Watson–Crick base pairing. Initially, all of these mononucleotides were randomly polymerized into short oligonucleotides of different lengths by peptide bonding [40]. The process was mediated by natural selection and RNA replication. Natural selection led to the elimination of useless random oligonucleotide sequences during base pairing. From these chaotic assemblages of oligonucleotides, only four bases, such as A, U, C, and G, were selected by exploiting the properties of the Watson–Crick base pairing, whereas hypothetical F and N bases were eliminated. The four standard bases are better than 2 or 6 based on estimates of arbitrary catalysis and the actual pairing energy of standard bases [45]. The four nucleotides were strung together to produce short pieces of oligonucleotide and RNA molecules, which could replicate with the aid of the peptide enzyme. Replication selected prebiotic RNA molecular bases from overwhelmingly large assortment of mononucleotides. These RNA molecules were random and noncoded, being a jumble assortment of nucleotide bases.

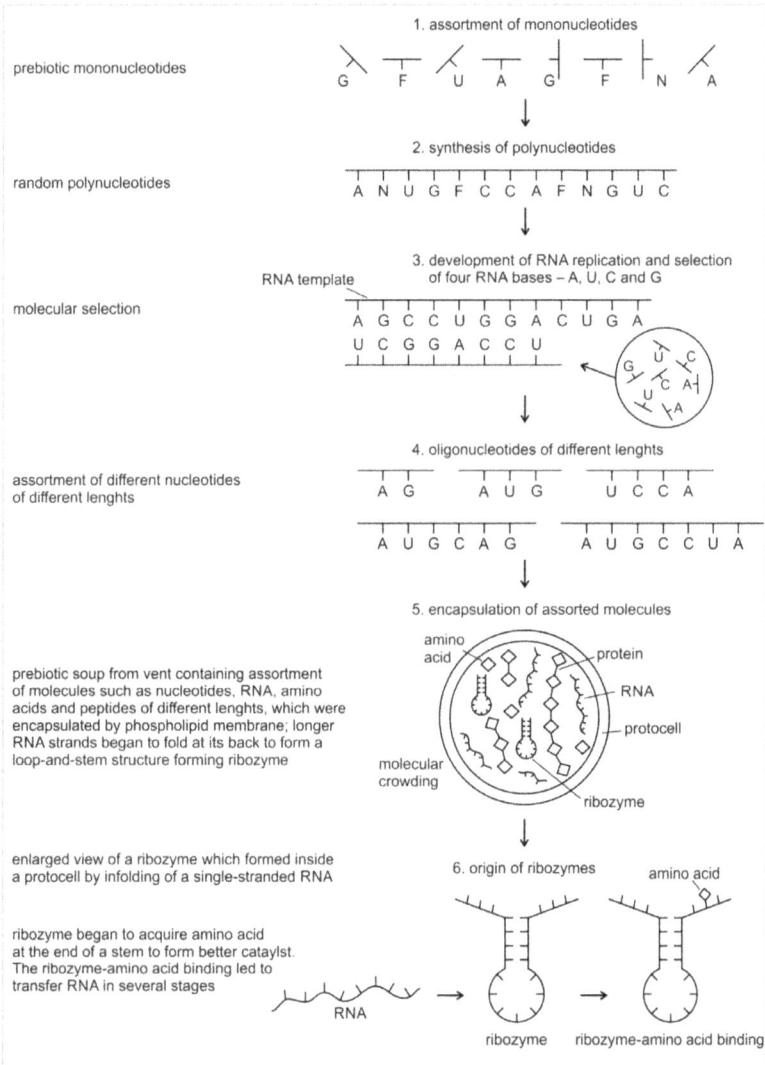

Figure 2. Six main steps represent the early evolution of non-coding RNA in the hydrothermal crater vent environment. In the first stage, there is an assortment of different nucleotides (including some not found in RNA). In the second stage, these nucleotides randomly assemble into polynucleotides by polymerization with the removal of water molecules. In the third stage, four nucleotides, A, U, G, and C were selected out during replication by the Watson–Crick base pairing. In the fourth stage, the nucleotides undergo polymerization to create a mixture of polynucleotides that are random in length and sequence. In the fifth stage, a variety of biomolecules from the vent environment, such as amino acids, mononucleotides, oligonucleotides, and peptides, are randomly encapsulated, creating molecular crowding. Because of crowding, the single-stranded RNA begins to fold, forming the double-stranded stem and single stranded loop that make the hairpin. In the sixth stage, this secondary structure of RNA is shown separately: it forms a ribozyme and begins to act as an enzyme. Stems are created by hydrogen bonding between complementary base pairs. The ribozyme acquires amino acids, at the CCA sequence of the stem, as "cofactors" increasing its catalytic efficiency. The opposite end of the loop consists of three, unpaired bases facing outward, forming a binding site for the attaching of three corresponding mononucleotides. This is the beginning of the emergence of the proto-tRNA.

Once some rudimentary template-dependent synthetic mechanism allowing for base-pairing was in place, molecules rich in A, U, C, and G were then progressively selected and amplified. These bases joined to form primordial RNA strands of different lengths, which began to self-replicate through a process of base pairing. Short sequences of nucleotides are normally better replicators than long sequences. Longer sequences suffer from an important evolutionary disadvantage; it takes longer to replicate a long sequence than a short once. If a pool of nucleotide sequences containing a range of length is left to code and replicate, then short sequences will dominate and long ones will become extinct [70]. The base pairing principle would later give rise to codon-anticodon hybridization, the origin of messenger RNA, transcription, and replication.

RNA is generally single-stranded and an informational molecule. The self-replication of RNA molecules occurs through a process of base pairing and dissociation. When one RNA strand is made in the vent environment, a second strand would automatically form through base pairing in such a way that cytosine always pairs with guanine, while adenine always pairs with uracil. Consequently, pairing is always between purine and pyrimidine. Because the hydrothermal vents in the crater basins were hot, double-stranded RNA, which formed by base-pairing, came apart through the dissociation of the two chains. When the strands separate, the cycle repeats with another round of base pairing, leading to two more double-stranded RNA molecules, one of which contains the original strand, containing its exact copy. By exploiting the properties of nucleotide base-pairing, coupled with the high temperatures of hydrothermal vent in the crater basin, short pieces of RNA replicated without the aid of any other molecules. Such complementary templating mechanisms lie at the heart of RNA replication, producing a large, more diverse population of RNA molecules (Figure 2).

Because RNA contains a sequence of bases that is analogous to the letters in a word, it can function as an information containing molecule. Moreover, RNA, being a single chain, is free to take any kind of shape; the structure that it can achieve by morphing its shape is wide-ranging, similar to protein. From this basic architecture of a single-stranded RNA molecule, different species of RNAs, such as ribozymes, tRNA, mRNA, and rRNA evolved inside protocells, with a supply of information, distinct in attribute and configuration in response to amino acids. There was a molecular choreography of different RNAs in the prebiotic world that led to the rudimentary translation. The advent and multifunction of different species of RNA molecules signal the transition from the age of *chemistry* to the age of *information*.

5.3. The Origin of Ribozyme

The RNA molecule has a secondary structure. It can form a localized double-stranded RNA stem by base pairing and a terminal loop to form a hairpin structure. In the stem, adenine forms a bond with uracil and cytosine pairs with guanine to form double-stranded RNA. The resulting hairpin structure is a key building block of many RNA secondary structures, such as ribozyme and tRNA (Figure 3). As an important secondary structure of RNA, an RNA hairpin can direct RNA folding, determine interactions in a ribozyme, protect structural stability for mRNA, provide recognition sites for RNA binding proteins, and serve as a substrate for enzymatic reaction [71]. Structurally, RNA hairpins can occur in different positions within different types of RNAs; they differ in the length of the stem, the size of the loop, the number and size of the bulges, and in the actual nucleotide sequence.

Ribozymes are RNA molecules that are capable of catalyzing specific biochemical reaction, similar to the action of protein enzymes. There are different classes of ribozymes, but they all appear to be associated with metal ions, such as potassium or magnesium. Different ribozymes catalyze different reactions, but almost all ribozymes are involved in catalyzing the cleavage of RNA chains in the formation of bonds between the RNA strands.

Most likely, the chemical bonding of a particular amino acid to a small RNA hairpin structure led to the origin of ribozyme. We assume that different kinds of RNA, protein enzymes, nucleotides, oligonucleotides, and amino acids were available in the prebiotic soup. The single-stranded nature of RNA molecule can be bent back on itself, in a hairpin loop, where the stems of the loops are maintained by base pairing to form a three-dimensional structure, just like a protein molecule to act as an enzyme. In some stem-loop configurations, two ends of the stem might remain free, containing the 3′ and 5′ ends. This 3′ end might function as an acceptor stem to form a covalent attachment to a specific amino acid (Figure 3). This small hairpin RNA molecule with specific terminal base sequences acquired the corresponding amino acid as a 'cofactor' to improve the catalytic range and efficiency to become initial ribozymes [46]. Many enzymes act with the help of one or more cofactors. The binding of amino acids to a ribozyme resulted in an enhancement of catalytic activity.

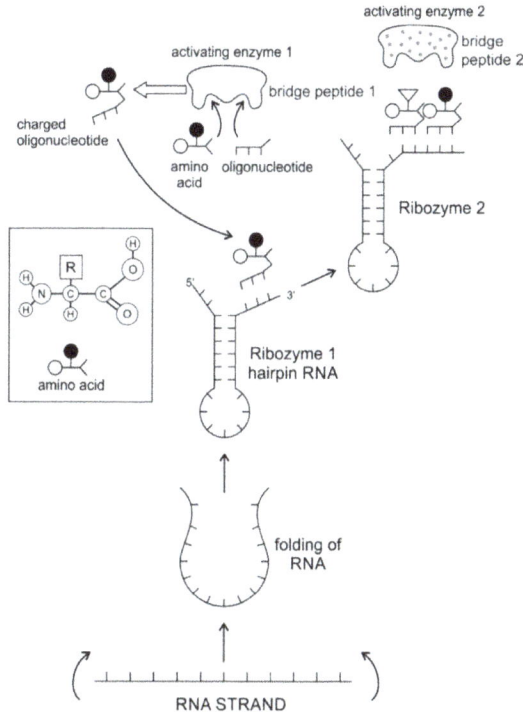

Figure 3. The origin of hairpin ribozyme and its chemical bonding with appropriate amino acid. A single-stranded RNA can develop secondary structure by infolding with double-stranded stem and single-stranded loop forming a hairpin ribozyme. The ribozyme acquired amino acid as cofactor to form a more efficient catalyst [46]. The amino acid is bound to an oligonucleotide (RNA molecule containing only three nucleotides) by an activation enzyme such as 'bridge peptide' [24], and the oligonucleotide is bound to the surface of the ribozyme by base pairing (ribozyme 1). The activating enzyme 2 would bind the next batch of amino acid and oligonucleotide is attached to ribozyme 2, forming the peptide bond.

Any specific binding between two molecules involves information, as if two molecules 'recognize' each other. An amino acid can be linked to an oligonucleotide with three bases by an activating enzyme; the charged oligonucleotide is then bound on the surface of a ribozyme by base pairing and delivers the appropriate amino acid (Figure 2). In this way, ribozymes are capable of producing short peptide chain. This de-novo peptide would play a role in stabilization, in order to become coded. Overtime, the original peptide forming ribozymes will specialize as amino acid specific adaptors. In the peptide/RNA world, different kinds of peptides were synthesized. Aminoacylation of ribozymes would be governed by the availability of amino acids. Most primordial amino acids in the prebiotic environment were alanine, glycine, valine, and aspartic acid. Initially, one kind of an amino acid and one kind of hairpin would be catalyzed by an activating enzyme, perhaps a precursor to the aminoacyl transfer tRNA synthetase, such as bridge peptide, which is a very short peptide that facilitated the emergence of self-sustained RNA-peptide complex supporting a primitive translation [24]. The BP was initially synthesized by chance as a result of the physical proximity of hybridized short, random aminoacylated ribozyme. Each BP is somewhat specific for an amino acid and for its corresponding ribozyme, but the specificity is low. Aminoacylated ribozyme would be involved in complex formation, bringing some of the aminoacylated ribozyme 3′-ends in close proximity. This would promote peptide bond formation between two adjacent amino acids (Figure 3). There was a feedback between ribozymes and bridge peptides. Later, a second amino acid, which is attached to a different hairpin by a different ribozyme, would be added, and so on to create a chain of polypeptide, supporting a primitive proto-translation. It is our contention that the interacting union of a hairpin ribozyme with a specific amino acid is cornerstone in the origin of information, transfer RNA, translation, genetic code, and protein synthesis. The ribozyme would give rise to tRNA and bridge peptide to pre-aaRS to aaRS. This is a combination of model between the one Koonin-developed model [72] describing a polymer transition out of the RNA world and the model of Carter [17] for the role of aaRS in the initial stages of code formation and translation.

A ribozyme has a well-defined tertiary structure that enables it to act like a protein enzyme in catalyzing biochemical and metabolic reactions. The relevance of ribozyme for the origin of tRNA is enormous. Ribozymes, being assembled in the prebiotic vent environment, could not only replicate themselves but would catalyze the formation of specific proteins. The adaptor ribozymes are the precursors of tRNA molecules and they play critical roles in the building of ribosomes. Ribosomal RNA functions as a peptidyl transferase in ribosomes to link the amino acids in protein synthesis, but the framework of transferase is provided by the ribosomal proteins.

5.4. The Origin of Transfer RNA

Any model for the development of protein synthesis must necessarily start with direct interactions between RNAs and amino acids. Chemical considerations suggested that direct interactions between the amino acids and the codons in mRNA were unlikely. The protein and mRNA languages seem to be unrelated. Amino acids do not read their codons. Some kind of an adaptor molecule must mediate the specification of amino acids by codons in mRNAs during protein synthesis [27]. The adaptor molecules were soon identified by other researchers as transfer RNAs (tRNAs), which serve as a reading device of mRNA through base pairing. The tRNA molecule binds to amino acids, associates with mRNA molecules, and also interacts with ribosomes to decipher and translate the code of mRNA.

It is generally believed that the first RNA gene, the *Ur-Gen*, was a precursor of modern tRNA [73]. tRNA is the ancestor of all RNAs. It is an ancient molecule that has evolved very little over time. The phylogeny of ribosomes suggests that tRNA is an ancient component of ribosomes that arose in the early prebiotic world [20].

A tRNA molecule is short, typically being 76 to 90 nucleotides in length, which serves as the physical link, a cipher, between the messenger RNA (mRNA) and the amino acid sequences of proteins [30]. Although the tRNA molecule is short, both its primary structure and its overall geometry are undoubtedly more complex than those of any other RNA species [74]. The translation of a message carried in mRNA into the amino acid language of proteins requires an interpreter. The amino acids themselves cannot recognize the codons in mRNA. The tRNA matches appropriate amino acids to the appropriate codons. To convert the three-letters words (codons) of nucleic acids to the one-letter, amino acids of proteins, tRNA molecules serves as the interpreters during translation. Each amino acid is joined to the correct tRNA by a special enzyme, aminoacyl-tRNA synthetase (aaRS).

tRNA participates in two clearly distinct steps in the translation process. The first step comprises the reactions that lead to the charging of the tRNA molecule with an amino acid. The second step comprises the complex reactions in which tRNA transfers its amino acids into a growing protein chain, in response to a specific codon. The chemical reaction catalyzed by the tRNA is simple—the joining of amino acids through peptide linkages. It performs the remarkable task of choosing the appropriate amino acid to be added to the growing protein chain by reading successive mRNA codons. The actual step of translation from mRNA into protein language occurs when amino acids and tRNAs are matched and joined. The translators that do this job are the aminoacyl-tRNA synthetases (aaRS). These enzymes are the only bilingual elements in the cell: they can recognize both the amino acid and the corresponding tRNAs. They are the key element of translation, being the links between the worlds of proteins and nucleic acids. The activation of tRNA occurs when a synthetase uses energy from ATP hydrolysis to attach an amino acid to a specific tRNA. There are twenty such synthetases, one for each amino acid. Together, they make up the complete dictionary for protein synthesis in a cryptic form that relies on tRNAs for decoding into the anticodon language. Each type of amino acid can be attached to only one type of tRNA, so each type of organism has many types of tRNA and more than 20 amino acids. There might be a coevolutionary process in which the anticodons and the corresponding amino acids were progressively mediated by natural selection. As ribosomes appear, tRNAs transport amino acids to ribosomes, where the amino acids are assembled into proteins.

Because of its molecular complexity, the origin of tRNA is controversial. The modern tRNA structure, with its complex configuration and multiple functions, might have originated from a simpler form, such as pre-tRNA molecules to select specific abiotic amino acids in the vent environment (Figure 4A–D). The pre-tRNA molecules with hairpin structures (stem and loop) might have evolved in some evolutionary stages of protein synthesis, originating from a linear chain of RNA [75]. The tRNA has a secondary and tertiary structure. In solution, the secondary structure of tRNA resembles a cloverleaf with three hairpin loops (Figure 4E,F). One of these hairpin loops contains a sequence of three nucleotides, called the anticodon, which forms base pairs with the mRNA codon. The other two loops of the cloverleaf form a D-arm and a T-arm. The unlooped stem contains the free 3′ and 5′ ends of the chain. The CCA sequence at the 3′ end of the acceptor stem forms a covalent attachment to the amino acid that corresponds to the anticodon sequence. The CCA sequence of the acceptor stem offered a binding site for the amino acid. The 5′ terminal contains a phosphate group. Both the anticodon and the acceptor stem sequence correlate with the role of amino acids in folded proteins [76]. The secondary structure tRNA molecule may provide some clue as to its ancestral molecular configuration. The cloverleaf-configuration of tRNA can be derived from a folded ribozyme with a single loop and an attachment site for the amino acid at the end of a stem (Figure 4E).

The most plausible scenario of the origin of the tRNA molecule is based on ribozymes. The chemical bonding of particular amino acids to small RNA molecules with specific base sequences was the crucial step. Perhaps the precursor of tRNA started as a simple ribozyme with a hairpin structure (Figure 4A,B). This ribozyme acquired amino acids at its 3′ end as a 'cofactor' (Figure 3): that is, an amino acid was attached to a ribozyme and made it a more efficient catalyst [77]. By using cofactors, the range of specificity of catalytic activity could be increased. One way of attaching an amino acid to a particular point on the surface of the ribozyme is at the end of a single-stranded unlooped stem of the hairpin, which is charged and begins to bind amino acid, which enhances the catalytic function of the ribozyme. With the stabilization of the catalytic reactions, these ribozymes began to participate in the first catalytic cycles. This configuration of a ribozyme linking an amino acid at the end may be the starting point for the origin of tRNA, where the unlooped stems contain the free 3′ and 5′ ends of the chain. This amino acid attachment to ribozymes by a specific assignment enzyme first occurred to make cofactors more efficient catalysts [46].

Aminoacylation of tRNA is an essential event in the translation system. Although, in the modern system, protein enzymes play the sole role in tRNA aminoacylation; in the primitive translation system, ribozymes could have catalyzed aminoacylation to tRNA or ancestral tRNA-like molecules. What was the catalytic function of ribozyme? If it was attaching an amino acid to its own end, it would not be logical that the substrate amino acid is the cofactor at the same time. It has been suggested that this attachment first occurred to make cofactors and it was carried by ribozymes. The RNA world hypothesis implies that the ribozyme functioned as an assignment enzyme to attach a particular amino acid to an ancestral tRNA for aminoacylation before the emergence of aaRS [77]. In the peptide/RNA world, we suggest that the ribozyme was not an aminoacylation catalyst; another molecule, such as bridge peptide, performed this function for the ligation of amino acid with ancestral tRNA [24]. In the early stage of aminoacylation, pre-aaRS, originally a protein enzyme, emerged as an assignment enzyme for charging ancestral tRNA [17–19]. In that case, the ribozyme should have another activity that is so advantageous as to help the molecule to survive. In our view, the cofactor function of ribozyme was utilized to form peptide bonds between adjacent amino acids before the emergence of the ribosome. This enzymatic activity may be precursor to that of the Peptidyl Transferase Center of the ribosome that is responsible for peptide bond formation. Another phenomenon in which the intervention of a ribozyme could have been of critical importance is RNA replication [68].

Many studies have suggested that the modern cloverleaf structure of tRNA may have arisen from a single ancestral gene by the duplication of half-sized hairpin-like RNAs by passing through some intermediate structures [76–83]. The linkage of an amino acid with a ribozyme at the end with a hairpin loop might be the starting point for the origin of tRNA, which is a quarter size of the modern tRNA molecule [47]. The relevance of ribozymes in the origin of tRNA is enormous. The equivalent effect of gene duplication might be accomplished by a simple ligation of two identical hairpins of folded ribozymes to create double hairpins, a D-hairpin and a T-hairpin, with an anticodon at the stem bases [82]. RNA ligation is a powerful driving force for the emergence of tRNA, joining two hairpin loops of ribozyme (Figure 4C). During the evolutionary transitions of the pre-tRNA molecule, the double hairpin structure with the D-hairpin and the T-hairpin formed in the ancient prebiotic world, with both the anticodon and the terminal CCA sequence adjacent to the D-hairpin (Figure 4D) [80].

The function of tRNA molecules depends on their precise three-dimensional structure. The cloverleaf tRNA folds into a more compact L-shaped tertiary structure, but each has a distinct anticodon and an attached amino acid (Figure 4H). One arm of the L-shaped tRNA structure has a minihelix with a single-stranded CCA end that is used for attaching a single amino acid; the other arm forms an anticodon loop, with three unpaired bases that may bind with the complementary codon of mRNA. Each tRNA molecule can carry one of the 20 different amino acids at its CCA minihelix end. Each type of amino acid has its own type of tRNA, which binds it and carries it to the growing end of a protein chain during the decoding of mRNA. The CCA end of the minihelix interacts with the large ribosomal subunit to form a peptide bond and the loop end interacts with the small ribosomal subunit for decoding mRNA triplets through codon-anticodon interactions [76].

We suggest that this half-sized hairpin structure of the pre-tRNA molecule acquired some functional capacity for translation before the emergence of tRNA (Figure 4C,D). The pre-tRNA molecule is the evolutionary precursor of the tRNA molecule. Direct duplication or the ligation of half-sized, hairpin-like structures—the pre-tRNA molecule—could have formed the contemporary full-length tRNA molecules, (Figure 4E). The acceptor stem bases and the anticodon stem/loop bases in tRNA in tRNA 5′-half and 3′-half fit together with the double-hairpin folding; this suggests that the primordial double-hairpin RNA molecules could have evolved to the structure of modern tRNA by gene duplication, with subsequent mutations to form the familiar overleaf structure [76,80]. In other words, two pre-tRNA molecules somehow fused together to form a tRNA molecule.

The half-sized pre-tRNA molecule with two loops (D-hairpin and T-hairpin) on one side, and anticodon and acceptor stem region of CCA end on the other side, is structurally and functionally independent and is more ancient than the other-half of the tRNA molecule [81]. This short, self-structured strand of the pre-tRNA molecule possesses a template domain, which is chargeable through interaction with specific amino acids, is probably the predecessor of tRNA (Figure 4C). This pre-tRNA molecule binds, with high specificity, to the amino acid corresponding to its anticodon; this reaction is catalyzed by a specific pre-aminoacyl-tRNA synthetase (pre-aaRS). tRNA evolution is closely linked to aminoacylation. There is a separate tRNA for each amino acid that carries a triplet sequences of nucleotides for anticodon. Later, the anticodon of pre-tRNA will guide the codon formation of the pre-mRNA.

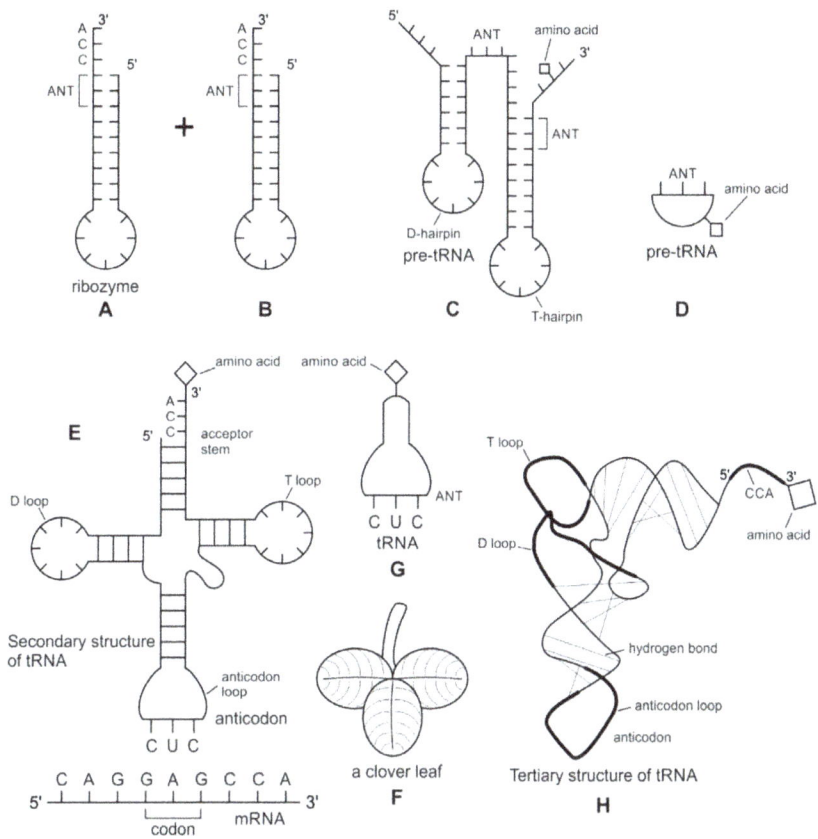

Figure 4. The double-hairpin model of the transfer RNAs (tRNA) formation, showing its evolutionary transitions [76,78,81]. (**A,B**), shows a secondary hairpin structure of two RNA molecules (such as ribozymes), each with a stem and a loop: the CCA sequence at the 3′ end of the stem offers a binding site for an amino acid, whereas the 5′ end offers a binding site for phosphorous; (**C**), the direct duplication or ligation of the hairpin structure may have generated a double hairpin structure, creating a D-hairpin and a T-hairpin. An anticodon (ANT) site forms between the two stems. In this newly conFigured pre-tRNA molecule, the acceptor site and anticodon site are now closer together, enabling it to decode a pre-messenger RNA (mRNA) molecule for protein synthesis (see Figure 20); (**D**), a schematic diagram showing the salient features of the pre-tRNA molecule, with the anticodon site; (**E**), the contemporary full-length tRNA molecule could have been formed by the ligation of two half-sized pre-tRNA structures. Its acceptor stem bases and anticodon stem/loop bases, at the tRNA 5′-half and the 3′-half, fit the double–hairpin folding. This suggests that the primordial double–hairpin RNA molecules could have evolved to modern tRNA. This new secondary structure of tRNA resembles a cloverleaf, its anticodon end forms a complementary base pair with the mRNA codon; (**F**), a cloverleaf from nature illustrates the structural similarity with the new tRNA molecule; (**G**), a schematic diagram showing the salient features of the tRNA molecule, emphasizing the anticodon. The tRNA serves a crucial role in matching an amino acid with a specific codon. When tRNA is bound to an amino acid it is called an aminoacyl tRNA. There is now a corresponding tRNA, with an appropriate anticodon, for each amino acid.; (**H**), the cloverleaf secondary structure of tRNA then folds to the L-shaped tertiary structure. At the CCA minihelix end, the aminoacylation site interacts with a large ribosomal unit for a peptide bond formation. The opposite end interacts with the small ribosomal subunit, to decode mRNA triplets through codon-anticodon interactions.

It should be apparent that tRNA molecules must contain a great deal of specificity, despite their small size. Not only do they (1) have the correct anticodon sequences, so as to respond to the right codons, but they must also (2) be recognized by the correct aaRS, to be activated by the correct amino acids, and (3) bind to the appropriate sites on the ribosomes to carry out their adaptor functions.

An important aspect of the specificity between amino acids and pre-tRNA is that, once this specificity is established, a mechanism for 'memorizing' or encoding variations in the sequence of pre-tRNA molecules becomes possible [73]. These pre-selected biomolecules of amino acids emerged from the existing prebiotic soup of the crater vent environments. Among the many essential components of the translation process, assignment enzymes, such as pre-aaRS, evolved to bind a specific amino acid to a pre-tRNA molecule (Figure 4).

5.5. The Origin of Metabolism

A prebiotic origin of metabolism is not fully understood. The core structure of the metabolic pathway is very similar across all organisms, which suggests the early origin of protometabolism in the prebiotic world [37]. Catalysts may have played an important role in establishing the early metabolism that ultimately led to the biosynthesis of protein. An intriguing possibility is that modern metabolic pathways emerged through a stepwise process of recruitment of ever more-effective catalysts to catalyze steps in primordial chemical-reaction networks. Metal ions of Fe, Mn, Zn, and Cu were also available in the vent environment, which help to mediate catalysis [10,11,13]. The synthesis of small organic molecules from inorganic precursors, including mineral-mediated synthesis, is probably the stimulus for the origin of metabolism. Large Hadean impacts may have made the atmosphere transiently rich in CO, which may have played a role in the origin of life and in fueling early biological metabolism. CO was an important trace gas on the prebiotic Earth, because it has high free energy and catalyzed the key reactions of prebiotic synthesis and in fueling early biological metabolisms [16].

Crystalline surfaces of common rock-forming minerals, such as pyrite and montmorillonite, are likely to have played several important roles in protometabolism [21]. Mineral surfaces with well-known catalytic properties might have promoted the polymerization of monomers, such as amino acids and nucleic acids. Many of life's essential macromolecules in the prebiotic world, including enzymes, carbohydrates, and RNA, form from water-soluble monomeric units—amino acids, sugars, and nucleic acid, respectively. Minerals surfaces provide a means to concentrate and assemble these bio-monomers. The polymerization of proteins from amino acids requires the dehydration and condensation mechanism that is precisely found in the fluctuating hydrothermal crater basins. It is well-known that amino acids concentrate and polymerize on clay minerals to form small, protein-like molecules [84]. Such reactions occur when a solution containing amino acids evaporate in the presence of clays. Subsequent studies have shown the adsorption and polymerization of amino acids on varied crystalline surfaces [23].

The problem of the origin of the evolution of metabolism has been recently advanced by the behavior of ZnS, which is capable of harvesting sunlight energy and converting this energy into the formation of chemical bonds of dicarboxylic acid from CO_2, thus providing the core reactions of universal metabolism before the existence of enzymes [85]. This paper has related how prebiotic metabolites available from simple sunlight promoted reactions can catalyze the synthesis of clay minerals (i.e., a zinc clay called sauconite). The work presents an excellent example of reproductive power of clay minerals and the mechanism by which prebiotic metabolites catalyze their formation. Clay minerals that act as sponges can retain water and polar organic molecules, and they might have played a key role in concentrating and catalyzing the polymerization of key organic molecules such as RNA and protein.

Small molecules—such as amino acids, short peptides, and cofactors—may have catalyzed reactions that are required to produce more complicated organic compounds. Although their catalytic abilities are known to be limited in both acceleration and specificity as compared with later molecular RNA or protein catalysts, some small molecules are remarkably effective catalysts.

The second stage, or metabolism, defined as the first set of reactions that are catalyzed by protein enzymes (and perhaps, ribozymes) prefiguring present-day metabolism, and perhaps already including certain central systems, such as the glycolytic chain and the Krebs cycle [68]. Centrally located within this network are the sugar phosphate reactions of glycolysis and the pentose pathway. This stage of metabolism appeared in the peptide/RNA world and it was modified and refined continuously during the origin of the first cells. As more enzymes were added and started to build their own network, new pathways could have developed.

5.6. The Origin of Aminoacyl-tRNA Synthetase

Aminoacyl-tRNA synthetases (aaRSs) are a superfamily of enzymes that are responsible for creating the pool of correctly charged aminoacyl-tRNAs, which are necessary for the translation of the genetic information (mRNA) through the ribosome. aaRSs are very ancient enzymes that are present in all organisms, and are one of the pioneer molecules that are formed by the polymerization of amino acids in incremental steps. Each enzyme catalyzes the activation of a specific amino acid and recognizes a specific tRNA for binding.

The unavailability of activated amino acids was the most critical barrier of protein synthesis. Aminoacylated ribozyme was the pioneer molecule to use the amino acid as cofactor and employed bridge peptide for activation. Later, with the development of tRNA, the activation reaction is catalyzed by specific aaRS, a derived product of bridge peptide. The first step is the formation of an aminoacyl adenylate with an amino acid and an ATP. The next step is the transfer of the aminoacyl group to a particular tRNA molecule to form aminoacyl-tRNA, or a charged tRNA. The mechanism of aaRS formation is well-known [77]. It reveals insight into how and why the tRNA molecule creates its own bilingual enzyme aaRS that can then connect it with the appropriate amino acid. It enhances the selection and sorting of the appropriate amino acids from the prebiotic soup for protein synthesis. Each aaRS is highly specific for a given amino acid. It has a highly discriminating amino acid activation site. Both amino acids and ATP were available in the hydrothermal vent, facilitating a reaction with tRNA to form aminoacyl-tRNA synthetase. Moreover, the proofreading ability by aaRS increases the fidelity of protein synthesis.

How do aaRS choose their tRNA partners? The aaRS recognize, on the one hand, individual amino acids, which they activate via conjunction with ATP; or, aaRS activate amino acids to generate its conjugate with AMP [77]. The synthetase first binds ATP and the corresponding amino acid to form an aminoacyl-adenylate, releasing inorganic pyrophosphate (PP_1). The next step is the transfer of the aminoacyl group of aminoacyl-AMP to a particular tRNA molecule to form aminoacyl-tRNA. The mechanism can be summarized in the following reaction series:

1. Amino acid + ATP $\rightarrow$ Aminoacyl-AMP + PP_1
2. Aminoacyl-AMP + tRNA $\rightarrow$ Aminoacyl-tRNA + AMP

Thus, the equivalent of two molecules of ATP are consumed in the synthesis of each aminoacyl-tRNA. One of them is consumed in the formation of the ester linkage of aminoacyl-tRNA, whereas the other is consumed in driving the reaction forward. The activation and transfer steps for a particular amino acid are catalyzed by the same aminoacyl-tRNA synthetase. Indeed, the aminoacyl-AMP intermediate does not dissociate from the synthetase. Aminoacyl-AMP is normally a transient intermediate in the synthesis of aminoacyl-tRNA. Synthetases can recognize the anticodon loops and acceptor stems of tRNA molecules. Their precise recognition of tRNAs is as important for high-fidelity protein synthesis, as is the accurate selection of amino acids.

aaRSs come in twenty flavors, with each one being specific to an amino acid and tRNA. These twenty enzymes are widely different, each being optimized to function with its own particular amino acid and the set tRNA molecules that are appropriate to that amino acid. They can be divided into two classes, termed class I and class II. The two aaRS superfamilies evenly divide translation into ten amino acids each. The initial activating enzyme was a bridge peptide that facilitated the aminoacylation of ribozyme (Figure 3). From bridge peptide, protozymes and then urzymes, and finally pre-aaRS and aaRS probably evolved [17–19,24]. We speculate that the precursor of aaRS was pre-aaRS, a hypothetical primordial ancestor that gave rise to two classes of aaRS, which are both multidomain proteins. Each aaRS uses different mechanisms of aminoacylation. In our model, the original aminoacylation enzymes were pre-aaRS, a simpler version of aaRS, which must have featured a strong linkage to the anticodon of a pre-tRNA molecule. This linkage must have featured a codon-like, trinucleotide binding site for the adaptor's anticodon, on the pre-aaRS. We propose that pre-aaRS is an enzyme, including an anticodon, plus a domain that is capable of binding and activating an amino acid and transferring to the pre-tRNA. Pre-aaRS is analogous to 'protozymes' and 'urzymes' [18,19], but is somewhat more advanced, because it would allow for tRNA/anticodon recognition. Protozymes retain about 40 percent of activity of the full-length of aaRS, even though they contain only about 10 percent as many amino acids. Next came 'urzymes', which retain about sixty percent of activity and have the same functional repertoire as the full-length enzymes. We speculate that pre-aaRS would be as long as the urzyme, but it has acquired additional anticodon binding function. The proposed evolutionary path from bridge peptide to protozyme to urzyme to pre-aaRS to aaRS documents increases the complexity of functions and would satisfy the rule of continuity [24].

5.7. The Origin of Messenger RNA and Translation

There are two haunting questions regarding the genesis of mRNA: (1) how mRNAs first appeared in the prebiotic environment, before the emergence of DNA and (2) how they evolved in the sequence of nucleotides, with the function of specifying amino acids as the fundamental components for the origin of the genetic code. The primordial mRNA was lost long ago in the *information* stage of biogenesis, leaving no trace of its origin. While existing evidence suggests that the genetic code was influenced by physico-chemical interactions between individual amino acids and strings of nucleic acids [69,86], researchers have yet to piece together the stepwise mechanisms by which it evolved over time.

In the prebiotic world, different species of RNA evolved through cooperation, each with a different function. Although random RNA strands grew during prebiotic synthesis by base pairing, in which some portions of the strand might show codon-like arrangement of nucleobases, they did not contain any genetic information (Figure 2). Moreover, the strings of nucleotide may be haphazardly interrupted by stop and start signals. A fundamental property of protein synthesis is that the amino acids are not added in a haphazard fashion. Their sequence is rigorously imposed by mRNA, which is itself is incrementally formed by tRNA. Each mRNA must be specially made allowing hybridization with tRNA, and specific to each protein.

Here, we propose a new model for the synthesis of custom-made mRNA by tRNA. The evolution of non-random coding mRNA served as the first medium for genetic information that coincided with the development of the genetic code and protein synthesis. As the tRNA molecules began to recognize and react with certain amino acids, they need a separate storage device for safe keeping the information of amino acid assignment. Because the selection of mRNA exclusively depends on codon-anticodon interaction, tRNA begins to make a specific strand of mRNA for the storage of amino acid information (otherwise, it is difficult to see how else mRNA molecules could have become involved with coding the strings of amino acids in a specific manner). We suggest the origin of a new generation of ancestral mRNAs—pre-mRNAs, were created by pre-tRNAs step-by-step. These newly synthesized pre-mRNAs have direct preferences for the amino acids that they tend to encode.

In our model, pre-tRNA molecules begin to select codons via base pairing with their anticodons; these short codon segments are linked to create a longer strand of pre-mRNA step-by-step for storing genetic information. In the pre-tRNA molecule, the site of attachment of the appropriate amino acid is proximate to the anticodon, making the communication between two active sites easier (Figure 5A,B). The physical proximity of the anticodon and the acceptor stem in ancestral pre-tRNA molecules is relevant to a long-sought goal-deriving amino acid/codon pairing rules from an ancestral nucleotide-based receptor-ligand recognition system [66]. A crucial aspect of the origin of pre-mRNA is that codon units are not just randomly added. Instead, the anticodon of pre-tRNA acts as a template to select the matching codon of a pre-mRNA strand. Using the base pairing mechanism, each anticodon of a charged pre-tRNA molecule begins to attract corresponding nucleotides from the prebiotic pool by base pairing (Figure 5D). After hybridization with anticodons, these triplet nucleotides begin to cluster and link together to form small chains of oligonucleotide with codon bases. Several small oligonucleotide chains begin to link to form a longer strand of a pre-mRNA molecule that becomes a database for storing the information of several amino acids (Figure 5E). This coded pre-mRNA became the binding partners for pre-tRNA, enhancing mutual stability and instant cognition. This is a turning point in the origin of translation when a pre-mRNA molecule becomes a digital strip for the storage of genetic information in a separate device in the nucleotide language. Translation is easier to evolve, logically as well as chemically, if there is already a triplet-amino acid assignment that is present. Eventually, several strands of pre-mRNA are joined to form a longer strand of pre-mRNA. These pre-mRNA genes are very short, no longer than 30 to 80 nucleotides. The main feature of pre-mRNA is its heterogeneity for information content. A triplet code sequence with a random codon assignment has very high information content in protein synthesis. With different combinations of codons and varied lengths of pre-mRNA strand, a wide range of amino acid information could be stored for the synthesis of longer protein chain (Figure 5F).

With the emergence of pre-mRNA, the information of anticodon assignment of large pre-tRNA populations can be transferred and stored in a codon message, along the strand of a pre-mRNA molecule. Along the linear strand of a pre-mRNA molecule, digital information for coding amino acids symbiotically emerged with the help of the anticodon of pre-tRNA molecules. Biological information was not only concentrated, but also specified along the strand of a pre-mRNA molecule. Charged pre-tRNA becomes the carrier of a specific amino acid that attached to the matching codon of pre-mRNA.

During the interaction of charged pre-tRNA with pre-mRNA, each aminoacyl pre-tRNA (aa-pre-tRNA) molecule transported and selected specific amino acids for protein synthesis. This is how information enters into the codon of the pre-mRNA molecule in a storage format for a specific amino acid via the anticodon. The information is laid down in the sequences of pre-mRNA, whose quantity is expressed by the lengths of those sequences. These base-pairing attachments between charged pre-tRNA and pre-mRNA provided the structural basis for translation.

The aa-pre-tRNA brings this specific amino acid to this pre-mRNA site during translation, where its anticodon binds to the complementary codon. Initially, four short oligonucleotides, each with a specific codon, were formed and joined in different combinations, specifying four amino acids, such as valine, alanine, aspartic acid, and glycine [88]. This is the first stage of the origin of translation along with the genetic code, involving four amino acids, in which a small number of amino acids were coded by a small number of triplets (Figure 5D). These four amino acids were readily available from the prebiotic vent environment. These oligonucleotides with codons are linked together by random combinations to form a pre-mRNA strand with a coded message (Figure 5E). Once the base sequence of pre-mRNA is stored for a number of amino acids, a rudimentary translation begins to initiate between pre-tRNA and pre-mRNA to synthesize the protein products that provide some modest catalytic, structural, and binding features in the peptide/RNA world. Most likely, the code assignments and the translation mechanism evolved together [75]. Pre-mRNA molecules, which were customized by pre-tRNA, multiplied in the vent environment and linked into longer strands of pre-mRNA to become a genetic reservoir, a digital recipe for proteins synthesis. However, at this stage, pre-mRNA can contain limited genetic information for four amino acids or their multiplied combinations.

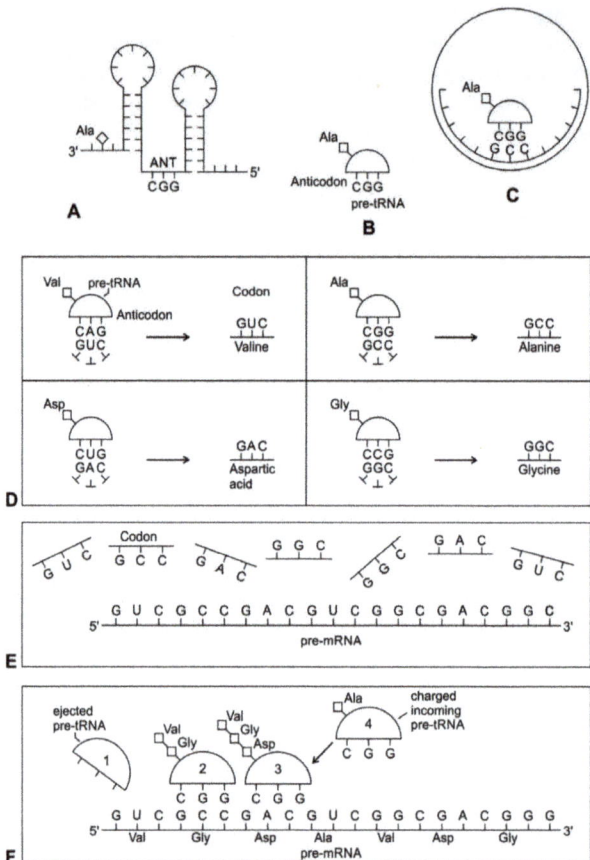

Figure 5. Primitive translation process began with interaction between pre-mRNA and pre-tRNA before the appearance of ribosomes. Pre-tRNA molecule serves as a crucial role in matching a prebiotic amino acid to a specific codon. (**A**), a pre-tRNA molecule with two hairpin loops of 3′ and 5′ terminals and an anticodon (ANT); the acceptor stem at the 3′ end forms a covalent attachment to a specific amino acid that corresponds to the anticodon sequence; (**B**), schematic representation of pre-tRNA emphasizing the 3′ end and corresponding anticodon; (**C**), encapsulated pre-tRNA and pre-mRNA molecule with codon-anticodon interaction; the inner cell membrane acts as a substrate to hold the pre-mRNA molecule in place. (**D**), the anticodon of a pre-tRNA molecule began to hybridize with corresponding nucleotide by base pairing; the triplet nucleotides were kinked to form a codon; In the abiotic stage, the primitive GNC code appeared, which codes four amino acids: valine, alanine, aspartic acid, and glycine [87,88]; (**E**), codons thus produced by pre-tRNAs, began to link in a strand to form a pre-mRNA with coding sequence; (**F**), pre-tRNA and pre-mRNA interactions to form rudimentary translation; the 3′ acceptor end of pre-tRNA gathers appropriate amino acid from the pool and binds it by activation enzyme; an aminoacetyl pre-tRNA with appropriate anticodon hybridizes with codon, ejecting the pre-tRNA; the next aminoacyl pre-tRNA then moves down another codon and repeats the process; amino acid released from the old pre-tRNA begins to join to form a protein chain.

During the initial translation process, each pre-tRNA carries its corresponding amino acid on its end (Figure 5F). When a charged pre-tRNA recognizes and binds to its corresponding codon of pre-mRNA, then the growing amino acid chain transfers to the single amino acid of the pre-tRNA. The pre-tRNA molecule begins to translate the codon of the pre-mRNA molecule in the 5′ to 3′

direction. The codon for the first amino acid in the chain (the amino end of the protein) is always at the 5´-end of the pre-mRNA. Likewise, the codon for the last amino acid in the chain is at the 3´-end of the pre-mRNA.

As the translation began along the strand of pre-mRNA, the triplet GUC coded for the amino acid valine. An aminoacyl pre-tRNA entered the site, where it then hybridized the codon. Here, a ribozyme, the precursor to peptidyl transferase of ribosome, performed two critical functions. First, it detached the valine from its pre-tRNA, which was ready to make a growing amino acid chain and released the pre-tRNA. Second, it catalyzed the formation of a peptide bond between that amino acid and the one that was attached to the next codon site. The first pre-tRNA, carrying the amino acid glycine, paired with the codon GCC. With the arrival of the second pre-tRNA, carrying valine, the first pre-tRNA, like a runner in a relay race, passed its glycine to the next, linking with valine and it was ejected. The third pre-tRNA with anticodon CUC hybridized with the next codon, GAC, bearing the aspartic acid, and picked up the link of glycine and valine. The next step repeats when a new aminoacyl pre-tRNA prepares to attach to the next codon site CGG for alanine. Here, it would receive the newly formed polypeptide link of valine-glycine-aspartic acids. To this link, alanine would be added. This is the way that a string of bases of pre-mRNA is translated into a sequence of amino acids. The released amino acids chain of valine, glycine, aspartic acid, and alanine are joined together by a peptide bond to form a newly synthesized protein (Figure 5F). Ribozymes functioned as a catalyst to break the acyl bond holding the growing amino acid chain on the pre-tRNA, and link the new incoming amino acid to the protein chain by a peptide bond. Those ribozymes that were involved in the protein synthesis were the precursors for the peptidyl transferase of the larger unit of the ribosomes.

The association between amino acids and codons—for example, between GUC and valine—is called the code. In this way, the genetic code begins to translate in a rudimentary form, as the short chain of proteins is built according to the instruction from the linear order of codons on the pre-mRNA. The process continues until the pre-tRNA molecule reaches the last codon in the pre-mRNA strand. It stops because there are no more codons to match. The ribozyme is clipped off by the completed protein chain. Once the complete protein is made, the pre-tRNA was discarded, and the pre-mRNA was broken down and its nucleotides recycled. The newly synthesized proteins functioned as enzymes for specific catalysis.

This initial code-programming and storage operation of the pre-mRNA by the pre-tRNA must have occurred within the protective environment of the protocells (Figure 5C). By pairing with the anticodons of the pre-tRNAs, the codons of the pre-mRNA not only selected the appropriate amino acids, but they also help to immobilize the pre-tRNAs. To initiate primitive translation, the pre-mRNA strand needed a substrate where pre-tRNA molecules would sequentially bind one codon after another.

How the primitive translation machinery maintains its proper reading frame is a question of primary importance. In the absence of the ribosomes, the inner surface of the protocell membrane would have served as a substrate for holding the pre-mRNA in position for pairing with the anticodon (Figure 5C). The spherical curved surface of the membrane probably facilitates the movement of pre-tRNA in downstream from 5´ to the 3´ ends of pre-mRNA during translation. This may be the beginning of the origin of reading frame, which is crucial for the reproducibility of translation; the codons of pre-mRNA should be read in a fixed direction with no gap between them.

The availability of several groups of new enzymes enlarged both the structural and the functional capabilities of the pre-mRNA and pre-tRNA molecules, evolving into the more efficient mRNA and tRNA. This evolutionary transformation was characterized by a progressive refinement of the translation system and an increase of the genetic code. As more and more pre-tRNA guided pre-mRNA molecules began to emerge, they were continuously replicated, increasing their population in the prebiotic pool, linking together in various combinations to form longer strands of mRNA molecules. tRNA and mRNA outnumber their precursors pre-tRNA and pre-mRNA through base pairing and replication. These longer mRNA genes arose as replication increased in accuracy. Each mRNA contained about 100 to 200 nucleotides (Figure 5E).

5.8. The Origin of Ribosomes

Translation needs one more piece of the molecular machine to continuously make protein in an assembly line—the ribosome. Ribosomes link amino acids together in the order that is specified by mRNA molecules. They provide the environment for controlling the interaction between codons of mRNA and anticodons of aminoacyl-tRNA in the creation of proteins. The translation of encoded information of mRNA and the linking of amino acids that were selected by tRNAs are at the heart of the protein production process. Ribosomes can link amino acids together at a rate of 200/min. Therefore, small proteins can be made fairly quickly. Once a new protein chain is manufactured, the ribosome is released from protein synthesis to enter a pool of free ribosomes that are in equilibrium with separate small and large subunits [77].

The ribosome is composed of two-thirds of RNA and one-third protein. It is made of about 50 ribosomal proteins (r-protein) that are wrapped up with four ribosomal RNAs (rRNA) and it is therefore a ribonucleoprotein (Figure 6). Although ribosomal proteins greatly outnumber ribosomal RNA, the rRNAs account for more than half the mass of the ribosome. A bacterial cell may contain as many as 20,000 ribosome complexes, which enable the continuous production of several thousand different proteins, both to replace degraded proteins and to make new ones for daughter cells during cell division. A ribosome physically moves along an mRNA strand, reads the codon sequences of mRNA, and catalyzes the assembly of amino acids into protein chains using the genetic code. It uses tRNAs to mediate the process of translation from the nucleotide language of mRNA into the amino acid language of proteins with the help of various accessory molecules. Each ribosome can bind one mRNA and up to three tRNAs. Central to the development of ribosomes are RNAs that spawn the tRNAs, and a symmetrical region that is deep within the large ribosomal RNA, where the peptidyl transferase reaction occurs [77,89,90].

Recent bacterial ribosomes shed light on the origin, evolution, morphology, and composition of primitive ribosome that emerged in the peptide/RNA world. The bacteria have smaller ribosomes, termed 70S ribosomes, which are composed of two major subunits of unequal size, which are called the large (50S) and the small (30S) subunits; each consists of one or two RNA chains and scores of proteins (Figure 6). The small subunit (SSU) is where mRNA and tRNA molecules interact to read the genetic code, and the large subunit (LSU) is where the growing protein chain is synthesized from the amino acids that are attached to tRNAs. Thus, the small subunit is mainly decoding mRNA, but the large subunit mainly has a catalytic function. In the large subunit, rRNA performs the function of an enzyme and it is termed as a ribozyme. In prokaryotic ribosomes, the small subunit, 30S, is made of one ribosomal RNA and 21 ribosomal proteins, while the large subunit, 50S, is made of two ribosomal RNAs and 31 ribosomal proteins. The two subunits fit snugly in a slot, through which a strand of the mRNA molecule runs between them, after the fashion of a tape through a cassette player. The ribosome glides through the mRNA tape, which then carries out its instructions bit by bit, linking the amino acids together, one by one in a specified sequence, until an entire protein has been synthesized. The ribosomal RNAs are programmed to recognize the codon as it appears on mRNA. When the production of a specific protein is finished, the two subunits of ribosome drift apart [89,90]. Ribosomes only have a temporary existence. The large and small subunits of the ribosome undergo a cycle of association and dissociation during each round of translation. Similarly, once the protein is made, mRNA is broken down and the nucleotides are recycled.

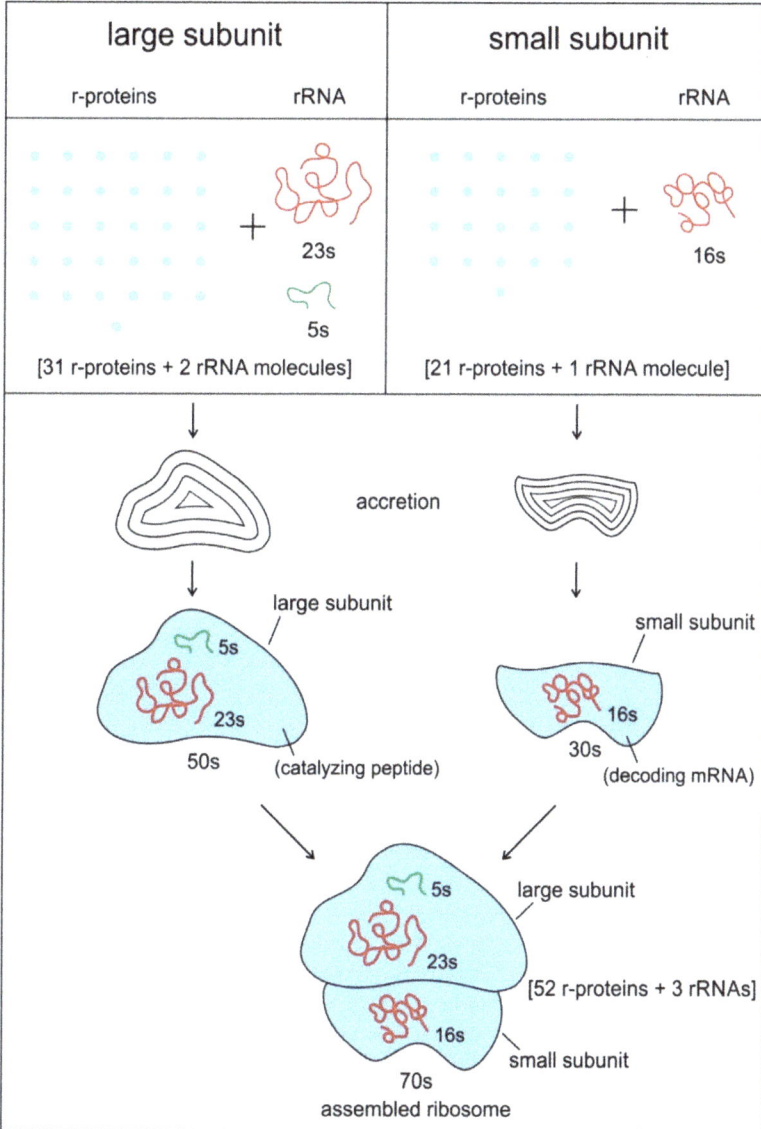

Figure 6. The origin of the ribosome. The ribosome consists of two subunits each with specific roles in protein synthesis. The basic form of the ribosome has been conserved in evolution. Perhaps, the early ribosome was similar to that of modern prokaryotes, which is a large ribonucleoprotein complex of three rRNAs and 52 r-protein molecules. Although ribosomal proteins greatly outnumber ribosomal RNAs, the rRNAs pervade both subunits. There is now evidence that rRNA interacts with mRNA or tRNA at each stage of translation, and that ribosomal proteins are necessary to maintain rRNA in a structure in which it can perform the catalytic functions. Most likely, the symbiotic interactions of ribosomal RNAs and ribosomal proteins gave rise to ribosomes, which grew by accretion. However, there is some controversy whether the small or large subunit appeared first. In our view, both units coevolved by the accretion of ribosomal RNAs and ribosomal proteins.

The ribosome evolved prior to the emergence of DNA and the cellular life in the peptide/RNA world. Ribosome evolution is intricately linked to the prior evolution of mRNA, tRNA, and primitive form of the genetic code and translation. The origins and evolution of ribosomes remain printed in the biochemistry of extant life and in the structure of the ribosome. Most theories propose that the ribosome was a functional takeover of a primitive RNA-based translation system in a coordinated series of chemical reactions. RNA is thought to be responsible for the bulk of the ribosome's work. Recent structures of ribosomes have unambiguously shown that the essential functions of the ribosome, such as decoding, peptidyl transfer, and translocation, all appear to be mediated by RNA [91]. Phylogeny of ribosome suggests that the origin of rRNA is linked to accretionary tRNA building blocks that gave rise to functional rRNA [20]. The decoding center where mRNA is located in the small subunit and it is primarily formed from 16S rRNA. The rRNAs are folded into highly compact and precise three-dimensional structures that form the core of the ribosome. The rRNAs give the ribosome its overall shape. Thus, the widely popular concept of 'the ribosome is a ribozyme' was born; the ribozymes must have preceded coded protein synthesis [39].

In recent times, the role of proteins in the origin of ribosomes is gaining currency, implying that the ribosome may have first originated in a peptide/RNA world, where both amino acids and a variety of enzymes were available [9,13–22,92]. Ribosomal proteins are not passive contributors to ribosome function. They are generally located on the surface, where they fill the gaps and crevices of the folded rRNA. The main role of the ribosomal proteins seems to fold and stabilize the rRNA core, while permitting the changes in rRNA conformation that are necessary for this RNA to catalyze efficient protein synthesis. The ribosomal proteins provide the structural framework for the 23S rRNA, which actually carries out the peptidyl transferase reaction. In the absence of ribosomal proteins, 23S rRNA is unable to serve as a peptidyl transferase activity. The assembly of large and small subunits that are dependent upon ribosomal proteins [17,92]. Several ribosomal proteins assist in the assembly of the large subunit by providing unstructured, highly positively charged protein sequences that bind amino RNA segments together and extend to the center of the subunit [92]. These extensions cooperatively fold with ribosomal proteins to produce the small subunit.

Why would an RNA structure evolve to make proteins if the protein did not already exist that would confer a selective advantage on the ribosomes capable of synthesizing them? The availability of even simple proteins could have significantly enlarged the otherwise limited catalytic function of RNA. Many prebiotic protein enzymes carried out several key functions in the primitive translation system. Moreover, the production of simple proteins had already commenced through the interactions of mRNA/tRNA/aaRS, before the origin of ribosome (Figure 5). Perhaps ribosomal proteins were synthesized during the primitive translation system, which were then recruited to build the ribosome step-by-step. RNAs and proteins developed a symbiotic relationship to create ribosomes in the peptide/RNA world [17–19]. These r-proteins took an active part in stabilizing the evolving ribosomes and in interacting with many rRNA sequences. Because the number of proteins greatly exceeded the number of RNA domains, it can hardly come as a surprise that every rRNA domain interacted with multiple proteins in ribosomes [91]. Ribosomes are not entirely ribozymes, but are more accurately ribonucleoprotein (RNP), a complex that can have as many as 62 r-proteins, with only three rRNA molecules (Figure 6). Virtually all r-proteins are in contact with the rRNA. Accordingly, it makes sense that this assemblage is a result of a long and complicated process of gradual coevolution of rRNAs and r-proteins. Both the assembly and synthesis of the ribosomal components must occur in a highly coordinated fashion [20]. Their phylogenetic analysis reveals that the ribosomal protein/rRNA coevolution manifested throughout the prebiotic synthesis process, but the oldest protein (S12, S17, S9, L3) appeared together with the oldest rRNA substructures that were responsible for both the decoding and ribosomal dynamics 3.3-3.4 Ga. Although protein synthesis is largely carried out by different kinds of RNA molecules within the ribosome, such as mRNA, tRNA, rRNA, and peptidyl transferase, aminoacyl synthetase (aaRS) played a crucial role as a protein enzyme that attached the appropriate amino acid onto its tRNA during protein synthesis. The synthetase, in terms of importance, is equal to the tRNAs in the decoding process, because it is the combined action of synthetases and tRNAs that allows each codon in the mRNA molecule to associate with its proper amino acid. Similarly, both rRNA and the 50S subunit proteins are necessary for the peptidyl transferase activity during peptide bond formation, but the actual act of catalysis is a property of the ribosomal RNA of the larger subunit (Figure 6). The cumulative conclusion that seems to be most in accord with biochemical evidence is that the peptide/RNA world preceded ribosome.

The accretion model describes the origin and evolution of ribosomes [20]. Given that the ribosome is quite ancient, it is likely that rRNAs and r-proteins coevolved to build this complex nanomachine. Ribosomes, like the rings of a tree, contain the record of their history, spanning four billion years. Like rings in the trunk of a tree, the ribosome contains components that functioned on in its early history. It accreted to grow bigger and bigger over time. However, the older parts froze after they accreted, like the rings of a tree (Figure 6). Recent phylogenetic work on ribosomal history suggests that both RNAs and proteins contributed to the formation of the ribosome core through accretion, recursively adding expanding segments [20,21]. Ribosomes contains life's most ancient and abundant polymers, the oldest fragments of RNA and protein molecules. It most likely a molecular relic of the peptide/RNA world [9].

Both ribosomal subunits have separate functions. Peptide bond formation occurs at the peptidyl transferase center (PTC) of the large subunit, whereas the mRNA sequences are decoded on the small subunit. mRNA decoding contributes to the specificity of protein synthesis on the ribosome. In isolation, both of the subunits can perform their respective functions (Figure 6). By itself, the large subunit will catalyze the formation of peptide bonds between aminoacyl-tRNA-like substrates. By itself, the small subunit binds mRNA, and when mRNA is bound, it will bind tRNAs in a codon-specific manner. In an RNA world scenario, the ribosome originated in the peptidyl transferase center of the large ribosomal subunit [93,94]. There are no r-proteins that are close to the reaction site for protein synthesis. This suggests that the protein components of the ribosome do not directly participate in the peptide bond formation catalysis, but rather the proteins act as a scaffold that may enhance the ability of rRNA to synthesize protein. Ribosomes themselves, although being fundamentally ribozymes in nature, still require r-proteins to fold their rRNAs into biologically active conformations and to optimize the speed and accuracy of their functions [85]. The ribosomal surface is an integrated patchwork of rRNAs and r-proteins.

Currently, there is a debate regarding the origin of the ribosomal subunits: which unit came first, the small or the large subunit? It is likely that the PTC of large ribosomal subunit evolved from pre-tRNA molecules by duplication of the minihelix [81]. In this view, the simple function of peptide bond formation at the PTC site came first, and the specifications that were based on the codon sequence came later. In other words, the large subunit of the ribosome came first, followed by the addition of the small unit. However, these proposals do not link the protein synthesis to RNA recognition and do not use a phylogenetic comparative framework to study ribosomal evolution.

Other authors who favor the small unit of ribosome as the first, deduced from the phylogeny of ribosome, offer a contrasting view of the origin of ribosomal subunits [20]. The study suggests that the components of the small ribosomal subunit evolved earlier than the catalytic peptidyl transferase center of the large ribosomal subunit. In this view, the ribosomal RNA and proteins coevolved tightly, starting with the oldest proteins (S12 and S17) and the oldest rRNA helix in the small subunit (the ribosomal ratchet responsible for ribosomal dynamics), ending with the modern multi-subunit ribosome. A major transition in the evolution of ribosomes at around 4 Ga brought independently evolving subunits together by infolding the inter-subunit contacts and interaction with full cloverleaf tRNA structures.

In our view, both the small subunit and the large subunit of the ribosome simultaneously appeared and worked together, because the decoding of mRNA and the peptide bond formation were both essential components during protein synthesis. These two subunits might have coevolved to join during translation and separate after protein synthesis. The rRNAs are folded into highly compact, precise three-dimensional structures to form the core of the ribosome, whereas the r-proteins are generally located on the surface, where they fill the gaps and crevices of the folded RNA and act to fold and stabilize the core [95]. As these two subunits expanded through accretion, eventually arriving at the size of the bacterial ribosome, the accretion stopped, they then bound together during protein synthesis, and finally spilt apart when the ribosome finished reading its mRNA molecule (Figure 6).

If the fundamental functions of the ribosome are based on rRNA, then why are there so many ribosomal proteins, some of which are highly conserved? One explanation is the rRNA does not fold into its functional state in the absence of r-proteins. Another reason for the presence of proteins in ribosomes is that they improve the efficiency and accuracy of the translation [93]. Both rRNAs and r-proteins work cooperatively in ribosomes to perform the multitask procedure of protein synthesis. Harish and Caetano-Anolles suggested that functionally important and conserved regions of the ribosome were recruited and could be relics of an ancient peptide/RNA world [20]. The corollary is that a fully functional biosynthetic mechanism that is responsible for primordial peptides and ancient r-proteins must have existed that in time was superseded by the ribosome.

According to this accretionary model, very early in ribosomal evolution, rRNA helices interacted with r-proteins to progressively form a core that mediated nucleotide interactions, which later served as the center for the coordinated and balanced RNP (ribonucleoprotein) accretion that evolved into our modern ribosomal function [20]. The early existence of smaller functional units of ribosome, which are capable of carrying out different translational steps, such as peptidyl transferase, decoding, and aminoacylation, along with the development of A, P, and E sites for the positioning of tRNA molecules, can be inferred from the phylogeny. These small functional RNA/protein units were incrementally accreted and then refined by the incorporation of additional rRNA and r-protein molecules. Similarly, the first atomic resolution of the larger of the two subunits of the ribosome suggests that the RNA components of the large subunit accomplish the key peptidyl transferase reaction [96]. Thus, rRNA does not exist as the framework to organize catalytic proteins. Instead, the proteins are the structural units and they help to organize the key ribozyme. A 'pure' RNA world is incompatible with the existence of the coevolutionary pattern that is proposed for ribosomal molecules.

Perhaps rRNA, such as noncoding ribozymes, acquired amino acids as cofactors, making them more efficient catalysts. By using cofactors, the range and specificity of catalytic activity can be increased. Ribozymes would have been in greater need of cofactors than protein enzymes, because, without them, the range of reactions that they can catalyze is much smaller [46].

In our endosymbiotic model, rRNAs and r-proteins were brought into close proximity within the plasma membrane to form the building block of the primordial ribosome. The origin of the ribosome precursor through fusion and the accretion of the key components of these ribosomal RNA and protein molecules is the likely scenario. The rRNA and r-protein molecules began to fuse because of a chiral preference and then formed the rudimentary ribosomes. Once the core of the ribosome formed, the mRNA and tRNA molecules were recruited to help in translation through a trial and error method. Once a true mRNA and the core small subunit of ribosome were in place, the ribosome would become increasingly complex by adding early conserved rRNA and r-proteins. Ribosomal proteins played an important role in supporting the ribosome structure and in promoting translation. With the onset of operational coding, tRNA began to assemble amino acids into long chains of proteins. Here, we suggest that a ribosome-like entity was one of the key intermediates between prebiotic and cellular evolution, which formed by endosymbiosis and the fusion of rRNA and r-protein molecules. Once ribosomes were installed inside the protocell membranes, the translation system was greatly improved.

In vitro constructions of ribosomes can shed new light on the mechanism of protein synthesis and provide deeper insights into the way that nature has assembled this complex machine. Working with *E. coli* cells, natural ribosomal proteins were combined with synthetically made rRNA, which self-assembled in vitro to create semi-synthetic, functional ribosomes [96–98]. Comprising 57 parts—three strands of rRNAs and 54 proteins—an artificial ribosome (termed Ribo-T), in which two subunits are tethered together by a short length of RNA, is able to carry out normal translation and pump out custom-made proteins. The ability to make ribosomes in vitro is a process that mimics nature and opens up new avenues for the study of ribosome synthesis, suggesting the coevolution of ribosomal RNAs and proteins.

5.9. Protein Synthesis

We have now reviewed the emergence of all major components of the translation machinery for protein synthesis. Translation of the mRNA template converts nucleotide-based genetic information into the 'language' of amino acids to create a protein product. Translation requires the input of an mRNA template, tRNAs, aminoacyl-tRNA synthetases, ribosomes, and various enzymatic factors. The tRNAs function as the adaptor molecules that transport amino acids to ribosomes in response to codons in mRNAs, where peptidyl transferase catalyzes the addition of amino acid residues to the growing protein chain in protein synthesis by means of peptide bonds. The ribosomes serve as the sites for protein synthesis and they link amino acids together in the order specified by mRNA. They always translate the mRNA from the 5′ to the 3′ direction, like a sliding machine.

Proteins have a modular chemical structure that allows for the construction of widely different molecular machines using the same basic set of amino acids, each with a different size and chemical character. Protein synthesis requires the concerted effort of dozens of different enzymes. 20 tRNA molecules, each with their own dedicated synthetase enzyme, are built for 20 amino acids. Modern protein synthesis proceeds with the participation of 20 amino acids, tRNA, mRNA, ribosomes, various enzymes, including aminoacyl-tRNA synthetase, ribozymes, peptidyl transferase, and a considerable number of proteinous factors, ATP, GTP, etc. More than 120 species of RNAs and proteins are involved in the process of protein synthesis [65]. These biomolecules were related, encapsulated, and interacted with each other in complex ways, like an autopoietic machine. Yet, the whole series of molecules in the translation process functioned with astounding precision, in a kind of molecular choreography, which gave birth to the universal genetic code.

The structure and function of the modern ribosome during translation are well-known in the literature and they will not be repeated here [77,99]. In the ribosome, there are three stages and three operational sites that are involved in the protein production line and all work in harmony. During the initiation stage, a small ribosome subunit links onto the 'start end' of an mRNA strand. Aminoacyl-tRNA also enters site A of the ribosome. The production of the protein has now been initiated. The second stage, elongation, consists of joining amino acids to the growing protein chain, according to the sequence that was specified by the message. The incorporation of each amino acid occurs by the same mechanism. In the termination stage, the ribosome reaches the end of the mRNA strand, a terminal, or 'end of the protein code' message. This registers the end of production for the particular protein that was coded by this strand of mRNA (Figure 7).

Translation is not the end of the protein synthesis process. Once released from the ribosome, the long chain of amino acids will spontaneously fold in intricate contortions into a unique three-dimensional configuration and proper characteristic shape: some parts form sheets, while others stack, curl, and twist into spirals. The sequence of amino acids determines the shape and conformation of a protein and, thereby, all of its physical and chemical properties. A protein molecule spontaneously folds during or after biosynthesis, but the folding process depends on the solvent, the concentration of salts, the temperature, the possible presence of cofactors, and the molecular chaperons [99]. Proteins must fold in specific ways to function properly.

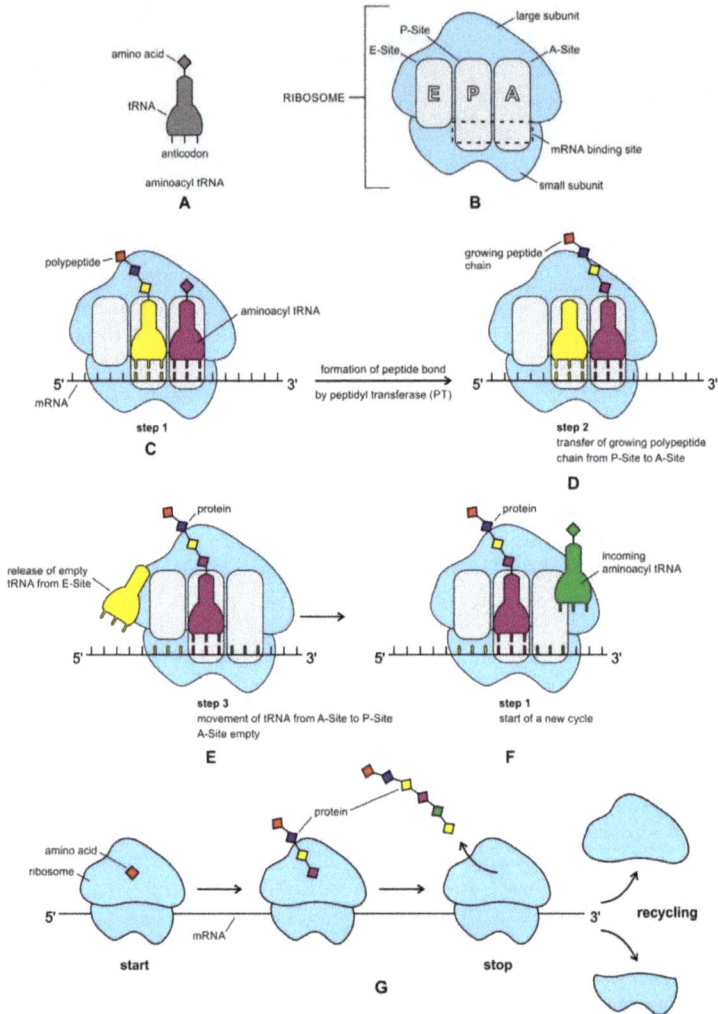

Figure 7. The translation machinery of the ribosome, where the mRNA message is decoded. The ribosome provides the substrate for controlling the interaction between mRNA and aminoacyl-tRNA. (**A**), an aminoacyl tRNA, with appropriate anticodon. (**B**), each ribosome has a binding site for mRNA, and three binding sites for tRNA. The tRNA binding sites are designated E-, P-, and A-sites (for exit, peptidyl-tRNA, aminoacyl-tRNA, respectively). The small subunit contains the binding site for mRNA. Translation takes place in a four-step cycle (**C–F**) that is repeated over and over during the synthesis of protein. (**C**), in step 1, an aminoacyl-tRNA, with appropriate anticodon, enters the vacant A-site on the ribosome where it hybridizes with a codon. (**D**), in step 2, the carboxyl end of the protein chain is uncoupled from the tRNA at the P-site, then joined by a peptide bond to the free amino group of the amino acid linked to the tRNA at the A-site. This reaction is catalyzed by an enzymatic site in the large subunit, called peptidyl transferase (PT). (**E**), in step 3, a shift in the large subunit (shown by arrow) relative to the small subunit in the 3′ direction, moves the two tRNAs into the E- and P-sites of the large unit, and then ejects the empty tRNA from E-site. (**F**), in step 4, the small subunit moves exactly three nucleotides along the mRNA molecule, bringing it back to its original position relative to the large subunit. This movement resets the ribosome with an empty A-site so that the next aminoacyl-tRNA molecule can bind. The cycle repeats when the incoming aminoacyl-tRNA binds to the codon of the A-site; (**G**), summarizes the life cycle of the ribosome during its translation [77,99].

6. The Origin and Evolution of the Genetic Code

A code is a set of rules that establish a correspondence between the objects of two independent entities. The genetic code is a correspondence between codons and amino acids. It is the universal language of life. It defines the rules by which information that is stored in mRNA sequences is translated into the corresponding amino acids sequences to proteins. The genetic code is universal; it is the same for all organisms, from simple bacteria to eukaryotes to animals to humans. The genetic code maps the three-letter words, in a four-letter alphabet, of the mRNA language (4^3 = 64 codons) to the protein language alphabet of 20 amino acids. Before focusing on the origin of the code, let us consider its most important properties. The universal genetic code consists of 64 codons that specify 20 amino acids, and the start and stop sites. The large number of codons is due to redundancy in the code; that is, several codons may specify the same amino acids. All but two of the amino acids (methionine and tryptophan) have more than one codon, many have two, one has three, several have four, and two of them have six codons (Table 1). The amino acids that are more often used in proteins are specified by a greater number of different codons. No codon goes unused. The genetic code has redundancy, but no ambiguity. For example, although codons GAU and GAC both specify aspartic acid (redundancy), neither of them specifies any other amino acid (ambiguity). The genetic code is nonoverlapping, meaning that the 'words' follow each other without gaps or overlaps. Each codon in mRNA specifies one amino acid in the protein product. The code is also comma-free. There are no commas or other forms of punctuation within the coding regions of mRNA molecules. During the translation, the codons are consecutively read. The code is ordered. Multiple codons for a given amino acid and codons for amino acids with similar chemical properties are closely related, which usually differ by a single nucleotide [65,66]. The arrangement of the genetic code is distinctly non-random and is such that neighboring codons are assigned to amino acids, with similar physical properties. Hence, the effects of translation error are minimized with respect to reshuffled codes. The digital information in the linear sequence of nucleotides in mRNA is translated into analog sequences of amino acids in proteins according to the genetic code [61]. The vast majority of living organisms follow the same universal genetic code. The most important exceptions to the universality of the code occur in the mitochondria of mammals, yeast, and several other species.

The table of universal (or standard) genetic code, showing the association of each three-letter code to its respective amino acid, is a little dictionary, a Rosetta stone, just as Morse code relates the language of dots and dashes to the twenty-six letters of the alphabet (Table 1). The very existence of two languages (with the code being a translational intermediary) implies a directional course of evolution. The table is not a random accident, but rather it is the result of very specific selection. The mRNA is a linear polymer of four different nucleotides and it is consecutively read in groups of three nucleotides (codons) to form the 'words' of the message without any comma. This is known as an 'open reading frame'. Every sequence in mRNA can be read in its 5′ → 3′ direction in three reading frames. Each three-base codon stands for a single amino acid, so there are 64 possible combinations of three nucleotides. The arrangement of the codons in the universal code is highly nonrandom. The code has been confirmed by several experimental methods [77,99].

The origin of the genetic code remains elusive, even though the full codon catalog was deciphered over 50 years ago [100]. It is still not clear why the genetic code might have originated in the prebiotic world leading to the *information* age. Perhaps some critical biomolecules in the prebiotic environment were cooperative and they began to attract each other. A stereochemical relation between some amino acids and cognate anticodons/codons is likely to have been an important factor in the origin of the genetic code. The biosynthetic relationships between amino acids and RNAs are closely linked to the organization of the genetic code. Different species of RNAs and proteins were manufactured by molecular machines during this stage, and all of the manufacturing processes require not only physical quantities, but also additional entities, like sequences and coding rules [101].

Table 1. The evolution of the Universal Genetic Code is seen in three distinct stages. The nucleotide sequence of an mRNA (in a digital format) is translated into the amino acid sequence of a protein (in an analog format) via the genetic code. Genetic information is encoded in mRNA using codons, comprised of four bases: uracil (U), cytosine (C), adenine (A), and guanine (G). The Figure shows the evolutionary pathway going from a GNC code (four codons) through a SNS code (16 codons), to the Universal Genetic Code. To decode a codon, the first letter is matched in the left column, the second letter on the top row, and the third letter on the right column. The 64 codons, along with the amino acid or stop signal that they specify, are shown in the boxes. All but two of the amino acids (methionine and tryptophan) have more than one codon. Note that in the mRNA, uracil replaces the thymine found in DNA. A (after [87]), B (after [88]); C, the Universal Genetic Code. Instead of a conventional representation, the modern genetic code is shown reflecting the order of codon occurrence from GNC, to SNS, to a modern code (columns G and U inverted).**Table 2.** Universal Genetic code showing numerical codons. **Table 3.** Universal Genetic code showing numerical codons **Table 4.** 20 Primary Amino Acids in the Genetic Code and their corresponding numerical codons.

The GNC primitive genetic code
[4 codons]

	U	C	A	G	
G	Alanine	Alanine	Aspartic acid	Serine or Glyine	C

	U	C	A	G	
G	Valine	Alanine	Aspartic acid	Serine or Glyine	C

The SNS primitive genetic code
[16 codons]

	U	C	A	G	
G	Valine	Alanine	Aspartic acid / Glumatic acid	Serine / Glumatic acid	C / G
C	Valine	Glumatic acid	Glumatic acid	Glumatic acid	C / G

	U	C	A	G	
G	Valine	Alanine	Aspartic acid / Glumatic acid	Glycine	C / G
C	Leucine	Proline	Histidine / Glutamine	Arginine	C / G

(Di Giulio 2008)   (Ikehara 2002)

The universal genetic code
[64 codons]

Second letter

		U	C	A	G	
	G	Valine GUU GUC GUA GUG	Alanine GCU GCC GCA GCG	Aspartic acid GAU GAC / Glutamic acid GAA GAG	Glycine GGU GGC GGA GGG	U C A G
First letter	C	Leucine CUU CUC CUA CUG	Proline CCU CCC CCA CCG	Histidine CAU CAC / Glutamine CAA CAG	Arginine CGU CGC CGA CGG	U C A G
	A	Isoleucine AUU AUC AUA / Methionine AUG (start codon)	Threonine ACU ACC ACA ACG	Asparagine AAU AAC / Lysine AAA AAG	Serine AGU AGC / Arginine AGA AGG	U C A G
	U	Phenylalanine UUU UUC / Leucine UUA UUG	Serine UCU UCC UCA UCG	Tyrosine UAU UAC / (stop codon) UAA UAG	Cysteine UGU UGC / (stop codon) UGA / Tryptophan UGG	U C A G

The accuracy of the genetic code translated depends on two-steps in protein synthesis: precise decoding of mRNAs and accurate synthesis of aminoacyl-tRNAs. aa-tRNAs are made by aaRSs, which match specific amino acids with the corresponding tRNAs, as defined by the genetic code. Thus, the crucial feature of the genetic code is the attachment of particular amino acids to tRNA molecules, a step that is carried out by assignment enzymes, such as aminoacyl-tRNA synthetase. We are suggesting this attachment first occurred between pre-tRNA and specific amino acid and it was carried out by pre-aaRS enzyme. Since various enzymes were available in the Peptide/RNA world, we propose that the functional enzyme for binding an amino acid to its corresponding tRNA was pre-aaRS from the beginning, not ribozyme, as previously suggested by other workers [46].

Although we know which codon encodes which amino acid, we do not know why the specific codon assignments take their actual form. Why are there exactly four nucleobases in mRNA? Why does life use 20 amino acids for making proteins, when 70 amino acids were available in the prebiotic environment from the cosmic source [4]? If the code evolved at a very early stage in the history of biosynthesis, perhaps during its prebiotic phase, the four nucleotides in mRNA and 20 amino acids in proteins may have been the most promising case for optimization by natural selection for chemical reactions that are relevant at that stage. Perhaps it is simply a 'frozen accident', a random choice that just locked itself in, and remained, mostly, unchanged once the optimal design was reached [27]. Any change of codon reassignment may be lethal, because it would trigger mutation, which would be dispersed throughout all proteins in the cell. This accounts for the fact that the code is universal in all organisms from bacteria to humans. To account for the uniform code in all organisms, one must assume that all life evolved from the last universal common ancestor (LUCA). Since then, the universal code remains unchanged for the last four billion years.

6.1. The Origin of the Genetic Code

Although multiple hypotheses have been proposed to explain why codons are selectively assigned to specific amino acids, empirical data is extremely rare and difficult to obtain, leaving many theories in the realm of conjecture. Three main concepts on the origin and evolution of the genetic code are:

(1) stereochemical theory, according to which codon assignments are dictated by physico-chemical affinity between amino acids and the cognate anticodons or codons; perhaps, tRNA molecules matched their corresponding amino acids by their stereochemical affinity [72,73,100,102–104]. Simply put, the hypothesis proposes that symbols in the genetic code (anticodons or codons) may directly bind to the objects (amino acids) that they stand for.

(2) Coevolutionary theory, which suggests that the code structure coevolved with the amino acid biosynthesis pathways [68,86,105]. This theory suggests that the genetic code is primarily an imprint of the biosynthetic pathways forming amino acids. There are two generations of amino acids: the ten primary or primitive amino acids were formed under prebiotic conditions; they serve as a starting point for the synthesis of the remaining ten amino acids that derived from the first set. What happened afterwards, is that some primitive systems evolved the ability to manufacture the secondary amino acids, and also eventually the primary amino acids.

(3) Adaptive theory, which postulates that the structure of the code was shaped under selective forces that made the code maximally robust, usually some kind of error minimization [76].

Many other models have emerged as addendums to one of the main models or as some form of hybrid. We believe that these three theories are not mutually exclusive and are compatible with the peptide/RNA world, because aaRS play a crucial role in translation. Without aaRS, however, tRNA molecules could not be matched with their corresponding amino acids.

It has long been conjectured that the universal genetic code (Table 1) evolved from a simpler primordial form that encoded fewer amino acids [27]. The earliest peptides started with 10 amino acids, which have been produced in prebiotic chemistry experiments [69,84]. Any model for the evolution of an early code and translation apparatus in the pre-DNA stage will have to provide conditions that allow for tRNA and mRNA of various enzyme factors not only to coexist, but also to coherently grow and to evolve optimal function. We have suggested that the first amino acid was incorporated as a cofactor by aminoacylated ribozyme via bridge peptide, thus initiating proto-translation and the genetic code. The ribozyme would give rise to tRNA and bridge peptide to aaRS [24]. Our proposed biochemical pathways for the origin of the translation favor three distinct phases for the origin of the genetic code. The early stage of coding might have been initiated in a peptide/RNA world by stereochemical interactions between pre-tRNA and amino acids, leading to the birth of pre-mRNA molecules for storing genetic information. The activating enzymes were pre-aaRS, precursors to modern aminoacyl-tRNA synthetases, which bound specific amino acid to corresponding pre-tRNA molecules. Subsequently, the code expanded with the involvement of more amino acid-tRNA ligations by aaRS and the progressive elongation of the mRNA strand for accommodating increasing genetic information. Finally, the code further expanded for redundancy and was optimized through codon reassignment with the emergence of ribosomes, which bound mRNA and tRNA to synthesize proteins (Table 1).

The stereochemical hypothesis postulates that a physico-chemical affinity between amino acids and cognate anticodons or codons is determined by the structure of the code [75,102–106]. The close linkage between the physical properties of amino acids and tRNA molecules was likely an essential step for the origin of code. A stereochemical relation between some amino acids and cognate anticodons/codons is likely to have been a significant influence in the earliest assignments [107]. It is possible that the chiral d-sugars in RNA attracted the chiral L-amino acids as a stereo pair. An exhaustive analysis of the stereochemical concept suggested that the genetic code originated before translation [108]. The stereochemical theory is supported by RNA aptamer experiments, in which RNA molecules evolved to bind specific amino acids [96]. Such experiments have provided critical empirical data, demonstrating the association of codon triplets with amino acids.

It is generally believed that specificities for some amino acids come from stereochemical interactions. However, the stereochemical hypothesis possesses its own set of problems. First of all, it is not preserved as a way of recognition today. If it was a factor for codon/amino acid recognition at the beginning, then it would be extremely important to be replaced later by protein type of recognition. Second, as demonstrated in experiments [104], for aptamer-amino acid interactions, five out of eight amino acids have close to random specificities, with just one exception—arginine—showing a significant specificity to recognize its own codon. However, arginine is not even in the list of initial codon formation (Table 1). As discussed earlier, bridge peptide is capable of stimulating interaction between specific RNAs and specific amino acids [24]. Bridge peptide seems to be an alternative to stereochemical way of recognition in the beginning.

Other experiments suggest that anticodons are selectively enriched near their respective amino acids in the ribosomal structure and such enrichment is correlated with the universal genetic code [109]. Ribosomal anticodon-amino acid enrichment reveals that specific codons were reassigned during code evolution. These authors concluded that anticodon-amino acid interactions shaped the evolution of the genetic code.

tRNA serves as the physical link between the mRNA and the amino acid, because mRNA could not make direct bond with an amino acid. It is a decoding device that reads the triplet genetic code of mRNA and it causes the insertion of codon specific amino acids in a growing protein chain during the process of translation. The specific coding between codon and amino acids takes place in a two-step process via tRNA. For each amino acid, there is a corresponding tRNA molecule for which it has the intrinsic affinity. tRNA molecules function as adaptors by mediating the incorporation of proper amino acids into proteins in response to specific nucleotide sequences in mRNA. The amino acids are attached to the correct tRNA molecules by a set of activating enzymes, aminoacyl-tRNA synthetases [110].

Since all tRNAs have similar structures, the identification must take place on a sequence level in combination with subtle structural variations. In most known cases, the anticodon bases are part of this set of identity elements. In *E. coli*, the tRNA species for 17 out of 20 amino acids are recognized by their anticodons [111]. Its matching codon is also recognized as data, the information of a specific amino acid. tRNA-codon recognition has always been assumed to be result of base-pairing. Anticodon-codon pairing might have initiated the first primitive translation.

It is generally believed that the linking of amino acids to tRNAs played a crucial role in the origin of coding and translation. The original amino acid-binding motifs could have been the actual anticodons of tRNAs. Several authors have proposed that abiotic tRNA molecules could have bound some abiotic amino acids to either improve stability or expand their functional capabilities, or both [89,107,112]. Without this initial amino acid binding site, it is difficult to see how else the tRNA molecules could have become involved with coding specific amino acids. tRNA-amino acid pairing interactions were a prelude to the code. Thus, our first clue to the origin of the code is to decipher how primordial tRNAs and amino acids were related by molecular recognition and chemical principles [113].

In contrast to stereochemical hypothesis, the coevolution theory suggests that the original genetic code specified a small number of abiotic simple amino acids, and that, as more complex amino acids were synthesized from these precursors, some codons that encoded a precursor were ceded to its more complex products. Wong [69,86,105,114] championed the coevolution theory, and further expanded by Di Giulio [87]. It proposes that primordial proteins consisted only of those amino acids that were readily obtainable from the prebiotic environment, representing about half of the twenty amino acids of today, and the missing amino acids entered the system as the code expanded along the pathways of amino acid biosynthesis. The coevolution theory postulates that prebiotic synthesis could not produce 20 modern amino acids, so a subset of the amino acids had to be produced through biosynthetic pathways before they could be opted for expanded genetic code and translation [75,87]. There are two types of amino acids, depending on whether they were supplied by the prebiotic environment (Phase 1) or were biosynthetically produced (Phase 2) [68,105]. The first phase of amino acids consists of glycine, alanine, serine, aspartic acid, glutamic acid, valine, leucine, isoleucine, proline, and threonine. Phase 2 amino acids include phenylalanine, tyrosine, arginine, histidine, tryptophan, asparagine, glutamine, lysine, cysteine, and methionine. The first phase of amino acids naturally emerged through prebiotic synthesis in the vent environment, before the emergence of ribosomes. They have been identified in meteorites. The ranks of amino acids in this list strongly correlate with the free energy that is available in the vent environment for their syntheses: the most thermodynamically efficient are on the top of the list. These 10 amino acids are considered as old and they were represented in the first stage of protein synthesis [69]. They would play important roles in the primitive GNC-SNS code (Figure 8). Phase 2, the amino acids entered the code by means of biosynthesis from the Phase 1 amino acids with the emergence of tRNA molecules, aminoacyl transferase enzyme, and ribosomes [68,82,105,108].

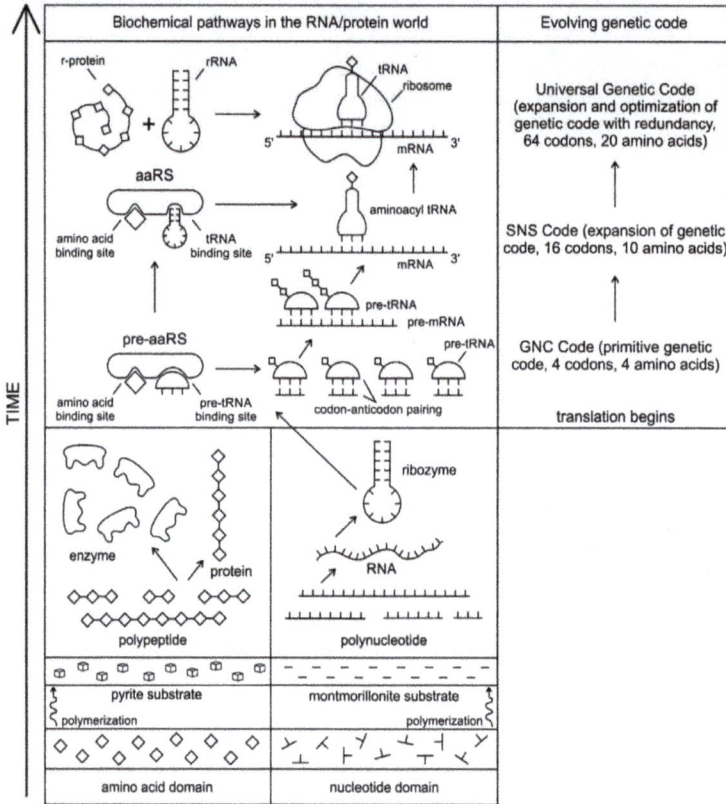

Figure 8. The inferred biochemical pathways for the origin of translation and the genetic code in the RNA/peptide world. The hydrothermal crater vent was crowded with several monomers such as amino acids and nucleotides, which were polymerized on the mineral substrate to form various peptides and RNAs. As ribozymes evolved into pre-tRNAs, each pre-tRNA molecule captures specific amino acid, assisted by pre-aaRS enzyme. Eventually, anticodons of pre-tRNAs created custom-made pre-mRNAs for the storage of genetic information. The interaction between pre-tRNA and pre-mRNA generated small protein chain by rudimentary translation and GNC primitive code with four codon and four amino acids. With the refinement of translation, pre-tRNA evolved in tRNA and pre-mRNA to mRNA with the expansion SNS code with 16 codons, and 10 amino acids. Finally, as ribosome appeared by fusion of ribosomal proteins and RNA, it facilitates high-fidelity translation, leading to universal genetic code with 64 codons and 20 amino acids.

6.2. Early Stage of Code Evolution: GNC Code

The early phase of the evolution of the genetic code is characterized by low fidelities of replication and translation, as well as by an initially low abundance of efficiently replicating units [115]. Hypercyclic organization offers multiple advantages over any other kind of structural organization. This hypercycle model can be built to provide realistic precursors, such as pre-tRNA and pre-mRNA. The interaction between pre-mRNA and pre-tRNA molecules is the beginning of the first stage of the biosynthesis of the templated protein chain, encoded in pre-mRNA. These new generations of amino acids are not only template directed but also sequence-directed [116]. Here, we propose that the interaction between pre-tRNA and amino acids led to the development of the pre-mRNA strand and the primitive GNC genetic code [88,116].

A primordial code must have a certain frame structure, a grammar of rules, otherwise message cannot be consistently read. The GNC hypothesis refers to the origin of genes. It suggests that the universal genetic code originated from a primitive four-amino acid system encoding primitive amino acids (glycine, alanine, valine, and aspartic acid) to create sequence-specific peptide chain [85,88]. The GNC codons include four codons (GGC, GCC, GAC, GUC), which code four primitive amino acids Each letter of GNC represents the following nucleotides: G = G; N = A, U, C, G; and, C = C. The GNC code defines the very earliest phases of the genetic code origin, reflecting the biosynthetic relationships between four amino acids and four codons (Table 1). Perhaps GNC code was promoted by the pre-tRNA/pre-mRNA interaction and coevolution [116,117] (Figure 5A). As amino acids overtook more and more catalytic duties, the genetic information that has been established so far had to be rewritten, a translation into the language of amino acids by specific interaction was inevitable. The translation required a mini dictionary of nucleotide-to-amino acid equivalence; hence, this was the inevitable moment for the genetic code to emerge. To perform protein translation an elaborate machinery of specialized enzymes is necessary. This machinery must be produced step-by-step before translation can take place at all (Figure 5D). It seems reasonable to start this process in simplified form while only using a restricted set of amino acids, such as glycine, alanine, aspartic acid, and valine, which were of prebiotic origin [71].

6.3. Transitional Stage of Code Evolution: SNS Code

GNC code evolved into the second generation of the genetic code, called an SNS type, where N arbitrarily denotes any four RNA bases and S denotes guanine (G) and cytosine (S) [80,108]. SNS is composed of 10 amino acids (glycine, alanine, aspartic acid, valine, glutamic acid, leucine, proline, histidine, glutamine, and arginine) and 16 codons (GGC, GGG, GCC, GCG, GAC, GAG, GUC, GUG, CUC, GUG, CCC, CGC, CAC, CAG, CGC, and CGG) [87,88]. The SNS type code shares similarity with the Phase 1 amino acids that were generated in prebiotic synthesis [64]. The remaining ten amino acids are derivatives of the first ten primitive amino acids. Support for the GNC-SNS primitive genetic code hypothesis comes from the following six indices: hydropathy, -helix, ß-sheet and ß-turn formabilities, acidic amino acid content, and basic amino acid content (Table 1). This early genetic code continued to evolve, maximizing its efficiency, until it arrived at its current state, the universal code. This universal code had the edge over the GNC-SNS primitive code, reliability wise, so natural selection would favor it and, through the process of successive refinement, an optimal code would be reached. The universal code is the optimization of functional efficiency to minimize error during translation.

6.4. Final Stage of Code Evolution: Universal Genetic Code

The present genetic code is most probably the outcome of a long selective process, in which many different codes were tested against each other. As more and more biotic amino acids were synthesized and are available in the vent environments, more complex molecules, such as tRNA, mRNA, and ribosomes emerged and produced Phase 2 amino acids. At this stage, the universal code began to appear (Table 1). A direct correlation has been found between the hydrophobicity ranking of most amino acids and their anticodons. In this stage, ribosomes emerged to facilitate high-fidelity translation. tRNA assigned more codons to mRNA and this led to the emergence of the universal genetic code with 64 codons specifying 20 amino acids [100]. The driving force during this process is not only to minimize translation error, but also positive selection for the increased diversity and functionality of proteins, which are made with a larger amino acid alphabet. With 64 codons, the strand of mRNA became longer, forming a continuous sequence with the start and stop sites for protein synthesis. In the 'codon capture theory', the number of encoded amino acids is kept constant and is equal to 20, and the coding codons change in the evolution, a key role that is played by the anticodon [104,113].

It has been suggested that the universal genetic code with 64 codons originated from the SNS code, which allowed redundancy [85]. Four codon assignments, corresponding to tyrosine, tryptophan, serine, and isoleucine, were newcomers from the SNS code, suggesting that these amino acids are later additions to the code [84]. This idea is consistent with the view that these four amino acids are later additions of code. Undoubtedly, there were many experiments with a variety of coding methods before adopting the current system, in which 61 codons specify 20 amino acids and three additional codons for the start and stop sites.

The code is obviously not the result of a random assignment of codons to amino acids. It has a structure. Synonyms are grouped. The large number of codons is due to redundancy in the code; that is, several codons may specify the same amino acids (Table 1). Some generalizations can be made regarding the redundancy of the code. For example, similar codons specify the same amino acids to reduce the harmful effects of mutation. For example, GUU, GUC, GUA, and GUG all specify valine. Similarly, amino acids that are used more often in proteins are specified by a greater number of different codons. For example, the most common amino acid, leucine, is coded by six codons (UUA, UUG, CUU, CUC, CUA, and CUG), and the relatively rare tryptophan by one codon (UGG) [46] (Table 1). The expanded genetic code is so universal that there is strong evidence that all life on Earth had a single origin in the universal code before the last universal common ancestor (LUCA) evolved.

Between the codon and anticodon, there is a paradox in the expanded genetic code. The 20 amino acids that were found in proteins are specified by 61 different mRNA codons. Instead of containing 61 different tRNAs with 61 different codons, though, most bacterial cells contain some 30-40 different tRNAs. Consequently, many amino acids have more than one tRNA to which they can attach; in addition, many tRNAs can pair with more than one codon [75].

Although some features of the expanded code may reflect the early version, there are others that appear to be adaptive. The genetic code has certain regularities and structures [115]. There is a strong correlation between the first bases of codons and the biosynthetic pathways of the amino acid that they encode. The first letter of the codon is allied to the precursor of the amino acid. The second letter signifies whether an amino acid is soluble or insoluble in water, its hydrophobicity. Amino acids that have U at the second position of the codon are hydrophobic, whereas those that have A at the second position are hydrophilic. Codons for the same amino acid typically only vary at the third position. The third letter is where redundancy lies with eight amino acids with a fourfold degeneracy, where all four bases are interchangeable. In all cases, U and C are interchangeable in the third position. In other words, the third position of the codon is information-free with much flexibility. Many amino acids are specified by more than one codon. Codons for the same amino acid tend to have same nucleotides at the first and second positions, but a different nucleotide in the third position. The relative lack of criticality is related to the fact that the pairing between the anticodons and codons often enjoy a certain flexibility, so those same anticodons can pair with more than one codon, a phenomenon that is known as wobble [118]. This is why the number of tRNAs and, therefore, of anticodons is smaller than the number of 64 codons, usually ranging between 35 to 45 [76]. Once the code was born, the need to minimize errors might have refined it. The code has been optimized over the eons and is not simply the product of chance, but of natural selection.

A key role of the universal genetic code is to maintain integrity and verify the specificity of each mRNA codon to a particular amino acid. There must be an accuracy strategy of cross checking that could reveal that the mRNA codons and amino acids will directly interact. Various authors have suggested that the original amino acid-binding motifs could have been the actual codons rather than anticodons [117,119]. However, contrawise, we believe the codon-amino acid pairing system might have evolved for code verification at a later stage of code evolution. Initially, anticodons developed between the interactions of pre-tRNA and pre-aaRS [111]. The anticodons selected the codons of pre-mRNA by base-pairing (Figure 5). As the genetic code was refined and optimized, verification on the strings of codon on the mRNA strand began, through quality control, to ensure that each tRNA successfully interprets the amino acid information for protein synthesis with a low error rate. Most likely, the amino acid-codon interaction, which is mediated by aptamers, evolved later for keeping the code error free.

7. Coevolution of Translation Machines and the Genetic Code

The contemporary genetic code of protein biosynthesis most likely evolved from a simpler code and process. It has been suggested that the present code is a random accident that is forever frozen in time [27,28], while others have argued that the code, like all other features of organisms, was shaped by natural selection. Both the stereochemical and the coevolutionary hypotheses provide possible mechanisms for the selection of amino acids by RNAs from a large pool of prebiotic soup, which are recruited for protein synthesis. During these processes of selection and recruitment of amino acids, the translation machine and the genetic code evolved. Natural selection has led to codon assignments of the genetic code that minimize the effects of translation errors and mutations during the evolution of the code. The adaptive hypothesis posits that the genetic code continued to evolve after its initial creation, so that the current code maximizes some of the functions.

We accept all three well-known hypotheses—the stereochemical, the coevolutionary, and the adaptive—for the origin and evolution of the genetic code at different stages. We concur with previous researchers that multi-generations of amino acids were sequentially produced [75,88,104,113] as the code expanded, along with the pathways of amino acid biosynthesis. The biosynthetic relationships between different generations of amino acids are closely linked to the evolution of the genetic code. We contend that information evolved along with the translation machines, and it played a vital role in the perfection of the translation process and genetic code. We concur with the view that the genetic code and the translation mechanism evolved together in the prebiotic world [64]. Here, we elaborate this concept of information-based coevolution of the translation machine and the genetic code that may provide a new window into the origins of translation and the genetic code.

The translation machines are an extremely complicated hierarchy of complex macromolecules that are symbiotically related to one another. Yet, the whole functions with remarkable precision. Once the translation machinery complex for protein synthesis is installed step-by-step, information enters into the system via symbiotic interactions of mRNA, tRNA, aaRS, and ribosome. This machinery implements the genetic code. Here, we summarize how such a complex translation machinery would evolve step-by-step into today's protein-synthesizing machinery, starting from the cosmic building blocks in hydrothermal crater-lake environment (Figure 8).

The origin of biomolecular machinery likely centered around the tRNA-amino acid alliance, both being ancient molecules in the prebiotic environment [74,117]. Various 'spare parts' of biomolecules for building translation machinery were available in the prebiotic soup during the *chemical* stage, from which some few were selected, based upon the chemical affinity between macromolecules. tRNA is the oldest and most central nucleic acid molecule. Its coevolutionary interactions with aaRSs define the specificities of the genetic code. The biochemical pathway that is outlined here for the emergence of the genetic code is the simplest and the most straightforward account of the development of RNA-dependent protein synthesis.

The *information* age emerged from a reciprocal partnership between small ancestral oligopeptides and oligonucleotides. They initially contributed to rudimentary information coding as well as catalytic rate accelerations. It begins with the molecular recognition, attraction, and communication between pre-tRNAs and amino acids, mediated by pre-aaRS. The role of tRNA synthetases in the origin of the genetic code is pivotal. It helps the anticodon of tRNA to pair with the right amino acid. It is the matchmaker between tRNA and its corresponding amino acid. Coevolution, the coordinated succession of structural changes that are mutually induced by the increasingly interacting and growing protein and nucleic acid molecules, played an important role during the origin of translation machinery and genetic code. aaRS coevolved with tRNA and tRNA coevolved with mRNA during the rise of the genetic code specificities. A novel mechanism of how tRNAs are recognized by certain aaRS has been suggested [19]. In this view, tRNAs carry two codes: the well-known anticodon and a second one in the acceptor stem (Figure 4E). These two codes are not arbitrary: the nucleic acid sequence of the acceptor stem and the anticodon code for distinct physical properties of amino acids. In other words, the codon/amino acid pairing reflects the different physical roles the different amino acids play in the structure of full, folded proteins. The genetic coding of three-dimensional (3D) protein structures evolved in distinct stages, initially based on the size of the amino acid and later on its compatibility with globular folding in water.

The genetic code and the translation mechanism evolved together in the prebiotic world [64]. Here, we discuss a simple but effective biological information system that works as a translation system. In Figure 8, we show the proposed biochemical pathways and coevolution of translation machinery and the genetic code in three stages. We outline how early RNAs and protein catalysts developed into the universal coding system that we have today. Our outline is necessarily speculative, but it suggests a series of transitional stages of symbiotic relationships between tRNAs and proteins that may have led to the origin and evolution of the genetic code. Since molecular evolution did not leave any fossil record, some of the transitional stages of the translation machinery are now erased by evolution as the final stage appeared. Thus, no record of code evolution has been detected so far.

Among different species of RNAs, tRNA has a very ancient history and it is more closely associated with protein during synthesis. Pre-tRNA, the tRNA's ancestor, likely played a central role of primitive translation early on. A stereochemical relation between some amino acids and cognate anticodons of pre-tRNA must have played an important informational role in the earliest assignments [67,104]. Bridge peptides also help in the aminoacylation of ribozymes with specific amino acids [24].

The origin of the code follows closely the biosynthetic pathways of refining the translation machinery complex in three successive stages in the peptide/RNA world (Figure 9):

- pre-tRNA/ pre-aaRS/pre-mRNA machine;
- tRNA/ aaRS/mRNA machine; and finally,
- tRNA/aaRS/mRNA/ribosome machine.

Here, we hypothesize that the genetic code evolved as pathways for synthesis of new amino acids became available with the progressive refinement of the translation machine. In the primitive translation machine, a symbiotic relationship is established among three components: pre-tRNA, pre-aaRS, and pre-mRNA, to create a short chain of amino acids, which form the biosynthetic protein. The protein chain grew through the addition of further residues of amino acids in the same manner. The result was a synthesis of the first protein, through the linking of the amino acids that were carried by the pre-tRNAs. At this stage of the GNC code, the translation machine began to form (Table 1, Figure 9). There are four codons in the GNC code that are assigned to four amino acids: valine, alanine, aspartic acid, and glycine, from which the first simple protein chain was created (Figure 9A, Figure 10).

In the next stage of translation, pre-tRNA evolved into tRNA through gene duplication. Pre-mRNA evolved into mRNA by linking several strands of pre-mRNA to increase the storage capacity. Pre-aaRS became aaRS through ligation to specific tRNA. These three modifications gave rise to the SNS code (Figure 9B). The superior information bearing qualities of mRNA, the superior catalytic potential of aaRS, and the better adaptor capacities of tRNA emerged from such complexes with gradual expansion of the genetic code. At this stage, tRNAs selected and recruited six more amino acids (glutamic acid, leucine, proline, histidine, glutamine, and arginine), in addition to primordial amino acids (valine, alanine, aspartic acid, and glycine) (Figure 11). These charged tRNAs then create 12 additional codons through base pairing and linking pre-mRNA strands, so that the newly synthesized mRNA strands were more information-rich for storage. mRNAs now possessed at least 16 (4 +12) codons, or combinations of these codons. The mechanisms of creating new strands of pre-mRNA are similar, as shown in Figure 5. Now, two sets of pre-RNA molecules are joined to form a new generation of mRNA. At this stage, the mRNA strands became longer, containing the digital information of 16 codons representing 10 amino acids, or combination thereof allowing for redundancy. The expanded SNS code was refined through the symbiotic interactions of the tRNA/mRNA/aaRS complex. The translation system was considerably improved from the GNC to the SNS stage, but the code remains only moderately robust and susceptible to errors because of the limitation of redundancy. The primitive GNC code expanded to an SNS code composed of

16 codons (GGC, GGG, GCC, GCG, GAC, GAG, GUC, GUG, CUC, GUG, CCC, CGC, CAC, CAG, CGC, and CGG) and 10 amino acids (glycine, alanine, aspartic acid, valine, glutamic acid, leucine, proline, histidine, glutamine, and arginine) [66,80]. The first 10 amino acids, found in the prebiotic environment, have been identified in carbonaceous chondrites [4]. The SNS genetic code is an imprint of the biosynthetic relationships between amino acids (Table 1). As the code expanded, aaRS began to evolve from the earlier pre-aaRS enzyme, and then displaced their less efficient precursors. Primordial class I and class II syntheses evolved from ancestral pre-aaRS. At this point, encoded proteins are longer and they possess enough amino acid diversity to take on some of the general features of contemporary proteins. The mRNA template provides the specifications for the amino acid sequences of the protein gene products. The recipe for the biogenic protein synthesis was inscribed in the codon sequences of mRNA (Figure 9B).

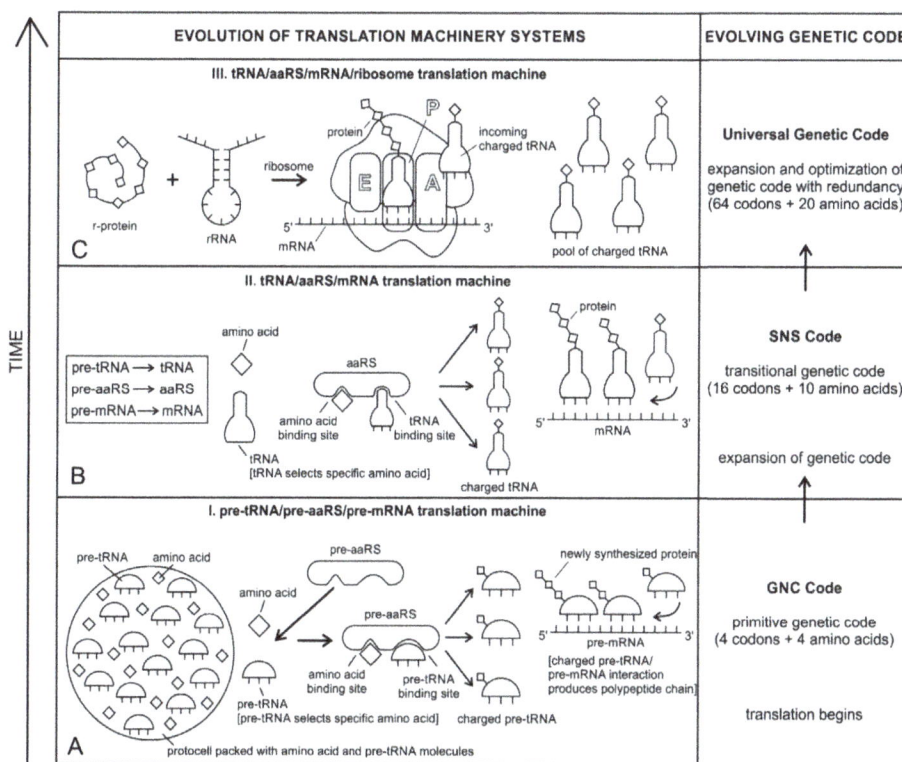

Figure 9. The inferred temporal order of evolution of translation machinery systems showing coevolution of translation machines and the genetic code. In our model, there are three stages of translation machinery systems: (**A**) pre-tRNA/pre-aaRS/pre-mRNA stage when GNC code evolved with the beginning of translation system; (**B**) tRNA/aaRS/mRNA stage when SNS code appeared; and finally, (**C**) tRNA/aaRS/mRNA/ ribosome stage when universal code evolved.

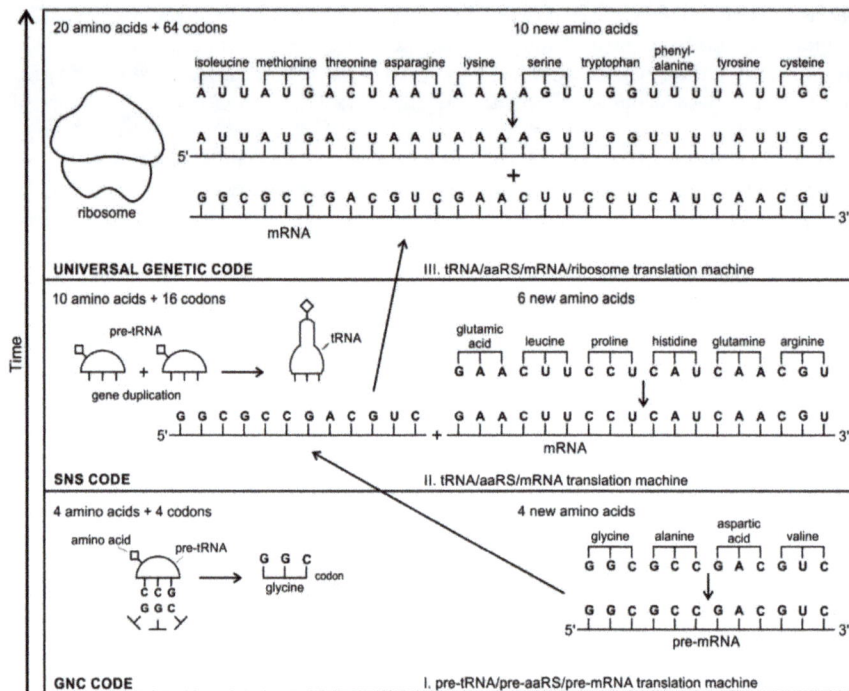

Figure 10. Three stages of the evolution of the genetic code corresponding to the evolution of the translation machines and the progressive addition of amino acids. Pre-tRNA molecule creates its custom-made pre-mRNA for storage of limited amino acid information in the beginning. Primitive translation process began with interaction between pre-mRNA and pre-tRNA. Pre-tRNA molecule in collaboration with pre-aaRS enzyme serve as crucial role in selecting and matching prebiotic amino acids from the prebiotic soup. At this stage, translation machine is simple, consisting of pre-tRNA/pre-aaRS/pre-mRNA. In the abiotic stage, the primitive GNC code appeared, which code four amino acids: valine, alanine, aspartic acid, and glycine [87,88]]. In the next stage, translation machine becomes modified and efficient with the evolution of the tRNA/aaRS/mRNA translation machine, when six new amino acids–glutamic acid, leucine, proline, and histidine were created. mRNA strand becomes more elongated and containing 16 codons and combination thereof with assignments of 10 amino acids with the emergence of the SNS code [87,88]. These 10 amino acids were readily available from the prebiotic environment [93]. Here, we see the beginning of degeneracy, where some the amino acids have more than one codon assignment. With the appearance of ribosome, the SNS code is modified to universal genetic code with 64 codons and 20 amino acids. The translation machine containing tRNA/aaRS/mRNA/ribosome becomes more robust with extensive degeneracy minimizing translation errors and mutation. Furthermore, amino acids with similar chemical properties seem to share similar codons. Ten more new amino acids were recruited at this stage from SNS stage: isoleucine, methionine, threonine, asparagine, lysine, serine, tryptophan, phenylalanine, tyrosine, and cysteine, totaling 20 amino acids. These new amino acids are derivatives of the first set of 10 primitive amino acids [86]. mRNA becomes independent storage device, and it can create its own strand by replication without the assistance of tRNA. mRNA strand becomes more elongated, containing information of 20 amino acids using 64 codons or combination thereof.

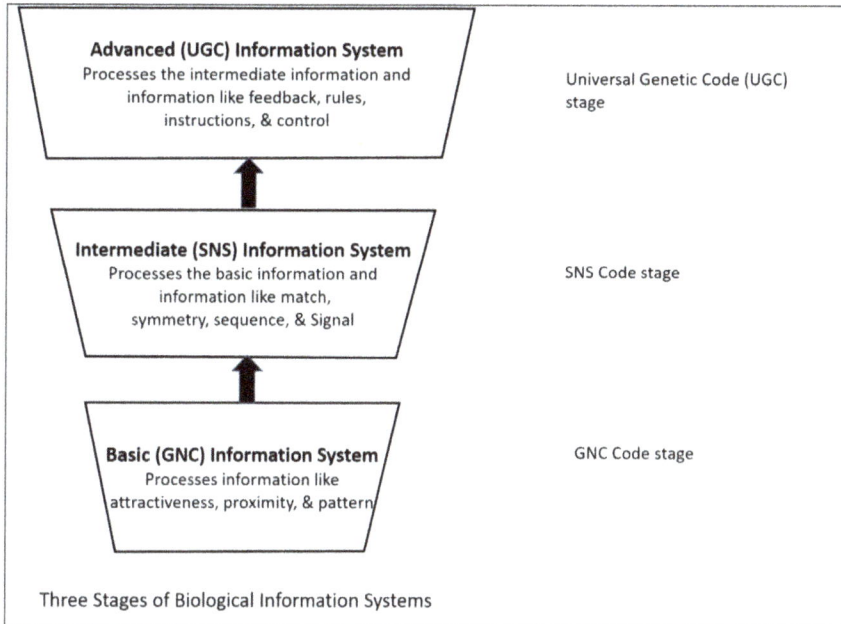

Figure 11. Evolution of Biological Information Systems. The basic biological system during the inception of the GNC code mainly processes the stereochemical properties of tRNA anticodons and primitive amino acids (alanine, aspartic acid, glycine, valine). The intermediate biological system during the origin of SNS code is able to process matching signals and signals etc. The advanced biological system during the origin of the universal genetic code is able to process rules, feedback, and instructions.

The final component of the translation machine, ribosome, is enormous, a hybrid of rRNAs and r-proteins. With the participation of the ribosome, the translation machinery became more elaborate with tRNA/aaRS/mRNA/ribosome complexes; this addition enabled higher specificity in the genetic coding. The ribosome was created through the symbiosis of the rRNA and r-protein, which increased the efficiency of translation, leading to the universal genetic code with its 20 amino acids and 64 codons. (Figure 9C) (Table 1). At this stage, tRNAs selected 10 additional amino acids (isoleucine, methionine, threonine, asparagine, lysine, serine, phenylalanine, tyrosine, cysteine, and tryptophan) (Figure 11). A variety of charged tRNAs then created the corresponding codons through base pairing, forming longer strands of mRNAs. Each mRNA at this stage has the potential of accommodating 64 codons or any combination thereof. The expanded universal code was stabilized with the symbiotic interactions of the tRNA/mRNA/aaRS complex. Once the ribosome appears in the scene, the translation is considerably refined to more efficiently facilitate protein synthesis. The key chemical step of protein synthesis on ribosomes is peptidyl transfer, in which the growing nascent peptide is transferred from one tRNA molecule to the amino acid and then bound to another tRNA. Amino acids are incorporated into the growing protein on the ribosome according to the sequence of the codons of the mRNA. When a ribosome finishes reading an mRNA molecule, the two subunits split apart. The structure of the universal code is highly robust against mutational and translational errors, because of its large allowance of redundancy. Although many deviations from the universal code exist, they are limited in scope and obviously secondary, and they would be introduced later in the evolutionary process. Viruses, bacteria, fungi, plants, animals, primates, and humans all use the same code.

7.1. Origin of the Prebiotic Information System

The prebiotic information system evolved along with the translational machines and the genetic code. The embedded prebiotic information system became more elaborated and advanced as the translation machines became increasingly complex. The information system evolved to process different kinds of information as it coped with the changing environment. The evolution of prebiotic information system can be broadly categorized as GNC (basic), SNS (intermediate), and Universal Genetic Code (advanced) levels of information. A GNC level of biological information has more of a physical nature and it includes things like attractiveness, proximity, and pattern. An SNS level of biological information includes match, symmetry, sequence, and signal, in addition to the information at the basic level. A Universal Genetic Code level of biological information adds rules, instruction, feedback, and algorithm to its repertoire (Figure 11). As implied above, these levels of biological information are cumulative. In other words, an advanced level of biological information also includes the basic and the intermediate levels of biological information. As protocells evolved their patterns (structures), by way of environmental necessities, the structure changed to handle specialized functions. Their structural components differentiated and elaborated to handle specific roles and functions. Protocells started to have a more modular structure, where each module played a specialized role(s). Several authors have found evidence of modular structures in organelles and cells [120,121]. A module is composed of one or many types of molecules. A modular structure requires a noise-free communication among its modules, in addition to the communication within a module. This scenario uses more information than a simple non-modular structure. The information system that is used by a protocell has to coevolve to handle a greater information demand as the modular structure of the protocell becomes increasingly elaborate and specialized.

The prebiotic information systems became increasingly sophisticated in order to process more and more advanced levels of biological information. Figure 11 shows the proposed co-evolution of the biological information systems in three stages. The GNC biological information system mainly dealt with physical, structural, and spatial type of information, whereas the UGC biological information system was a sophisticated system that was capable of handling rules, feedback, and instructions, etc., in order to support the various functions of a translation process.

It is instructive to view information systems at three levels—basic (GNC), intermediate (SNS), and advanced (UGC) level, as shown in Figure 11. The basic biological information system can be compared with early man-made information systems, such as the Turing machine and the computer systems of early 1950s. The intermediate biological information system can be compared to a system that has more elaborate parts, such as memory, data storage, processor, and logic. The computer systems of the 60s and 70's can be used as an illustration of the intermediate biological information systems. The advanced biological information system is very modular, distributed, and has a sophisticated memory structure and communication mechanism seen today in our man-made information systems that are based on embedded and distributed architectures. The pre-tRNA/pre-aaRS/pre-mRNA stage used a basic information system to process basic types of information. The tRNA/aaRS/mRNA stage used an intermediate information system to process intermediate levels information. The aaRS/tRNA/mRNA/ribosome stage used a more advanced information system that was able to process advanced types of information.

All of the signal processing devices, both analog and digital, have traits that make them susceptible to noise. Noise reduction is a goal of all communication systems. Biological information systems are of no exception. Biological processes, such as protein synthesis, undergo random fluctuations—'noise' or errors that are often detrimental to reliable information transfer. With the evolution of the code, denoising methods were implemented through the redundancy of codons. A practical consequence of redundancy is that errors in the third positions of the triplet codon only caused silent mutations or an error that would not affect the protein, because the hydrophilicity or hydrophobicity was maintained by the equivalent substitution of amino acids [69]. The biological information system model includes the process of translating the genetic code into corresponding amino acids as an error-prone information channel [63]. In this scenario, evolution drives the emergence of a genetic code as amino acids map that minimizes the impact of error. The codon to amino acid assignment is treated as a noisy information channel, when the mapping of codons to amino acids becomes nonrandom. The inherent noise (i.e., error in translation) in the channel poses a problem: how can a genetic code be constructed to withstand noises while accurately and efficiently translating information? The answer is redundancy: several codons can specify a single amino acid. This redundancy either implies that there is more than one tRNA for many of the amino acids or that some tRNA molecules can base pair with more than one codon. In fact, both situations occur. Redundancy explains why so many alternative codons for an amino acid differ only in their third nucleotide (Table 2). In an RNA genome, the genes and messenger are one and the same molecule, usually present in numbers of copies [68]. In such a system, innumerable mutations may take place without lethal effects. If one gene molecule and its translation product are disabled, many other unharmed molecules remain to carry on the function involved. Redundancy has been considerably increased from GNC code to SNS code and it has become extreme in the universal code for optimization, perhaps to minimize the noise or translation errors. We show a plausible correlation between stepwise modifications in the translation machinery and the evolution of the genetic code.

Table 2. Universal Genetic code showing numerical codons.

Second letter

First letter	U (1)	C (2)	A (3)	G (4)	Third letter
U (1)	UUU 111, UUC 112 — Phe F; UUA 113, UUG 114 — Leu L	UCU 121, UCC 122, UCA 123, UCG 124 — Ser S	UAU 131, UAC 132 — Tyr Y; UAA 133 STOP Ochre, UAG 134 STOP Amber	UGU 141, UGC 142 — Cys C; UGA 143 STOP Opal, UGG 144 Trp W	U (1), C (2), A (3), G (4)
C (2)	CUU 211, CUC 212, CUA 213, CUG 214 — Leu L	CCU 221, CCC 222, CCA 223, CCG 224 — Pro P	CAU 231, CAC 232 — His H; CAA 233, CAG 234 — Gln Q	CGU 241, CGC 242, CGA 243, CGG 244 — Arg R	U (1), C (2), A (3), G (4)
A (3)	AUU 311, AUC 312, AUA 313 — Ile I; AUG 314 Met M	ACU 321, ACC 322, ACA 323, ACG 324 — Thr T	AAU 331, AAC 332 — Asn N; AAA 333, AAG 334 — Lys K	AGU 341, AGC 342 — Ser S; AGA 343, AGG 344 — Arg R	U (1), C (2), A (3), G (4)
G (4)	GUU 411, GUC 412, GUA 413, GUG 414 — Val V	GCU 421, GCC 422, GCA 423, GCG 424 — Ala A	GAU 431, GAC 432 — Asp D; GAA 433, GAG 434 — Glu E	GGU 441, GGC 442, GGA 443, GGG 444 — Gly G	U (1), C (2), A (3), G (4)

7.2. The Beginnings of the Darwinian Evolution

A key question for the origin of life is how the Darwinian evolution comes to be in the peptide/RNA world. Several mechanisms establishing the correspondence between codons/anticodons and their cognate amino acids have been suggested, either directly or via tRNAs. All of the hypotheses for the origin of life incorporate peptides as stabilizing factor, including the differences in the role that they played in code formation. In the peptide/RNA world, the peptides should have been part of the very first steps in the establishment of the genetic code [19,21,23]. Recently, Kunnev and Gospodinov [24] proposed that hybridization-induced proximity of short aminoacylated RNAs (ribozymes) led to the synthesis of peptides of random sequence. Among these, emerged a type of peptide (named bridge peptide) that was capable of interaction between specific RNAs and specific amino acids. Most likely, the ribozyme-amino acid complex would improve/stabilize a particular pair of specific mRNAs.

Here, we accept the concept of 'bridge peptide' [24] in the context of the evolution of the translation machine and propose a hypothesis that could have led to the RNA-encoded protein synthesis. We suggest that aminoacylated ribozymes use amino acid as cofactor in the process of selection after stereochemical interactions between amino acids and their cognate codons/anticodons. A three-nucleotide long RNAs could be charged with a proper amino acid by an enzyme (bridge peptide) and this RNA will interact with the selected ribozyme by Watson–Crick base pairing and subsequently participate in the formation of short peptides. The selection would favor one amino acid in combination with one ribozyme and one bridge peptide that would become specific to each other. Most likely, a feedback loop from the information source (ribozyme interacting with three bases of RNA) and the stabilizing factor (bridge peptide) was established, perhaps by chance. As a result, interplay between information and its supporting structures/functions emerged. These events led to the coevolution of translation and the genetic code and also triggered the Darwinian evolution. The ribozyme would give rise to tRNA, and bridge peptide to aaRS. This transformation would result the emergence of the translation machine.

The three major events of the code evolution by translation machine—GNC, SNS, and UGC—appeared to be sequential, happening one after the other. This was the beginning of Darwinian evolution, when gradual complexity occurred and more advanced structures/functions emerged. The selection would give rise from pre-mRNA to mRNA, pre-tRNA to tRNA, pre-aaRS to aaRS, and finally the emergence of ribosome. In this evolutionary scenario, the genome size proportionally increased, covering the information for all proteins and ribozymes.

8. Design of Translation Machines and the Genetic Code

Life is characterized and sustained by a number of information rich biological processes that govern cellular functions and greatly contribute to its overall complexity. A biological process may involve the use of one or more modules within a cell. This involves the communication of different types of information, such as signals and connectivity etc., between and within a module [122].

8.1. Simulation of Translation Machines and Cells

The interest of computer scientists regarding the question of origin of life dates back to the origins of computer science. Several attempts have been made to simulate the functions of a molecular translation machine. Von Neumann pioneered the field of bio-inspired digital software/hardware [32]. His self-reproducing automata is now regarded as one of the greatest theoretical achievements in the early stages of artificial life research. He found striking parallel between artificial automata (such as a computer) and natural automata, such as various nanobots in the cells. He introduced the concept of Universal Constructor (UC), a self-constructing machine, which is capable of building any other machine, provided that it can access its description or information tape. This approach was maintained in the design of his cellular automata, which is much more than a self-replicating machine. The UC is more like a Turing machine with a tape control that could store and execute instructions. There are three components of von Neumann's UC machine:

- a memory tape, containing the description (a one-dimensional string of elements;
- the constructor itself, a machine that is capable of reading the memory tape and interpreting its contents; and,
- a constructing arm, directed by the constructor used to build the offspring (the machine described in the memory tape).

A universal constructor with its own description could build a machine like itself. To complete the task, the universal constructor needs to copy its description and insert the copy into an offspring machine. Von Neumann noted that, if the copying machine made errors, these mutations would provide inheritable changes in the property, like the evolutionary process. He realized that the biological machine is much more sophisticated than his UC. Unlike mindless automata, which must be told exactly what to do in order to build the correct objects, a biological machine plays a dual role: it contains instructions—an algorithm—to make a certain kind of translation machine and related enzymes (e.g., mRNA, tRNA, ribosome, aaRS, and other enzymes), but additionally it can be blindly copied as a merely physical structure without reference to the instructions. Another major difference between UC and evolving natural organisms is the lack of feedback in the fitness channel of the former. Cellular automata have been useful artificial models for exploring how relatively simple rules, when combined with spatial memory, can give rise to complex emergent patterns; it may be relevant for understanding new questions regarding the cell division and its relation to information. Subsequently, Von Neumann's UC has been modified by several workers to create Artificial life. However, von Neumann's UC has the information system outside the machine. Accordingly, when UC reproduces its offspring, it lacks the information tape. It has to be added each time during reproduction. It is analogous to vesicle division—a mechanical division of an empty protocell, devoid of instruction.

8.2. Genetic Code Vs. Binary Code

Genetic code is often compared with the binary code that is used by the computers. Significant similarities and differences exist between the two types of the code. Each system has its advantages and limitations. The primary or source alphabet that is used in computers and electronic communication is the binary digit (0, 1), or a bit, a contraction for 'binary digit'. [56]. The bit is the smallest unit of information on a computer. Binary information is grouped into sets of eight bits, which are called bytes; each byte thus has one of 2^8 or 256 possible configurations of zeroes and ones. A byte is just eight bits and it is the smallest unit of memory that can be addressed in many computer systems.

A binary source alphabet could be extended by forming ordered pairs, ordered triplets, ordered quadruplets, and so forth to form receiving alphabets that are larger than two [56]. In molecular biology, these extensions are called codons. The simplest unit of mRNA, on the other hand, is the nucleotide, which can have one of four bases—A, U, C, and G, the quaternary 'bit' [123]. However, we think that the use of the word bit in a quaternary system of mRNA is a misnomer. Here, we choose a new name in the genetic code, called qit, or quaternary digit instead of bit. The quits are A, U, C, and G. This increased variation means that each nucleotide of mRNA can hold twice as much information as each digit of a binary program. The qit creates more algorithmic randomness than bit and it is more information rich. Shannon's great insight in information theory is entropy. Entropy measures the degree of uncertainty or randomness in a system. Entropy is the opposite of information. It destroys information. We can reduce the entropy to the point where the stored information becomes maximal and transmission is highly reliable. Genetic information is low in entropy and high in information content. Entropy measures the degree of randomness that is introduced by errors. This is why the genetic codes is evolved in three stages to incrementally minimize the errors during the translation.

In mRNA, the genetic information comes in triplets of nucleotides or codons, which represent different amino acids, meaning that each codon in mRNA has only 4^3, or 64, possibilities. Each codon is thus an extension, or 'byte', and it has exactly as much information as a six-bit byte, or in computer terminology a code word, since 2^6 is 64 possible sequences for codons. Here, we use a new terminology for representing genetic information, called 'qyte' instead of 'byte'. In our terminology, each qyte is three-qit long, giving 4^3 possibilities.

Both binary and genetic codes contain signals that indicate where to begin and end the reading of their messages. Computers use start and stop bits for this purpose, while the genetic code contains one start codon and three stop codons. In a binary code, a single inaccurate bit causes its byte to have a different value, which can cause significant errors. However, mRNA exhibits greater flexibility and it is more resilient in comparison, as many nucleotide changes do not result in changes to the value of the amino acids that are coded by a codon.

However, information that is contained in life exists in two forms, digital (genetic) and analog (metabolism), and both appeared concurrently in the peptide/RNA world [48]. Digital information is encoded in linear polymers, such as DNA and RNA in discrete codons, analog information is manifest in the differing concentrations of biomolecules, especially proteins that get passed from generation to generation. Analog information systems dominate in the early prebiotic stages, but digital information systems dominate the information age [6]. Recognizing that there are two sequential events, first the origin of an analog chemical system that is capable of adaptive evolution and then a digital revolution, the origin of life problem becomes much more tractable [124]. Acceptance of this dichotomy and this progression helps to resolve the question of dual roles of RNA and peptides in generating information systems.

8.3. Conversion of Three Letter Codons into Numerical Codons

The genetic code is obviously not the result of a random assignment of codons to amino acids. It has a structure; the synonyms are grouped. The language-based terminology of the genetic code reflects the fact that both genes and proteins are essentially one-dimensional arrays of chemical letters. The nucleic acid alphabet comprises of four chemical letters, A, U, C, and G, whereas proteins are built from twenty different amino acids, represented by 20 abbreviated letters. In order to better visualize the codon distribution in the universal genetic code table, we substitute nucleobase alphabets of mRNA with numbers, as follows: 1 for U, 2 for C, 3 for A, and 4 for G. [In case of DNA codons, 1 represents thymine (T).] We have now created a universal numerical codon matrix in a structural format consisting of 64 numerical codons that specify 20 amino acids, and the start and stop codons (Table 2).

In Table 3, the abbreviation of the universal genetic code table is shown in numerical codons with redundancy. Each matrix cell displays information in numerical codon and its corresponding amino acid. Because of numerical distribution of codons in rows and columns, one can easily visualize the distribution of codons and their redundancy in the matrix cells; it was less obvious in standard genetic code using combinations of four letters. Looking at Table 3, we can say that codons beginning with 4 formed first, followed by codons with 2. Codons with prefix 1 and 3 were added last at the genetic code table.

Table 3. Universal Genetic code showing numerical codons.

		1		2		3		4	
1	111		121		131		141		1
	112	F (Phe)	122		132	Y (Tyr)	142	C (Cys)	2
			123						3
	113		124	S (Ser)	133	X (Stop)	143	J (Stop)	4
	114	L (Leu)			134	Z (Stop)	144	W (Trp)	
2	211		221		231		241		1
	212		222		232	H (His)	242		2
	213		223				243		3
	214	L (Leu)	224	P (Pro)	233		244	R (Arg)	4
					234	Q (Gln)			
3	311		321		331		341		1
	312		322		332	N (Asn)	342	S (Ser)	2
	313	I (Ile)	323						3
			324	T (Thr)	333		343		4
	314	M (Met) (Start)			334	K (Lys)	344	(R) Arg	
4	411		421		431		441		1
	412		422		432	D (Asp)	442		2
	413		423				443		3
	414	V (Val)	424	A (Ala)	433		444	G (Gly)	4
					434	E (Glu)			

In Table 4, we have shown a one-letter abbreviation of 20 amino acids, and its corresponding numerical codons. We have used three additional letters, J, X, and Z (shown in bold font) to signify three stop codons, namely opal, ochre, and amber, respectively.

Table 4. 20 Primary Amino Acids in the Genetic Code and their corresponding numerical codons.

1-Letter Abbreviation	3-Letter Abbreviation	Amino Acid	Numerical Codons
A	Ala	Alanine	421, 422, 423, 424
B	—	—	—
C	Cys	Cysteine	141, 142
D	Asp	Aspartic acid	431, 432
E	Glu	Glutamic acid	433, 434
F	Phe	Phenylalanine	111,112
G	Gly	Glycine	441, 442, 443, 444
H	His	Histidine	231, 232
I	Ile	Isoleucine	311, 312, 313
J	**Stop**	**Opal**	143
K	Lys	Lysine	333, 334
L	Leu	Leucine	113, 114, 211, 212, 213, 214
M	**Met (Start)**	**Methionine**	314
N	Asn	Asparagine	331, 332
O	—	—	—
P	Pro	Proline	221, 222, 223, 224
Q	Gln	Glutamine	233, 234
R	Arg	Arginine	241, 242, 243, 244, 343, 344
S	Ser	Serine	121, 122, 123, 124, 341, 342
T	Thr	Threonine	321, 322, 323, 324
U	—	—	—
V	Val	Valine	411, 412, 413, 414
W	Trp	Tryotophan	144
X	**Stop**	**Ochre**	133
Y	Tyr	Tyrosine	131, 132
Z	**Stop**	**Amber**	134

Using these three tables as guides, we have developed software to simulate the translation of the numerical codon sequence of mRNAs to produce its corresponding amino acid sequence. We name this software 'Codon-Amino Acid-Translator-Imitator' or (CATI) that mimics the process of reading a sequence of codons and translating it into a sequence of the corresponding amino acids and vice versa. The CATI software can also handle the reverse process, where a sequence of amino acids is translated into a sequence of corresponding codons.

Table 5 shows some sample outputs of CATI. Table 5 is made up of several sections. In the first section, column one shows a given set of numerical codon sequences (read from a spreadsheet). Column 2 shows the corresponding amino acid sequences. Table 5 also shows the translation of randomly generated numerical codons. The second section of Table 5 shows the translation of randomly generated numerical codons. Table 5 also shows the output of the reverse process—translating a given sequence of amino acids into the corresponding sequence of numerical codons. The third, fourth, and fifth sections of Table 5 show the translation of given amino sequences into the corresponding numerical codon sequences. Since several possible codon sequences can form a given amino-sequence, we show the count of all possible codon sequences and just a few actual codon sequences in the table. The last section in Table 5 shows the conversion of DNA codon sequences into the corresponding numerical codon sequences. While using the distribution of numerical codons, we can visualize at least some of the steps by which nature might have invented the code.

Table 5. Conversion of numerical codons into corresponding amino acids and vice versa using Codon–Amino Acid-Translator-Imitator (CATI) software.

Numerical Codon Sequences	Corresponding Amino Sequences
142343311141334	CRICK
431424243144312332	DARWIN
433312332122324434313331	EINSTEIN
22113124442231431343l	PYRAMID
144423321434242	WATER
43342324132323l	EARTH
42431443324431314142333l	AMERICAN
3144243134324333312314232312344	MAIDENHAIR
141244423131111312124231	CRAYFISH
144423321434244321421331144243324331333243311244131	WATERITANGERIINEFRY

The Randomly Generated Numerical Codon Sequences of Up to Length: 99	Corresponding Amino Acid Sequences
31423424111224112221442112341214242312434424441141211242342441113212442322412 4334433133	MQRFRSLASVCASSGGCFAAVYSAPSKEX
3141212222214212224234214434212411221324232134112421234144132121412242334312ll 3221222432111132243134	MSPPAPQLELVPTTSLRQWYSVRKILPPDFYRZ
31411112144442124344211443423221424444232132433112211441222243334442122413224 4221413143	MFSGLEAWSTCGHYESLVPARGLRYRPVJ

Amino Acid Sequences	Corresponding Numerical Codon Sequences
The Count of all Possible Codon Sequences for the following Amino Sequence: 221,184	Only the First 6 Codon Sequences Generated
SDSYDPCTGL	342432342132432224231423234423l3
SDSYDPCTGL	341432342132432223142232443213
SDSYDPCTGL	123432342132432223142232443213

Table 5. *Cont.*

	Only the First 6 Codon Sequences Generated
SDSYDPCTGL	12243234213243222314232344321 3
SDSYDPCTGL	12443234213243222314232344321 3
SDSYDPCTGL	12143234213243222314232344321 3
The Count of all Possible Codon Sequences for the following Amino Sequence: 1.591208761965867 8 × 10⁵¹	
SDSYDPCTGLLQKSPQCCNTDILGVANLDCHGPPSVPTSPSQFQASCVADGGRSA RCCTLSLLGLALVCTDPVGI	34243234213243222314232344321321323333342223314214233232432313214341342 33322134321422324432232334241322332334222334223311223342331424134232432434 43343342233431421423232133422134421321344321342213413142323432223413443313
SDSYDPCTGLLQKSPQCCNTDILGVANLDCHGPPSVPTSPSQFQASCVADGGRSA RCCTLSLLGLALVCTDPVGI	34143234213243222314232344321321323333342223314214233232432313214341342 33322134321422324432232334241322332334222334223311223342331424134232432434 43343342233431421423232133422134421321344321342213413142323432223413443313
SDSYDPCTGLLQKSPQCCNTDILGVANLDCHGPPSVPTSPSQFQASCVADGGRSA RCCTLSLLGLALVCTDPVGI	12243234213243222314232344321321323333342223314214233232432313214341342 33322134321422324432232334241322332334222334223311223342331424134232432434 43343342233431421423232133422134421321344321342213413142323432223413443313
SDSYDPCTGLLQKSPQCCNTDILGVANLDCHGPPSVPTSPSQFQASCVADGGRSA RCCTLSLLGLALVCTDPVGI	12243234213243222314232344321321323333342223314214233232432313214341342 33322134321422324432232334241322332334222334223311223342331424134232432434 43343342233431421423232133422134421321344321342213413142323432223413443313
SDSYDPCTGLLQKSPQCCNTDILGVANLDCHGPPSVPTSPSQFQASCVADGGRSA RCCTLSLLGLALVCTDPVGI	12443234213243222314232344321321323333342223314214233232432313214341342 33322134321422324432232334241322332334222334223311223342331424134232432434 43343342233431421423232133422134421321344321342213413142323432223413443313
SDSYDPCTGLLQKSPQCCNTDILGVANLDCHGPPSVPTSPSQFQASCVADGGRSA RCCTLSLLGLALVCTDPVGI	12143234213243222314232344321321323333342223314214233232432313214341342 33322134321422324432232334241322332334222334223311223342331424134232432434 43343342233431421423232133422134421321344321342213413142323432223413443313
The Count of all Possible Codon Sequences for the following Amino Sequence: 3,538,944	Only the First 4 Codon Sequences Generated
DPCTGLLGLAV	43222314223244321321344321342 3413
DPCTGLLGLAV	43122314223244321321344321342 3413
DPCTGLLGLAV	43222214223244321321344321342 3413
DPCTGLLGLAV	43122214223244321321344321342 3413

Table 5. *Cont.*

DNA Codon Sequences	The Corresponding Numerical Codon Sequences
TTGCAGAATTTGTAAG	14234331114334
GATGCGCGATGGATCAAC	43142423144312332
GAAATCAACTCCACGGAGATAAAT	43331233212232443431331
CCTTATCGGGCCATGATAGAT	22113124442231431431
TGGGCAACTGAGCGC	14442332143424
GAAGCACGTACACAT	43342324132323
GCGATGGAACGGATATGTGCAAAT	42431443324313141423331
ATGGCGATAGACGAAAATCATGCAATCAGG	31442431343243333312314231231442312344
TGTCGGGCATATTTTATCTCGCAT	14124442311131121242231
TCGGCAACTGAGCGGACTGCTAATGGGCGAACGAATAAACGAATTTCGGTAT	14442332143424321421331442433243313332431124341231

CATI accepts inputs two ways—from an excel spreadsheet and from a set of randomly generated sequences. A user can create a set of numerical codon sequences while using a spreadsheet. CATI can also generate a random sequence of numerical codons of an arbitrary length for translation.

Application potential carries great potential for numerical codons in bioinformatics. For example, it can be used in translating codon sequences in DNA sequencing in numerical forms, which is the process of determining the precise order of nucleotide bases within a DNA molecule. The simultaneous quantification of mRNA and protein in a single translation process highlights the increasing importance of numerical codons in various analysis tools. During protein synthesis, we do not have to translate nucleotide language to amino acid languages. Both nucleotides and amino acids can be expressed in numerical formats. The advent of rapid DNA sequencing methods has greatly accelerated biological and medical research and discovery. The use of numbers rather than letters may expedite similarity searches between two strands of DNA. Thus, numerical codons can be used for the DNA barcoding of a species, or DNA profiling of a person as a parallel system. The Genome Sequence Data Base (GSBD), operated by the National Center for Genome Resources (NCGR), is a national database of publicly available nucleotide sequences and associated biological and bibliographic annotation. As a pilot study, the data of a small gene can be converted to numerical codon sequences by our CATI software, for a feasibility study to see whether it affords better DNA mining, alignment of two DNA sequences, and for searching methods, storage, and data retrieval systems in the future.

CATI uses numerical codes to represent and manipulate codon sequences. This enables CATI to provide better performance in terms of speed and memory when compared to most of the other software when it comes to processing large sequences of codons and amino acids. This allows for us to undertake faster translation and sequence alignments. The randomly generated numerical codons can also be subjected to some constraints during the sequence generation in order to have a desirable amino acid content.

CATI, when fully developed and implemented, can help to perform various types of analysis of codon and amino acid sequences. It can also help to identify similarity between two or more sequences of DNA. We envision CATI as an effective tool in analyzing and synthesizing non-coding as well as coding mRNAs under different constraints and conditions and performing various types of sequence alignments. In the computer, manipulation numbers have advantages over the manipulation of letters. For example, with an appropriate internal representation of numbers, bit operations can be performed giving a higher speed of computation. We believe that CATI is advantageous over many multiple DNA-protein or RNA-protein translation tools that are available online, which are based on the manipulation of letters. DNA has the potential to provide large-capacity information storage. CATI may provide new insight for developing the storage and retrieval of large data sets in DNA in the future. We have not developed this idea in this paper, and but in a subsequent paper, we want to explore the application of CATI in various translation algorithms.

The CATI software is based on the Model View Controller (MVC) design pattern [33,125]. The MVC facilitates the design idea by segregating functional/task responsibilities and assigning them to different components/modules of software. This leads to an architecture with components that are relatively independent of each other. The three major components are (Figure 12A):

- The Model is the central component of the pattern. It manages the data (information), its associated logic, and the rules of application.
- The View is a (visual) representation of the model, the user interface. It relates to the logic (code) that produces the output. A view can be a form of any output representation of information.
- The Controller accepts input and acts a monitor to mediate (i.e., coordinate) between the tasks of view and model. It handles events that are generated by the user and communicates those changes to the Model, which updates its state accordingly and communicates any changes back to the Controller. The Controller then updates the view to reflect those changes.

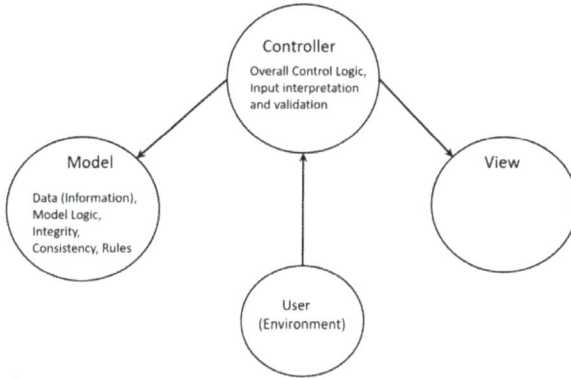

A

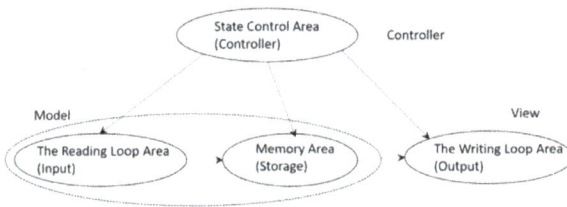

B

Figure 12. (**A**), a Model-View-Controller (MVC) Architecture pattern, and (**B**), an implementation Architecture of von Neumann's Universal Constructor (UC). The solid arrows in the Figure show the flow of control among the components. For example, the solid arrow between the controller and model implies that the controller directs the actions of the model. A dotted arrow indicates a flow of data (information).

These three architectural components of the system enable us to imitate the biological information process in a flexible and modular way. It is important to point out that the MVC design pattern is very similar to the implementation architecture of John von Neumann's Universal Constructor of the self-reproducing automata. Given a description, the Universal Constructor theoretically produces any automata from available parts. The universal constructor has not yet been manufactured physically. However, several attempts have been made to computationally implement the universal constructor. A very good implementation of John von Neumann's self-reproducing UC machine has been recently developed [126]. The overall architecture of this implementation is shown in Figure 12B. It has four components—the State Control Area, the Reading Loop Area, the Memory Area, and the Writing Loop Area. The State Control Area is the overall coordinator of all the other components. The Reading Loop Area reads the tape information and temporarily stores them in the memory area. The information in the memory area is used by the Writing Loop Area to produce the output. We suggest that there is a very close similarity between the Universal Constructor's architecture, as shown in Figure 12B and the MVC Design Pattern. The Controller of the MVC design pattern corresponds to the State Control Area, the Model corresponds to the combination of the Reading Loop Area and the Memory Area, and the View corresponds to the Wring Loop Area. In fact, the MVC design pattern can be interpreted as a more modern operationalization of John von Neumann's Universal Constructor.

We have found that the MVC design pattern is very good at abstracting the natural protein translation machine into an architecture that is modular and neatly separates the various aspects of information processing into three roles—model, view, and controller.

8.4. Algorithmic Design of CATI

We now present the algorithm that is used by CATI. For simplicity, the algorithm shows the overall logic without dividing it into the MVC components that are shown in Figure 13. CATI uses a codon chart in the form of a codon-amino mapping shown in Table 5. CATI is given a sequence of numerical codons. CATI can also generate a random sequence numerical codon on its own. This sequence of numerical codons is then translated into the corresponding sequence of amino acids.

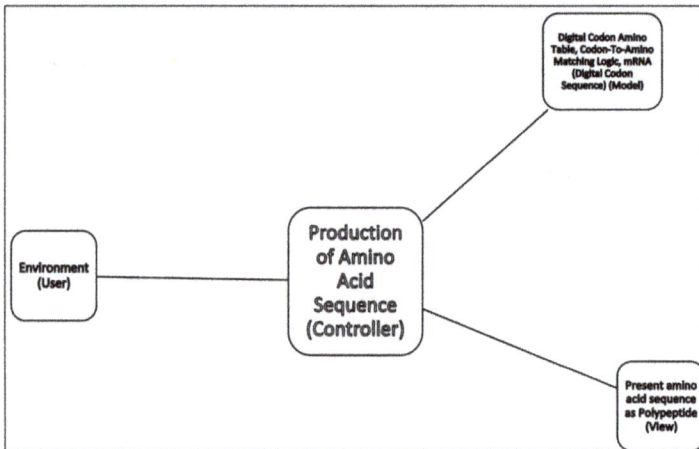

Figure 13. An overall architecture of CATI based on the MVC pattern. It shows the controller, model, and view aspects of the logic.

Figure 14 shows the algorithm. The main step of the algorithm in Figure 14 is the third step of CATI. It uses the ideas of a pattern recognizer (step 3.a), an adapter (step 3.b), and a sequence builder (step 3.c). CATI, in essence, plays the combined role of ribosome, aaRS, and tRNA taken together.

Produce-Amino-Acid-Sequence (Given: a digital-codon-sequence)

// Given a sequence of digital codons, the algorithm uses table 2 in the translation
// of mRNA type sequence into the corresponding sequence of amino acids

1. Check the digital-codon-sequence for a proper sequence.
2. Start with an empty amino-acid-sequence to hold the amino acids.
3. While there is a digital-codon remaining in the digital-codon-sequence
    a. Find a match for the digital-codon in table 2
    b. Get the corresponding amino acid from table 2
    c. Add the amino acid to the growing amino-acid-sequence
4. Display (release) the completed amino-acid-sequence

Figure 14. The overall Logic (Algorithm) of CATI. The logic combines the role of ribosomes, aaRS, tRNA, and mRNA into a single overall process.

9. Simulation and Visualization of the Translation Pathways

Computer simulations are useful in evolutionary biology for hypothesis testing, for verifying analytical methods, for analyzing interactions among evolutionary processes, and they are widely used in different disciplines. In general, computer simulations allow for the study of complex systems, including those analytically intractable [127]. Here, we use forward simulation models from primitive machines to advance translation machines, mimicking the biosynthetic processes for the origin of the genetic code, and for testing our hypothesis of the coevolution of the translation machines and the genetic code.

In this section, we visualize the early stages of translation machine evolution in three stages. We use a simulation and modeling software, called AnyLogic, which is commercially available (www. anylogic.com), to simulate and visualize the translation machines. We simulate translation machines at three levels of evolution:

- pre-tRNA/pre-aaRS/pre-mRNA
- tRNA/aaRS/mRNA
- tRNA/aaRS/mRNA/ribosome

Although the molecular organization of the genetic code is now known in detail, how the code came into being has not been satisfactorily addressed. We have already discussed the coevolution of translation machines and the genetic code in chapter 7; this offers a simple relation between the codon reading efficiency and the accuracy of the codon translation machine. Here, we highlight some of the features of this coevolution for visualization. The rapid and accurate translation of genetic code into proteins is the hallmark of the *information* stage; it evolved in three distinct stages through the availability of amino acids and the improvement of the translation machine. It is now widely accepted that the earliest genetic code did not encode all 20 amino acids that were found in the universal genetic code, as some amino acids have complex biochemical pathways and were probably not available in the prebiotic environment. Therefore, the genetic code evolved as pathways for the synthesis of new amino acids became available [73,92,102]. In our view, the code evolved in step with the amino acid biochemistry and the refinement of the translation machine (Figure 11).

Currently, the simulation simply visualizes the translation process without any provision for parameter changes. In the future, we plan to parameterize the simulation of the translation processes and its visualization.

9.1. Stage I. Visualization—pre-aaRS-pre-tRNA-pre-mRNA Machinery

The first information system emerged in the prebiotic world as a primordial version of the translation machine and the genetic code. The most primitive translation machine consists of pre-aaRS/pre-tRNA/Pre-mRNA molecules. Four primordial amino acids were specific to four pre-tRNAs, and four pre-aaRS enzymes began to translate the genetic information from pre-mRNA, and to synthesize short polymer chains of protein (Figure 15). The code that evolved at this stage was the primitive GNC code [88,115], involving four codons (GGC, GCC, GAC, and GUC), which created four primordial amino acids (glycine, alanine, aspartic acid, and valine). Since there was no redundancy at this stage, the translation errors are high.

The informational associations among these biomolecules are shown in the form of an information structure. This information structure showing macromolecules and their association with each other can be captured in the form of a class diagram (Figure 16). In a class diagram, a rectangular shape represents an informational object. The solid lines connecting these objects reflect an associative relationship between the objects. It is important to note that only a few attributes of each object are shown in the class diagrams. This is for illustration purposes only. Here, our main focus is on the interaction among the objects and not on providing a comprehensive list of attributes for each object. Figure 16 shows the information structure during the first stage of the genetic code, the GNC code. Pre-tRNA, pre-AARS, pre-mRNA, amino acid, codon, anticodon, as well as protein, and nucleotide are shown as objects in this Figure. The relationships that are shown in a class diagram are static, structural, and associative. A class diagram does not show any dynamic or temporal relationship.

A pre-AARS attaches the appropriate amino acid to its pre-RNA with the correct anticodon. A pre-mRNA has a sequence of codons. These are generally short-length sequences that deal with the GNC genetic codes. An amino acid carrying pre-RNA was able to base pair with codons in a pre-mRNA and helped to produce a protein as per the information in the pre-mRNA. At this stage, the types of information used are generally in the form of attractiveness, proximity, and pattern. The right combination of a catalyst, information, and the material acts as a translation machine to produce a new biological artifact. A pre-aaRS/pre-tRNA/pre-mRNA machine's MVC architecture shows a collaboration among these three machine parts that control the formation of a biosynthetic protein chain (Figure 17). The controller uses pre-mRNA, amino acid, anticodon, and other parts of the translation machine as information to translate (convert) a codon into the corresponding amino acid with the help of a charged pre-tRNA that acts as an adaptor. The charged pre-tRNA is shown as a view. It produces the amino acid, as an output, based upon its anticodon matching with the codon in pre-mRNA. These amino acids become part of a sequence in the form of a protein chain.

A visualization of the stage I translation machine has been created using Anylogic software. Appendix A provides instructions on how to run the visualization model in the AnyLogic cloud. The visualization model shows the overall translation process regarding how various molecules dynamically interact to produce a protein.

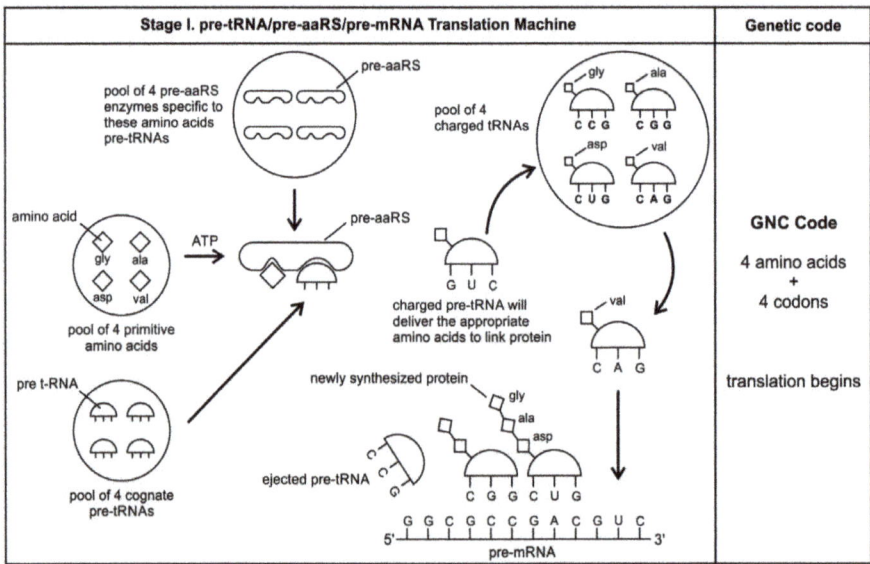

Figure 15. The pre-aaRS/pre-tRNA/pre-mRNA translation machine. Pre-aaRS is the matchmaker between pre-tRNA and amino acid. Four primitive amino acids and their cognate four pre-tRNAs and pre-aaRS molecules were selected from the prebiotic soup. Each amino acid with its specific pre-tRNA molecules was catalyzed by pre-aaRS enzyme in the presence of ATP to create a charged pre-tRNA molecule. In a similar way, four charged molecules were available to decode the short string mRNA one at a time. During the hybridization of anticodon of pre-tRNA with codon of pre-mRNA, each pre-tRNA delivers the appropriate amino acid, which is linked to form a chain of biosynthetic protein for the first time, containing four amino acids. This is the first stage of translation, when primitive GNC code evolves.

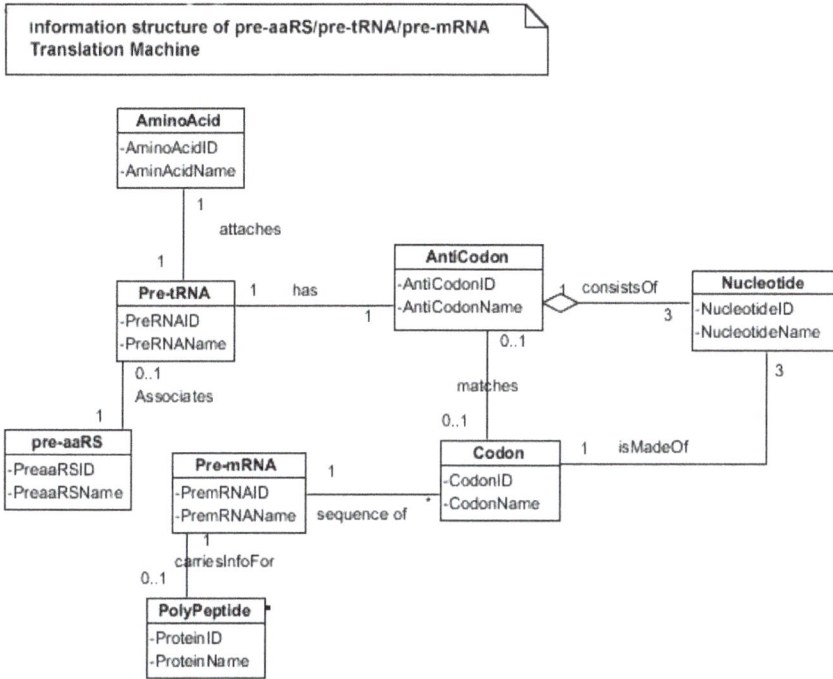

Figure 16. A class diagram showing an information structure during the first stage of translation system. This diagram shows relationships among various parts of the primitive translation machine. Pre-tRNA attaches to a specific amino acid with help of pre-aaRS molecules. The charged pre-tRNA molecule has an anticodon that hybridizes with the corresponding codon of pre-mRNA. As pre-tRNA begins to decode pre-mRNA molecules, short chain of protein is synthesized for the first time in a prebiotic environment. The linkage of an amino acid to a pre-tRNA established the primitive GNC genetic code.

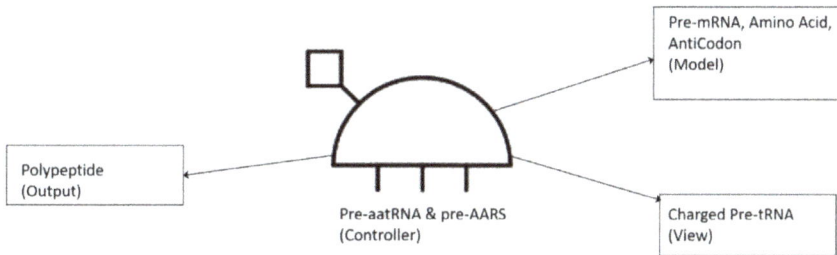

Figure 17. An MVC architecture of a pre-aaRS/pre-tRNA/pre-mRNA machine. Pre-aaRS and pre-tRNA direct charged pre-tRNA that will decode pre-mRNA to a growing protein.

9.2. Stage II. Visualization—aaRS-tRNA-mRNA Machinery

The translation machine is refined to the second stage (Figure 18) with the development of the aaRS/tRNA/mRNA machine, which increases the efficiency and decreases translation errors. At this stage, the GNC code evolved into transitional SNS code with 16 codons (GGC, GGG, GCC, GCG, GAC, GAG, GUC, GUG, CUC, GUG, CCC, CGC, CAC, CAG, CGC, and CGG), which code 10 amino acids (glycine, alanine, aspartic acid, valine, glutamic acid, leucine, proline, histidine, glutamine, and arginine) [88,115]. Because of the redundancy of codons, the translation error is minimized.

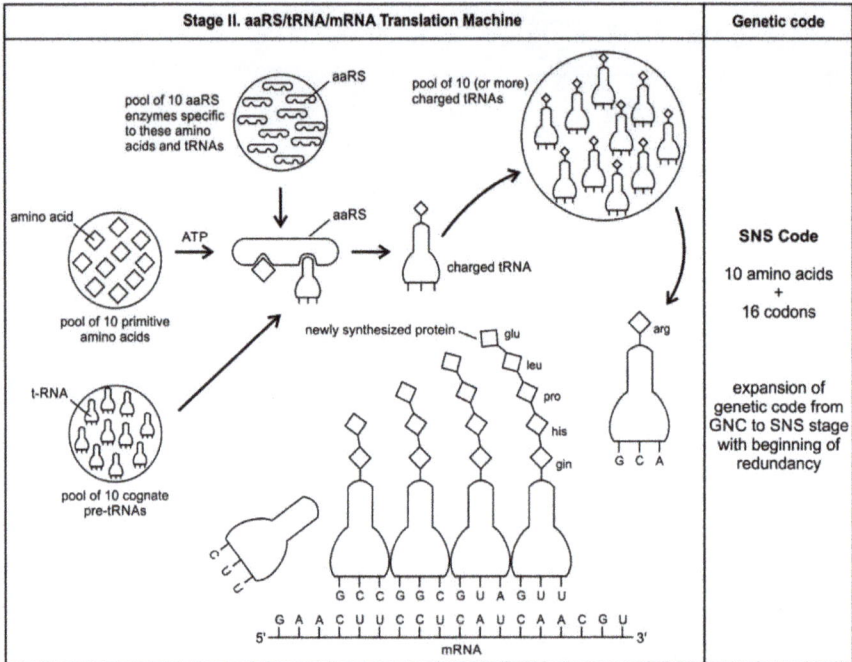

Figure 18. The aaRS/tRNA/mRNA translation machine. Ten primitive amino acids joined with specific tRNA molecules by aaRS enzymes to form a pool of 10 charged tRNA molecules. These charged tRNA molecules begin to decode mRNA, creating a chain of longer, biosynthesized protein molecule. At this stage, SNC code appears with 10 amino acids for 16 codons. The translation is moderately efficient with the appearance of redundancy to minimize the translation errors.

Figure 19 shows the information structure of the aaRS/tRNA/mRNA translation machine during the second stage of the universal genetic code. At this stage, additional information in the form of match, symmetry, and sequence are also available. A tRNA is transformed into a charged tRNA by the aaRS, as shown in Figure 19. The anticodon of the charged tRNA matches with the corresponding codon in mRNA. A protein chain is formed by the decoding of mRNA by tRNA.

Using the MVC framework, we suggest that, in the case of an aaRS machine, proteins represent the 'output', the corresponding charged tRNAs and amino acid ligation (aa-tRNA) as a 'view', and a combination of aminoacyl tRNA and aaRS as a 'controller', and mRNA as a 'model' that holds codons as information (Figure 20). The directional arrows represent the control and 'communication' between the various parts of the machine. For example, an aaRS coordinates and facilitates the activities of mRNA, tRNA, and amino acid ligation in the formation of protein chain. The aaRS acts as a facilitator and helps to produce (select) the amino acid that matches with the codon in mRNA. Briefly, the overall logic of the second stage machine is as follows: specific tRNA binds with a particular amino acid, tRNA, and then incorporates the amino acid into a growing protein at a position is that determined by the anticodon, the anticodon matches with a codon in mRNA, and the codon acts as an information carrier that matches with the specific tRNA. The final result is the release of the linked amino acids, which are the protein chain.

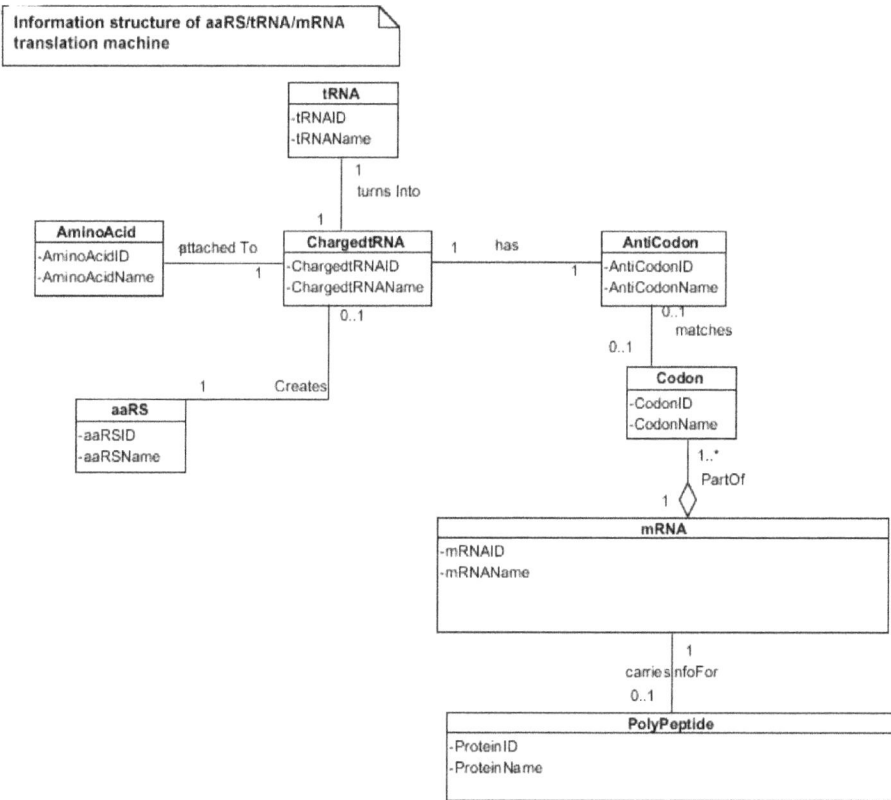

Figure 19. A class diagram showing the interactions of aaRS/tRNA/mRNA translation machine showing the information structure during the origin of the SNS code. At this stage, 10 primitive amino acids are available to create 10 or more charged tRNAs for decoding mRNA. An amino acid in the charged tRNA will be incorporated into a growing protein chain, at a position that is dictated by the anticodon of the tRNA.

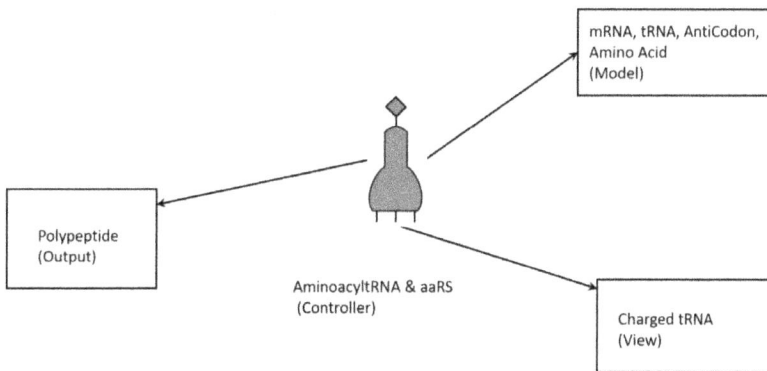

Figure 20. An MVC architecture of aaRS/tRNA/mRNA translation machine. aaRS and tRNA facilitate the interaction between a charged tRNA and a mRNA. A charged tRNA is an adaptor that acts as a view and helps to release the amino acid to form a chain of protein.

In the Appendix A, we show the instructions on how to run the visualization model for the second stage of translation machine.

9.3. Stage III. Visualization—aaRS-tRNA-mRNA-Ribosome Machine Complex

By the third stage, the translation machine has fully evolved, now consisting of the aaRS/tRNA/mRNA/ribosome machine that brings forth the universal genetic code (Figure 21). The translation of the universal genetic code into protein by ribosomes requires precise mRNA decoding by tRNA. At this stage, ribosomes emerged to facilitate high-fidelity translation. About 31 tRNAs and 20 aaRS enzymes assigned 64 codons specifying 20 amino acids. Of these 64 codons, 61 represent amino acids and three are start and stop signals. Although each codon is specific to only one amino acid, the code is degenerate, because a single amino acid may be coded for more than one codon. The redundancy of the universal genetic code optimized translation errors and mutations [108]. Codons for the same amino acids tended to bundle together. Perhaps the organization of the amino acids with particular sequences of the code minimized the errors that crept into the proteins. Among the 20 amino acids in the universal code, approximately half came from the prebiotic soup; as we see in the SNS code, the remaining half of amino acids were derivatives of the first set of 10 primitive amino acids by biosynthesis [78].

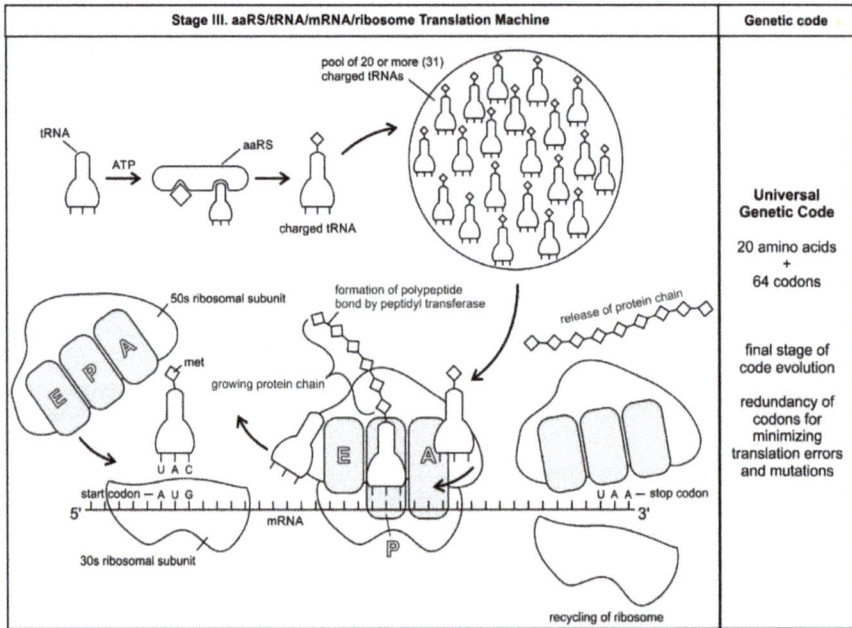

Figure 21. The aaRS/tRNA/mRNA/ribosome translation machine. tRNA delivers amino acid to ribosome that serves as the site of protein synthesis. Each ribosome has a large 50S subunit and a small 30S subunit that join together at the beginning of decoding of mRNA to synthesize a protein chain from amino acids carried by a tRNA. The correct tRNA enters the A site of the ribosome and appropriate amino acid is incorporated into the growing peptide chain, which transfers from tRNA in the P site to the tRNA of A site. As the ribosome moves, both tRNAs and mRNA then shit to the E site. Each newly translated amino acid is then added to a growing protein chain until ribosome completes the protein synthesis. At this stage, universal genetic code is optimized with 20 amino acids for 64 codons, including start and stop codons. The translation is highly efficient with start and stop codons; redundancy minimizes the translation errors and mutations.

Figure 22 shows the information structure that was available during the third stage of the genetic code. During this stage, a ribosome uses rules, and feedback types of information, in addition to the other types of information during translation. A ribosome acts like a biological assembly machine in the translation of mRNA into protein. A ribosome performs the protein synthesis with the assistance of two other kinds of molecules—mRNA and tRNA.

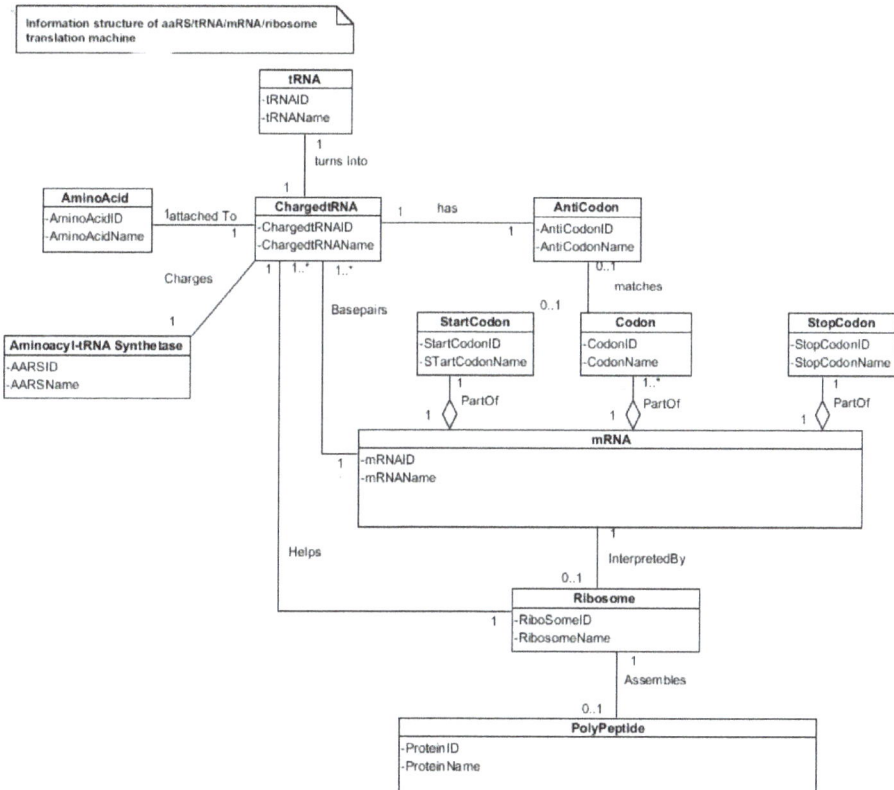

Figure 22. Class diagram showing the interactions of aaRS/tRNA/mRNA/ribosome translation machine. The diagram is similar to that of Figure 22. In addition, it shows the introduction of a ribosome, which decodes mRNA with the help of charged tRNA.

Figure 23 shows an MVC model of the ribosome machine. A ribosome machine is sometimes equated with a factory with several machines. It uses other machines, such as an aaRS machine, to complete the translation process. A ribosome plays the role of the controller. mRNA is a model containing the information in the form of a sequence of codons. An aa-tRNA machine plays the role of a view that supplies an amino acid. Note that this machine is depicted in Figure 21. Here, the ribosome machine uses the aa-tRNA machine as its submachine (sub part), signifying a functional hierarchy among macromolecules. The ribosome produces the peptide chain (protein) by establishing the proper match (fit) between an aa-tRNA and the corresponding codon in the mRNA.

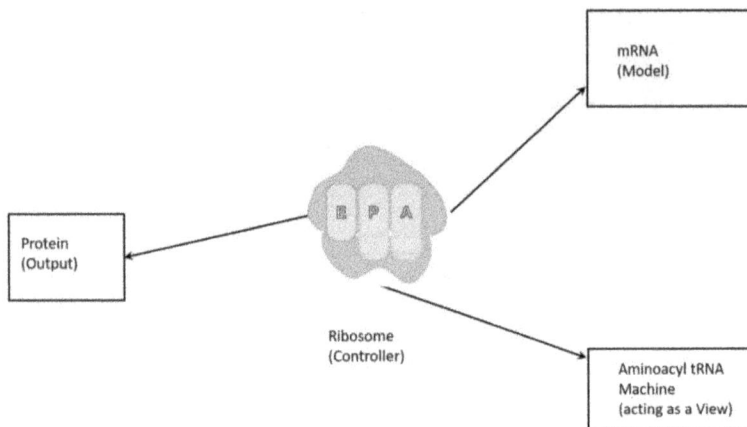

Figure 23. An MVC structure of aaRS/tRNA/mRNA/ribosome translation machine. A ribosome is a part of a bigger machine that uses the aaRS machine to decode the mRNA information into a corresponding sequence of amino acids to link a protein chain.

In the translation process of the third stage, the translation machine, with the assistance of a ribosome, is visualized using an AnyLogic model. We show the instructions in the Appendix A how to run the third stage of the visualization model.

10. Discussion and Conclusions

Although the origin of the prebiotic information is not fully understood, the manufacturing processes of different species of RNAs and proteins by molecular machines in the peptide/RNA world require not only physical quantities, but also additional entities, like sequences and coding rules. The demand for a wide range of specific enzymes to catalyze complex prebiotic chemistry was the prime selective pressure for the origin of the information systems for creating programmed protein synthesis. These coded proteins are specific and quite different from the random peptides that are generated by linking amino acids in the vent environment. There is a great potential in the application of numerical codons in bioinformatics, such as barcoding, DNA mining, or DNA fingerprinting.

We have reviewed the bottom-up pathways of prebiotic synthesis that address several hallmarks in living systems, such as the encapsulation and protocell division, peptide/RNA world, origin of mRNA, origin of aaRS, information processing, energy transduction, and adaptability. The scenarios for the origin of the translation machinery and the genetic code that are outlined here are both sketchy and speculative, but follow those biosynthetic pathways. It is the informational role of RNAs, aided by a series of enzymes, which is key to transforming nonliving chemistry into translation machines and the genetic code.

There are several novel ideas in the origin of prebiotic information that are presented in this paper:

1. The peptide/RNA world was more parsimonious in the vent environment than the popular RNA world hypothesis. It is easier to make proteins than RNAs in the vent environment. The duality of replication and metabolism is the intrinsic property of life and it must have appeared simultaneously before the origin of the first cells. Both RNAs and proteins worked in tandem to jumpstart the life assembly.

2. The *Information* stage is a crucial step in the origin of life prior to the origin of DNA and the first cell. We emphasize that reproduction is not possible without information. Life is information stored in a symbiotic genetic language. Information is an emergent property in the peptide/RNA world. The molecular attraction between tRNA and amino acid led to the translation machinery

and the genetic code. The demand for specific protein enzymes over peptides in the peptide/RNA world was the selective agent for the emergence of the *information* age.

3. Both mRNAs and proteins were invariably manufactured by molecular machines that required sequences and coding rules. The crucial step was the ligation of a specific amino acid to its corresponding pre-tRNA molecule that created a repertoire of complex machinery parts for translation. tRNA is an ancient molecule that created custom-made mRNA for the storage of amino acid assignment. During this stage, the translation and the genetic code coevolved.

4. The molecular basis of the genetic code manifests itself in the interaction of aaRSs and their cognate tRNAs. Aminoacylated ribozymes used amino acids as cofactor with the help of bridge peptides as a process for selection between amino acids and their cognate codons/anticodons; this self-sustained RNA-peptide complex may trigger proto-translation. As bridge peptide evolved to pre-aaRS and ribozyme to pre-tRNA, many of their structures were modified, but their functions were continued and elaborated. Eventually, the interaction of aaRS and tRNA was established.

5. The piecemeal buildup of translation machines consisting of tRNAs, mRNAs, aaRS, and ribosomes are proposed.

6. The existing theories on the origin and evolution of the genetic code are compatible with our coevolution model of translation machines and the genetic code. We suggest that there were three stages in the evolution of the genetic code—GNC, SNS, and finally the universal genetic code [88,115]. The code evolved through the progressive refinery of translation machines, from the pre-tRNA/pre-aaRS machine, to the tRNA/aaRS/mRNA machine, and finally the tRNA/aaRS/mRNA/ ribosome machine. This was the beginning of the Darwinian evolution that exhibited an interplay between information and its supporting structure. The evolution of the translation machine reflects the incremental enrichment of information content in the genetic bank of mRNA. An evolutionary path from bridge peptide to protozymes to urzymes to pre-aaRS to aaRS suggests increasing the complexity of functions and satisfying the rule of continuity.

7. Using a computer simulation, and a visualization model of the possible biosynthetic pathways that led to the origin of the information system, we show the step-by-step evolution of the translation machines and the genetic code.

An mRNA strand with strings of codons is a relatively small and simple molecule possessing limited storage capacity. It can store small amounts of genetic information; the capability of a ribozyme as an enzyme is severely limited. However, this catalytic deficiency is compensated by a variety of enzymes that are available in the hydrothermal vent environment. The short life of mRNA makes the protocell very responsive to changing conditions in the environment. Later, more durable DNA would emerge to become the molecule of choice for the large storage of genetic information regarding protein synthesis, replacing mRNA from its main function. The new generation of mRNA is created by the transcription of DNA. mRNA becomes a daughter of DNA to carry out its specific instruction of translation and protein synthesis.

The *information* age, with the origin of translation and the genetic code, was a watershed event in biogenesis triggering the origin of DNA and the first cells. The information age is quite distinct and more derived than the prebiotic chemical stage, and it is a necessary prelude to the biological age. However, it lacks the one crucial attribute of life: cell division. In the prebiotic information stage, each mRNA became a gene that contained the recipe for a specific protein. However, the information system would be fully developed with the appearance of DNA that contained a permanent storage for both hereditary information and the transcription capability. With the emergence of DNA, the central dogma is established; information flows from DNA to mRNA to proteins.

The new information paradigm suggests that life is organic chemistry, plus information, plus code, plus cell division, where replication, sequencing, coding, transcription, and reproduction become important attributes. The advent of cell division defines the emergence of the first cells from their protocell precursors. Life began when a cell was capable of dividing into two identical daughter cells. A protocell in the prebiotic information age did not acquire the capability of identical cell division.

Author Contributions: The concept was developed and designed by S.C.; computer simulation was done by S.Y. Both authors contributed in writing of this paper.

Funding: This research received no external funding.

Acknowledgments: We thank Mavis Liang for inviting us to contribute this article in the special issue of *The Origin of and Early Evolution of Life* volume and Gabriel Wang for guiding the manuscript through the editorial process. We owe an immense debt of gratitude to the many authors, who have helped us with their thoughtful, well-documented, and enlightening expositions in the origin of life research. We thank Oliver McRae for reading the manuscript for clarity and brevity. We thank five anonymous reviewers and the editor for their helpful suggestions and constructive input. We thank Volkan Sarigul for illustrations. This work was supported by Texas Tech University.

Conflicts of Interest: The authors declare no conflict of interest.

Appendix A

This appendix is a user guide for the visualization model developed in the AnyLogic software. It describes as to how to run the visualization model for each stage of the translation machines. The visualization models are hosted on the AnyLogic cloud, a service that allows simulation models to be available online on Internet. These hosted models can be run under a browser such as Chrome, Internet Explorer, and Fire Fox, etc. In the following sections, it is assumed that an Internet browser is up and running either on a laptop or on a PC.

Instructions for Stage I Visualization

This section shows the steps to run the stage I model. Please follow these steps:

Step 1. Paste the following URL link in your browser's address line and go to that link: https://cloud.anylogic.com/model/5a297e73-af36-4af5-a92c-a416021a9bda?mode=DASHBOARD&experiment=82e7de3a-b6df-4ab2-9304-eeba402cc447

Step 2. You should see a web page that looks like the one shown in Figure A1.

Figure A1. A screen shot of the web page showing the button to be pushed.

Step 3. Press the button as indicated in Figure A1. The visualization model should start running in a few seconds.

Step 4. The model can be stopped any time by closing the webpage.

Instructions for Stage II Visualization

This section shows the steps to run the stage II model. Please follow these steps:

Step 1. Paste the following URL link in your browser's address line and go to that link: https://cloud.anylogic.com/model/c19bfa43-d338-4a1b-93c3-4071d738add7?mode=DASHBOARD

Please follow the steps 2 to 4 as described above in the stage I visualization.

Instructions for Stage III Visualization

This section shows the steps to run the stage III model. Please follow these steps:

Step 1. Paste the following URL link in your browser's address line and go to that link: https://cloud.anylogic.com/model/4c291275-7bc4-4510-8fbd-e349bb59141e?mode=DASHBOARD

Please follow the steps 2 to 4 as described above in the stage I visualization.

Note: Please note that the visualization models are hosted in AnyLogic Cloud. After running the model, the screen can be closed by clicking on the "X" button on top right-hand side. These models have an 'Animated Run Time Limit' in the cloud. The model can be closed and re-run if you get a message as shown below:"Animated run time limitThe model has run for the maximum time allowed … "Simply close the message box, then close the model by pressing on the "X" button and run the model by pressing on the play button as shown in Stage I instructions.

References

1. Deamer, D.W. *First Life: Discovering the Connections between Stars, Cells, and How Life Began*; University of California Press: Berkeley, CA, USA, 2012.
2. Bernstein, M.P.; Sandford, S.A.; Allamonda, L.J. Life's fur flung raw material. *Sci. Am.* **1999**, *263*, 42–49. [CrossRef]
3. Deamer, D.W.; Dworkin, J.P.; Sandford, S.A.; Bernstein, M.P.; Allamandola, L.J. The first cell membranes. *Astrobiology* **2002**, *2*, 371–381. [CrossRef] [PubMed]
4. Pizzarello, J.R.; Cronin, J.R. Non-racemic amino acids in the Murchison and Murray meteorites. *Geochem. Cosmochem. Acta* **2000**, *64*, 329–338. [CrossRef]
5. Marchi, S.; Bottke, W.F.; Elkins-Tanton, L.T.; Bierhaus, M.; Wuennemann, K.; Morbidelli, A.; Kring, D.A. Widespread mixing and burial of Earth's Hadean crust by asteroid impacts. *Nature* **2014**, *511*, 578–582. [CrossRef] [PubMed]
6. Chatterjee, S. The hydrothermal impact crater lakes: The crucibles of life's origin. In *Handbook of Astrobiology*; Kolb, V.M., Ed.; CRC Press, Taylor & Francis: Boca Raton, FL, USA, 2018; pp. 265–295.
7. Chyba, C.; Sagan, C. Endogenous production, exogenous delivery and impact-shock synthesis of organic molecules: An inventory for the origin of life. *Nature* **1992**, *355*, 125–132. [CrossRef] [PubMed]
8. Chatterjee, S. The RNA/protein world and the endoprebiotic origin of life. In *Earth, Life, and System*; Clarke, B., Ed.; Fordham University Press: New York, NY, USA, 2015; pp. 39–79.
9. Chatterjee, S. A symbiotic view of the origin of life at hydrothermal impact crater lakes. *Phys. Chem. Chem. Phys.* **2016**, *18*, 20033–20046. [CrossRef] [PubMed]
10. Cockell, C.S. The origin and emergence of life under impact bombardment. *Phil. Trans. R. Soc.* **2006**, *B361*, 1845–1856. [CrossRef] [PubMed]
11. Osiniski, G.R.; Tornabene, L.L.; Banerjee, N.R.; Cockell, C.S.; Flemming, R.; Izawa, M.R.M.; McCutcheon, J.; Parnell, J.; Preston, L.J.; Pickersgill, A.E.; et al. Impact-generated hydrothermal systems on Earth and Mars. *Icarus* **2013**, *224*, 347–363. [CrossRef]
12. Kring, D.A. Impact events and their effect on the origin, evolution, and distribution of life. *GSA Today* **2000**, *10*, 1–7.
13. Martin, W.F.; Sousa, F.L.; Lane, N. Energy at life's origin. *Science* **2014**, *344*, 1092–1093. [CrossRef] [PubMed]
14. Boehnke, P.; Harrison, T.M. Illusory Late Heavy Bombardments. *Proc. Nat. Acad. Sci. USA* **2016**, *113*, 10802–10805. [CrossRef] [PubMed]
15. Mojzsis, S.J.; Harrison, T.M.; Pidgeon, T.T. Oxygen-isotope evidence from ancient zircons for liquid water at the Earth's surface 4300 Myr ago. *Nature* **2001**, *409*, 78–181.
16. Kasting, J.F. Atmospheric composition of Hadean-Early Archean Earth: Theimportance of CO. *Geol. Soc. Spec. Pap.* **2014**, *504*, 19–28.

17. Carter, C.W., Jr. What RNA world? Why a peptide/RNA partnership merits renewed experimental attention. *Life* **2015**, *5*, 294–320. [CrossRef] [PubMed]
18. Carter, C.W., Jr. An alternative to the RNA world. *Nat. Hist.* **2016**, *125*, 28–33. [PubMed]
19. Carter, C.W., Jr.; Wills, P.R. Interdependence, reflexivity, impedance matching, and the evolution of genetic coding. *Mol. Biol. Evol.* **2017**. [CrossRef] [PubMed]
20. Harish, A.; Caetano-Anolles, G. Ribosomal history reveals origin of modern protein synthesis. *PLoS ONE* **2001**. [CrossRef] [PubMed]
21. Bowman, J.C.; Hud, N.V.; Williams, J.D. The ribosome challenge to the RNA world. *J. Mol. Evol.* **2015**, *80*, 143–161. [CrossRef] [PubMed]
22. Hazen, R.M. *Genesis: The Scientific Quest for Life*; Joseph Henry Press: Washington, DC, USA, 2005.
23. Caetano-Anolles, K.; Caetano-Anolles, G. Piecemeal buildup of the genetic code, ribosome, and genomes from primordial tRNA building blocks. *Life* **2016**, *6*, 43. [CrossRef] [PubMed]
24. Kunnev, D.; Gospodinov, A. Possible emergence of sequence specific RNA aminoacylation via peptide intermediary to initiate Darwinian evolution and code through origin of life. *Life* **2018**, *8*, 44. [CrossRef] [PubMed]
25. Farias, S.T.; Rego, T.H.; Jose, M.V. tRNA core hypothesis for the transition from the RNA world to the ribonucleoprotein world. *Life* **2016**, *6*, 15. [CrossRef] [PubMed]
26. Lupus, A.N.; Alva, V. Ribosomal proteins as documents of the transition from unstructured (poly)peptides to folded proteins. *J. Struct. Biol.* **2017**, *198*, 74–81. [CrossRef] [PubMed]
27. Crick, F.H.C. The origin of genetic code. *J. Mol. Biol.* **1968**, *38*, 367–379. [CrossRef]
28. Crick, F.H.C. Central dogma in molecular biology. *Nature* **1970**, *227*, 561–563. [CrossRef] [PubMed]
29. Walker, S.I.; Davies, P.C.W. The algorithmic origins of life. *J. R. Soc. Interface* **2012**. [CrossRef] [PubMed]
30. Küppers, B.O. *Information and the Origin of Life*; MIT Press: Cambridge, MA, USA, 1990.
31. Rosen, R. Complexity and information. *J. Comput. Appl. Math.* **1988**, *22*, 211–218. [CrossRef]
32. Von Neumann, J. *Theory of Self-Reproducing Automata*; Burks, A.W., Ed.; University of Illinois Press: Chicago, IL, USA, 1966.
33. Reenskaug, T. MODELS-VIEWS-CONTROLLERS. Technical Note, Xerox PARC. December 1979. Available online: http://heim.ifi.uio.no/~{}trygver/mvc/index.html (accessed on 20 July 2018).
34. Cech, T.R. RNA as an enzyme. *Sci. Am.* **1986**, *255*, 64–75. [CrossRef] [PubMed]
35. Cech, T.R. Crawling out of the RNA world. *Cell* **2009**, *136*, 599–602. [CrossRef] [PubMed]
36. Gilbert, W. The RNA world. *Nature* **1986**, *319*, 618. [CrossRef]
37. Orgel, L.E. The origin of life. *Sci. Am.* **1994**, *271*, 77–83. [CrossRef]
38. Robertson, M.P.; Joyce, G.F. The origins of the RNA world. *Cold Spring Harb. Perspect. Biol.* **2012**, *4*, a003608. [CrossRef] [PubMed]
39. Cech, T.R. The ribosome is a ribozyme. *Science* **2000**, *280*, 878–879. [CrossRef]
40. Shapiro, R. A simpler origin of life. *Sci. Am.* **2007**, *296*, 24–31. [CrossRef]
41. De Duve, C. The beginnings of life on earth. *Am. Sci.* **1995**, *83*, 428–437.
42. Wachterhäuser, G. The cradle of chemistry of life: On the origin of natural products in a pyrite-pulled chemoautotrophic origin of life. *Pure Appl. Chem.* **1993**, *65*, 1343–1348. [CrossRef]
43. Damer, B.; Deamer, D.W. Coupled phases and combinatorial selection in fluctuating hydrothermal pools: A scenario to guide experimental approaches to the origin of cellular life. *Life* **2015**, *5*, 872–887. [CrossRef] [PubMed]
44. Lan, P.; Tan, M.; Zhang, Y.; Niu, S.; Chen, J.; Shi, S.; Qiu, S.; Peng, X.; Cai, G.; Cehng, H.; et al. Structural insight into precursor tRNA processing by yeast ribonuclease P. *Science* **2018**, *362*. [CrossRef] [PubMed]
45. Dyson, F. *Origins of Life*; Cambridge University Press: Cambridge, UK, 2004.
46. Maynard Smith, J.; Szathmary, E. *The Origins of Life*; Oxford University Press: New York, NY, USA, 1999.
47. Kampfner, R.R. Biological information processing: The use of information for the support of function. *Biosystems* **1989**, *22*, 223–230. [CrossRef]
48. Moreno, A.; Ruiz-Mirazo, K. The Informational nature of biological causality. In *Information and Living Systems: Philosophical and Scientific Perspectives*; Terzis, G., Arp, G.R., Eds.; MIT Press: Cambridge, MA, USA, 2011; pp. 157–175.

49. Shanks, N.; Pyles, R.A. Problem solving in the life cycles of multicellular organisms: Immunology and Cancer. In *Information and Living Systems: Philosophical and Scientific Perspectives*; Terzis, G., Arp, G.R., Eds.; MIT Press: Cambridge, MA, USA, 2011; pp. 157–175.

50. Miller, W.B. Biological information systems: Evolution as cognition-based information management. *Prog. Biophys. Mol. Biol.* **2018**, *134*, 1–26. [CrossRef] [PubMed]

51. Zwass, V. Information System. 2016. Available online: https://www.britannica.com/topic/information-system (accessed on 30 June 2018).

52. Turing, A.M. On computable numbers, with an application to the Entscheidungsproblem: A correction. *Proc. Lond. Math. Soc.* **1937**, *43*, 544–546.

53. Turing, A.M. The chemical basis of morphogenesis. *Phil. Trans. R. Soc. Lond.* **1952**, *237*, 37–72.

54. Shanon, C.E. A mathematical theory of communication. *Bell Syst. Tech. J.* **1948**, *27*, 379–423, 623–656. [CrossRef]

55. Davies, P. *The Fifth Miracle*; Simon & Schuster: New York, NY, USA, 1999.

56. Biro, J.H. Biological information—Definitions from a biological perspective. *Information* **2011**, *2*, 117–139. [CrossRef]

57. Wiener, N. *Cybernetics, Second Edition: Or the Control and Communication in the Animal and the Machine*; The MIT Press: Cambridge, UK, 1965.

58. Von Bertalanffy, L. *General System Theory: Foundations, Development, Applications*; George Braziller Publisher: New York, NY, USA, 1968.

59. Adriaans, P. Information. In *The Stanford Encyclopedia of Philosophy*, 2018 ed.; Zalta, E.N., Ed.; Stanford University: Stanford, CA, USA, 2012; Available online: https://plato.stanford.edu/archives/fall2018/entries/information (accessed on 5 August 2018).

60. Buckland, M. Information as a thing. *J. Am. Soc. Inf. Sci.* **1991**, *42*, 351–360. [CrossRef]

61. Floridi, L. The logic of being informed. *Log. Anal.* **2006**, *49*, 433–460.

62. Price, J.R. *An Introduction to Information Theory: Symbols, Signals, and Noise*; Dover: New York, NY, USA, 1980.

63. Tlusty, T. A model for the emergence of the genetic code as a transition in the noisy information channel. *J. Theor. Biol.* **2007**, *292*, 331–342. [CrossRef] [PubMed]

64. Woese, C. Evolution of the genetic code. *Naturwissen* **1973**, *60*, 447. [CrossRef]

65. Osawa, S. *Evolution of the Genetic Code*; Oxford University Press: Oxford, UK, 1995.

66. Woese, C.R. *The Genetic Code: The Molecular Basis for Genetic Expression*; Harper & Row: New York, USA, 1967.

67. Koonin, E.V. Why the central dogma: On the nature of biological exclusion principle. *Biol. Dir.* **2015**, *10*, 52. [CrossRef] [PubMed]

68. De Duve, C. *Singularities: Landmarks on the Pathways of Life*; Cambridge University Press: New York, USA, 2005.

69. Wong, J.T.F. A co-evolution theory of the genetic code. *Proc. Nat. Acad. Sci. USA* **1975**, *72*, 1909–1912. [CrossRef] [PubMed]

70. Bada, J.L. New insights into prebiotic chemistry from Stanley Miller's spark discharge experiments. *Chem. Soc. Rev.* **2013**, *42*, 2186–2196. [CrossRef] [PubMed]

71. Chen, I.A. Prebiotic chemistry: Replicating towards complexity. *Nat. Chem.* **2015**, *7*, 101–102. [CrossRef] [PubMed]

72. Wolf, Y.I.; Koonin, E.V. On the origin of the translation system and the genetic code world by means of natural selection, exaptation, and subfunctionalization. *Biol. Dir.* **2007**, *2*, 14. [CrossRef] [PubMed]

73. Svoboda, P.; Cara, A. Hairpin RNA: A secondary structure of primary importance. *Cell. Mol. Life Sci.* **2006**, *63*, 901–908. [CrossRef] [PubMed]

74. Eigen, M.; Winkler-Oswatitsch, R. Transfer-RNA, an early gene. *Naturwissen* **1981**, *68*, 282–292. [CrossRef]

75. Woese, C.R. On the evolution of the genetic code. *Proc. Nat. Acad. Sci. USA* **1965**, *54*, 1546–1552. [CrossRef] [PubMed]

76. Di Giulio, M. The origin of tRNA molecule: Implications for the origin of protein synthesis. *J. Theor. Biol.* **2004**, *226*, 89–93. [CrossRef] [PubMed]

77. Freeman, S. *Biological Science*, 2nd ed.; Pearson Prentice Hall: Upper Saddle River, NJ, USA, 2005.

78. Di Giulio, M. On the origin of transfer RNA molecule. *J. Theor. Biol.* **1992**, *159*, 199–209. [CrossRef]

79. Di Giulio, M. Was it an ancient gene codifying for a hairpin RNA that, by means of direct duplication, gave rise to the primitive tRNA molecule? *J. Theor. Biol.* **1995**, *177*, 95–101. [CrossRef]

80. Tamura, K. Origins and early evolution of the tRNA molecule. *Life* **2015**, *5*, 1687–1699. [CrossRef] [PubMed]
81. Tanaka, T.; Kikuchi, Y. Origin of cloverleaf shape of transfer RNA-the double hairpin model: Implication for the role of tRNA intro and long extra loop. *Viva Orig.* **2001**, *29*, 134–142.
82. Widmann, J.; Di Giulio, M.; Yarus, M.; Knight, R. tRNA creation by hairpin duplication. *J. Morphol. Evol.* **2005**, *61*, 524–530. [CrossRef] [PubMed]
83. Nagaswamy, U.; Fox, G.F. RNA ligation and the origin of tRNA. *Orig. Life Evol. Biosph.* **2003**, *33*, 199–209. [CrossRef] [PubMed]
84. Lahav, N. *Biogenesis: Theories of Life's Origins*; Oxford University Press: New York, NY, USA, 1999.
85. Zhou, R.; Basu, K.; Hartman, H.; Matocha, C.J.; Sears, S.K.; Vali, H.; Guzman, M.I. Catalyzed synthesis of zinc clays by prebiotic central metabolites. *Sci. Rep.* **2017**. [CrossRef] [PubMed]
86. Wong, J.T.F. Coevolution of genetic code and amino acid biosynthesis. *Trends Biochem. Sci.* **1981**, *6*, 33–36. [CrossRef]
87. Ikehara, K. Origins of gene, genetic code, protein and life: Comprehensive view of life systems from a GNC-SNS primitive code hypothesis. *J. Biosci.* **2002**, *27*, 165–186. [CrossRef] [PubMed]
88. Di Giulio, M. An extension of the coevolution theory of the origin of the genetic code. *Biol. Dir.* **2008**, *3*, 37. [CrossRef] [PubMed]
89. Noller, H.F. The driving force for molecular evolution of translation. *RNA* **2004**, *10*, 1833–1837. [CrossRef] [PubMed]
90. Noller, H.F. Evolution of protein synthesis from an RNA world. *Cold Spring Harb. Perspect. Biol.* **2012**, *4*, a003681. [CrossRef] [PubMed]
91. Ramakrishnan, V. The ribosome: Some hard facts about its structure and hot air about its evolution. *Cold Spring Harb. Symp. Quant. Biol.* **2010**, *74*, 155–179. [CrossRef] [PubMed]
92. Altstein, A.D. The progene hypothesis: The nucleoprotein world and how life began. *Biol. Dir.* **2015**, *10*, 67. [CrossRef] [PubMed]
93. Ban, N.; Nissen, P.; Hansen, J.; Moore, P.B.; Steitz, T.A. The complete atomic structure of the large ribosomal unit at 2.4 A resolution. *Science* **2000**, *289*, 905–920. [CrossRef] [PubMed]
94. Noller, H.F. On the origin of ribosome: Coevolution of subdomains of tRNA and rRNA. In *The RNA World*; Cold Spring Harbor Laboratory Press: Plainview, NY, USA, 1993; pp. 137–156.
95. Schimmel, P.; Alexander, R.W. All you need is RNA. *Science* **1998**, *281*, 658–659. [CrossRef] [PubMed]
96. Nissen, P.; Hansen, J.; Ban, N.; Moore, P.B.; Steitz, T.A. The structural basis of ribosome activity in peptide bond synthesis. *Science* **2000**, *289*, 920–929. [CrossRef] [PubMed]
97. Orelle, C.; Carlson, E.D.; Szal, T.; Florin, T.; Jewett, M.C.; Mankin, A.S. Protein synthesis by ribosomes with tethered subunits. *Nature* **2015**, *524*, 119–124. [CrossRef] [PubMed]
98. Savir, Y.; Tlusty, T. The ribosome as an optimal decoder: A lesson in molecular recognition. *Cell* **2013**, *153*, 471–479. [CrossRef] [PubMed]
99. Alberts, B.; Johnson, A.; Lewis, J.; Raff, M.; Roberts, K.; Walter, P. *Molecular Biology of the Cell*, 4th ed.; Garland Science: New York, NY, USA, 2002.
100. Koonin, E.V.; Novozhilov, A.S. Origin and evolution of the genetic code: The universal enigma. *Life* **2008**, *61*, 99–111.
101. Barbieri, M. What is information? *Phil. Trans. R. Soc.* **2016**, *A374*, 20150060. [CrossRef] [PubMed]
102. Woese, C.R. On the evolution of cells. *Proc. Nat. Acad. Sci. USA* **2002**, *99*, 8742–8747. [CrossRef] [PubMed]
103. Dunhill, P. Triplet nucleotide-amino acid pairing: A stereochemical division between protein and non-protein amino acids. *Nature* **1966**, *210*, 25–26. [CrossRef]
104. Yarus, M.; Widmann, J.J.; Knight, R. RNA-amino acid binding: A stereochemical era for the genetic code. *J. Mol. Evol.* **2009**, *69*, 406–429. [CrossRef] [PubMed]
105. Wong, J.T.F. Emergence of life: From functional RNA selection to natural selection and beyond. *Front. Biosci.* **2014**, *19*, 1117–1150. [CrossRef]
106. Freeland, S.J.; Knight, R.D.; Landweber, L.F.; Hurst, L.D. Early fixation of an optimal genetic code. *Mol. Biol. Evol.* **2000**, *17*, 511–518. [CrossRef] [PubMed]
107. Szathmary, E. The origin of the genetic code: Amino acids as cofactors in an RNA world. *Trends Genet.* **1999**, *15*, 223–229. [CrossRef]
108. Rodin, A.S.; Szathmary, E.; Rodin, S.N. On the origin of genetic code and tRNA before translation. *Biol. Dir.* **2011**, *6*, 14. [CrossRef] [PubMed]

109. Johnson, D.B.F.; Wang, L. Imprints of the genetic code in the ribosome. *Proc. Nat. Acad. Sci. USA* **2010**, *107*, 8298–8303. [CrossRef] [PubMed]

110. Ling, J.; Reynolds, N.; Ibba, M. Aminoacyl-tRNA synthesis and translation quality control. *Ann. Rev. Microbiol.* **2009**, *63*, 61–78. [CrossRef] [PubMed]

111. Giege, R.; Sissler, M.; Florentz, C. Universal rules and idiosyncratic features in tRNA identity. *Nucleic Acid Res.* **1998**, *16*, 5017–5035. [CrossRef]

112. Poole, A.M.; Jeffares, D.C.; Penny, D. The path from the RNA worlds. *J. Mol. Evol.* **1998**, *46*, 1–17. [CrossRef] [PubMed]

113. Yarus, M. RNA-ligand chemistry: A testable source of genetic code. *RNA* **2000**, *6*, 475–484. [CrossRef] [PubMed]

114. Wong, J.T.F.; Ng, S.K.; Mat, W.K.; Hu, T.; Xue, H. Coevolution theory of the genetic code at age forty: Pathway to translation and synthetic life. *Life* **2016**, *6*, 12. [CrossRef] [PubMed]

115. Eigen, M.; Schuster, P. The hypercycle: A principle of natural self-organization. Part A: Emergence of the hypercycle. *Naturwissens* **1997**, *64*, 541–565. [CrossRef]

116. Lahav, N. Prebiotic co-evolution of self-replication and translation in RNA world? *J. Theor. Biol.* **2001**, *151*, 531–539. [CrossRef]

117. Copley, S.D.; Smith, E.; Morowitz, H.J. A mechanism for the association of amino acids with their codons and the origin of genetic code. *Proc. Nat. Acad. Sci. USA* **2005**, *102*, 4442–4447. [CrossRef] [PubMed]

118. Crick, F.H.C. Codon-anticodon pairing: The wobble hypothesis. *J. Mol. Biol.* **1966**, *19*, 548–555. [CrossRef]

119. Knight, R.D.; Landweber, L.F. Rhyme or reason: RNA-arginine interactions and the genetic code. *Chem. Biol.* **1998**, *5*, R215–R220. [CrossRef]

120. Hartwell, L.H.; Hopfield, J.J.; Leibler, S.; Murray, A.W. From molecular to modular cell biology. *Nature* **1991**, *402*, C47–C52. [CrossRef] [PubMed]

121. Dagley, M.J.; Lithgow, T. TOM and SAM machineries in mitochondrial import and outer membrane biogenesis. In *The Enzymes*; Dalbey, R.E., Koehler, C.M., Tamanoi, F., Eds.; Academic Press: New York, NY, USA, 2007; Volume 25, pp. 309–343.

122. Goodsell, D.S. *The Machinery of Life*; Springer: New York, NY, USA, 2010.

123. Yockey, H.P. *Information Theory, Evolution, and the Origin of Life*; Cambridge University Press: Cambridge, UK, 2005.

124. Baum, D.A.; Lehman, N. Life's late digital revolution and why it matters for the study of the origins of life. *Life* **2017**, *7*, 34. [CrossRef] [PubMed]

125. Reenskaug, T. The Model-View-Controller (MVC) Its Past and Present. 2003. Available online: http://heim.ifi.uio.no/~{}trygver/2003/javazone-jaoo/MVC_pattern.pdf (accessed on 22 July 2018).

126. Pesavento, U. An Implementation of von Neumann's Self-Reproducing Machine. *Artif. Life* **1995**, *2*, 337–354. [CrossRef] [PubMed]

127. Arenas, M. Computer programs and methodologies for the simulation of DNA sequence data with recombination. *Front. Genet.* **2013**, *4*, 9. [CrossRef] [PubMed]

Article

Plausible Emergence of Autocatalytic Cycles under Prebiotic Conditions

Stefano Piotto [1,*]**, Lucia Sessa** [1]**, Andrea Piotto** [1]**, Anna Maria Nardiello** [1] **and Simona Concilio** [2]

[1] Department of Pharmacy, University of Salerno, Via Giovanni Paolo II, 132, 84084 Fisciano SA, Italy; lucsessa@unisa.it (L.S.); apiotto@hotmail.com (A.P.); annardiello@unisa.it (A.M.N.)

[2] Department of Industrial Engineering, University of Salerno, Via Giovanni Paolo II, 132, 84084 Fisciano SA, Italy; sconcilio@unisa.it

* Correspondence: piotto@unisa.it; Tel.: +39-320-4230068

Received: 7 January 2019; Accepted: 2 April 2019; Published: 4 April 2019

Abstract: The emergence of life in a prebiotic world is an enormous scientific question of paramount philosophical importance. Even when life (in any sense we can define it) can be observed and replicated in the laboratory, it is only an indication of one possible pathway for life emergence, and is by no means be a demonstration of how life really emerged. The best we can hope for is to indicate plausible chemical–physical conditions and mechanisms that might lead to self-organizing and autopoietic systems. Here we present a stochastic simulation, based on chemical reactions already observed in prebiotic environments, that might help in the design of new experiments. We will show how the definition of simple rules for the synthesis of random peptides may lead to the appearance of networks of autocatalytic cycles and the emergence of memory.

Keywords: origin of life; hypercycle; Monte Carlo

---

1. Introduction

The realization of a unified concept of life, incorporating all known biological systems separated from physical and chemical systems, is a long and complex process that is far from conclusion. Addy Pross [1] argues that the question about how life emerges is strongly connected with the definition of life. According to the synthetic theory of evolution, "life is a self-sustained chemical system capable of undergoing Darwinian evolution" [2]. This definition can be considered a derivation of another definition, formulated by M. Perret and elaborated upon by J. D. Bernal [3], that states: "Life is a potentially self-perpetuating system of linked organic reactions, catalyzed stepwise and almost isothermally by complex and specific organic catalysts which are themselves produced by the system."

A persuasive model of the origin and evolution of life came from the work of Eigen and Schuster in 1977. In their work [4] they suggested that self-replicative macromolecules, such as RNA and DNA in a suitable environment, exhibit behavior which we may call "Darwinian" and which can be formally represented by the concept of the "quasi-species." Though sophisticated, the hypercycle model requires other requisites to realize the condition of life. Compartmentalization is necessary to prevent dilution of chemicals and the stability of hypercycles but does not take into account the emergence of homochirality and the memory of the system [5].

The present work originates in this theoretical framework. We aim to address the problem of the emergence of a stable order in the spontaneous and prebiotic synthesis of peptides.

A system is considered "alive" when it is capable of producing its constitutive elements, though with some errors, and when it is capable of replicating itself, keeping roughly the same composition. These conditions define a system as autopoietic. The synthesis of constitutive materials is a necessary, though not sufficient, condition. It is essential to propose a way in which a system could keep the memory of its composition.

In living beings, the information about what must be synthesized, and when, is stored in DNA or RNA molecules. The genetic code is the map that translates the codon into an amino acid. In modern-day biology, this translation is accomplished by means of the ribosome, a complex molecular machine which was not available for the first emergent life. Let us assume that, before the appearance of a genetic code, it was necessary to have a system to maintain the memory of the formed molecules.

According to the RNA-world scenario [6], RNA molecules could provide both the information storage (in their sequences) and the catalytic activity. In the RNA scenario, proteins are not required in the early stage of life development. Nevertheless, the RNA-world was fiercely opposed [7] because of two orders of reasons: The prebiotic synthesis of RNA is very difficult, and RNA molecules catalyze only a few reactions, mainly the cleavage of RNA, despite the observation that all types of reactions necessary for nucleotide and peptide synthesis can be catalyzed by ribozymes [8]. RNA-world requires the prebiotic synthesis of some molecules, written information (in this case RNA molecules), and machinery to read and execute those data. The information must be copied and shared among the progeny.

In our opinion, the most severe critique of the RNA-world scenario is that the information can be stored in RNA molecules only with the help of a complex (protein) machinery. In more general terms, memory is hard to fit in any model without storage.

Many research works demonstrated that amino acids could be formed in prebiotic times, along with other organic compounds. The famous Miller–Urey experiment, for example, produced a mixture of amino acids [9]. Several amino acids have been found even in meteorites [10]. A consensus list of prebiotic amino acids includes Ala, Asp, Glu, Gly, Ile, Leu, Pro, Ser, Thr, and Val [11]. The oligomerization of amino acids to form new peptides is extremely hard to predict.

Several experiments demonstrated that the presence of small peptides in early Earth was plausible [12]. Peptide growing and degradation was considered in the pioneering work of Kauffman [13] and Eigen [14], for a curious case in the same year. In 1971, Eigen proposed the hypercycle, based on Orgel-type replicating "Watson–Crick" RNA strands, but with the additional idea that "a replicating pair 1 would catalyze or help the replication of 2, 2 would help 3 and so on, until N closed the 'hypercycle' and helped replicate 1".

Kauffman made the radical simple assumptions that the early molecules were polymers of two types of monomers, A and B, say two amino acids, or later, two nucleotides. These could undergo only cleavage and ligation reactions. Monomers A and B; and dimers AA, AB, BA, and BB, might serve as sustained food inputs to the system. Given this, the reaction network among the molecules could be determined. In 1971 the formation of medium or long peptides by amino acid polymerization was not yet wholly investigated experimentally. Many researchers share the idea that some random peptides could have been formed spontaneously and thus helped the onset of protobiological mechanisms (like autocatalytic and cross-catalytic cycles, complex formation with metals or other non-peptide molecules, membrane functionalization and control of trans-membrane traffic, new peptide formation by chain elongation or fragment condensation, synthesis and stabilization of nucleic acids, and so on) [15–21].

Since the work of Kaufmann and Eigen, several attempts have been made to simulate the emergence of autopoietic structures on a computer, but it very soon appeared clear that the number of required structures was too high to be calculated. A hundred amino acids can be combined in circa 10^{130} possibilities. The chemical space is too vast to be investigated either experimentally or computationally. The number of options is so vast that a particular sequence of 100 amino acids could likely have never appeared on our planet. Not only that, it must be considered that a single copy of a protein is not enough to sustain reproduction. It is important to have the accumulation of a particular protein for an organism to have an evolutionary advantage.

Nature could not have verified all possible amino acid combinations and, therefore, we face the problem of how the "few" existent proteins were produced and/or selected during prebiotic molecular evolution and, subsequently, how through a series of spontaneous steps of increasing molecular complexity they could produce life as self-reproducing protocells. Do existing proteins have some specific features that make them eligible for selection, for example in terms of particular thermodynamic, kinetic, or other physical properties?

Let us assume a system similar to the one proposed by Urey in 1953 [22]. We can hypothesize the spontaneous synthesis of amino acids and the regular production of fresh material. Let us assume that some kind of polymerization can take place on a mineral surface (or clay [23]) to catalyze the elongation of short peptides. The presence of a surface offers the advantage of spatial confinement of the amino acids that may lead to high peptide concentrations. Let us assume a continuous formation and degradation of peptide taking place onto the surface.

We can imagine the rain washing away the newly synthesized materials from the mineral surface, and the sun evaporating water and increasing the peptides' concentrations in these ponds. At high concentrations, these peptides can interact with each other. Any theoretical model should consider the presence of competing chemicals or parasites. Parasites of hypercycles may interrupt the cooperation of replicators [24].

The study of the emergence of autocatalytic cycles from short peptides presents some other critical problems. The first one is related to the number of possible sequences that can be generated. The astronomically large number precludes the possibility of exhaustive analysis. The second problem is the identification of the onset of such a cycle. Instead of using the twenty natural amino acids, we decided to limit their number to those present in prebiotic conditions only.

Within this theoretical framework, we aimed to evaluate the condition for the onset of hypercycles, while addressing the stability against parasite sequences and the establishment of a kind of dynamic memory. We have written a computer program (Genesys) to reproduce the chemical synthesis of random peptides. As starting pool, we chose the amino acids present in the Murchison meteorite [25] (A, V, G, E, D, and S) with concentrations inversely proportional to the formation energy, and a reservoir of amino acids to fish from.

With these six amino acids in large quantities, we have defined some rules that, once set, should permit the emergence of new properties. We began to catalyze the first reactions among amino acids, and we observed the formation of dimers; these dimers can, in turn, lengthen or degrade and return to their initial state. If the speed of synthesis is slightly higher than the speed of degradation, gradually longer peptides will accumulate. Shorter peptides will be slightly more stable than longer ones because the latter can break into many pieces. Following this approach, after a few generations of synthesis and degradation, an enormous number of different sequences will appear. Some sequences could be "parasitic" and compete with emerging hypercycles.

We have explored the possibility of exploiting the similarity among sequences to drastically reduce the number of synthesized peptides and to protect hypercycles from parasites. The similarity among peptides could favor the aggregation, similar to what happens in plaque formation. Even the simplest form of peptide activity, proteolysis, can direct the cut of an elongated peptide in a position that would lead to the accumulation of similar peptides.

Of course, a peptide α may interact with a peptide β that shares only a portion of the entire sequence, and still will lead to the accumulation of α or β, or more, peptides. The idea of similarity increasing the concentration of different peptides can be misleading. One may think that proteolysis would dramatically increase the number of possible sequences, but similarity can play a more sophisticated, and still spontaneous, role. For example, in the case of elongation, it may lead to the propagation of subsequences across the population. This step permits an increase in the number of available three-dimensional structures, and, at the same time, limitation of the number of sequences. Rather than having the accumulation of a myriad of different peptides, one can observe an increase in the structural complexity without an explosion of the number of peptides. With the onset of new kinds of sequences, we may find new three-dimensional structures, and new properties may emerge.

As the number of possible combinations for analysis increases with the square power of the number of sequences, it is clear that this method would not lead to any peptide. It is well known that two similar sequences generally have similar three-dimensional structures and biological functions.

We assumed that similar peptides could more easily aggregate to very different peptides and, in this way, have a lower probability of degradation. This hypothesis favors the persistence of similar sequences but does not favor their development. To have an accumulation of a particular sequence, it is necessary to have a sort of "memory" of the system. In biological times, memory is offered by specialized molecules that preserve the information of the sequences (DNA or RNA), together with a code for their translation. In prebiotic times, memory can emerge dynamically with autocatalytic cycles exploiting the similarity between two sequences to favor a template effect. Let us imagine four peptides (A, B, X, Δ) as these reported:

[A] GGGAAAVGVGAA
[B] GGGAAAAGAGVVVVAAGVAV
[X] AAVGVGAVVAAVVA
[Δ] VAAGVAVGGGVGAA

We can hypothesize that, with the help of similarity-driven catalysis, a cycle of this type may emerge:

$$
\begin{array}{ccc}
A & \text{---} & B \\
[& &] \\
\Delta & \text{\_} & X
\end{array}
$$

In the presence of autocatalytic cycles, the concentration of peptides {A, B, X, Δ} may vary cyclically. This means that the number of occurrences of particular sequences will not change in monotonous ways, but cyclically. More precisely, the periodic variation in concentration of a specific peptide is a necessary, but not sufficient, condition for the appearance of a cycle. It is important to note that we are considering the concentration oscillations arising from an autocatalytic activity, rather than fluctuation induced by external factors (temperature, water concentration, or lightning).

The primary objective of the present work is to set up the conditions by which such an autocatalytic cycle can emerge spontaneously from the random synthesis of linear peptides. Here we want to test if the introduction of similarity may lead to the onset of autocatalytic cycles. The onset of self-sustaining autocatalytic cycles would represent a rudimental and primordial form of memory in the system without the need of DNA or RNA.

2. Materials and Methods

Genesys is a program written in C++ that allows the user to build random libraries of peptides. From an initial set of amino acids, Genesys uses a Monte Carlo method to bind two amino acids or to lengthen and shorten the present peptides. At each generation, Genesys randomly chooses between present peptides and amino acids and proceeds with the operations of synthesis and fragmentation.

The calculation is iterative. Each "generation" consists of the following steps: starting pool, synthesis, and fragmentation. Starting pool represents the set of all the amino acids and all the starting oriented polypeptides. The method takes into account the orientation of the polypeptides. For example, the polypeptide GENESYS is different from the polypeptide SYSENEG because each sequence is oriented in the N-C direction.

The synthesis starts randomly from the first molecule of the pool that is randomly coupled with another molecule. "Randomly" represents the analog of the immediate proximity of two molecules in the real world. Once the pairs are made, the algorithm proceeds to calculate the probability P that the two polypeptides can combine. Let us define two strings: S1 and S2. P is calculated in terms of their length and on the basis of their similarity (*simil*) as defined in the following:

$$\text{If } (length\ S1 \leq 3\ AND\ length\ S2 \leq 3)$$
$$P = c1 * c2$$
$$\text{else}$$
$$P = c1 * c2 * simil$$

(1)

Whereas

$$simil = \frac{\max H}{\min(length\ S1;\ length\ S2)}; \leq 1$$

(2)

Length S1 and S2 represent the number of amino acids constituting the two peptides of the pair; c1 and c2 are the interaction coefficients as arbitrarily defined in the program to favor the interaction of longer peptides (listed in Table 1).

Table 1. The interaction coefficients.

Length	Interaction Coefficient
1	0.083
2	0.168
3	0.310
4	0.500
5	0.690
6	0.832
7	0.917
8	0.968
9	0.982
10	0.992

MaxH is the maximum of the substitution matrix *H*. The matrix is built following the same criteria of the BLOSUM matrix [26]. It is used to score alignments between different sequences. *H* is built by progressively finding the matrix elements, starting at $H_{1,1}$ and proceeding in the directions of increasing *i* and *j*. Each item is set according to:

$$H_{i,j} = max \begin{cases} 0 \\ H_{i-1,j-1} + S_{i,j} \\ H_{i-1,j} - d \\ H_{i,j-1} - d \end{cases}$$

(3)

The $S_{i,j}$ is the similarity score of comparing two amino acids (obtained from the similarity Table 2 as defined in Equation (4)), and *d* is the penalty for a single gap. The penalty gap *d* is set to 2.

$$S_{i,j} = \frac{10}{10 + |S_i - S_j|}$$

(4)

The values of $S_{i,j}$ range from 0 to 1. The similarity table (Table 2) is derived from chemical–physical properties such as polarity and ΔG of solvation, where the polarity (polar) is taken from [27], and the amino acid energy of solvation ΔG_w is taken from [28].

The similarity coefficients (S) are obtained by rounding the results of the following equation [28]:

$$S = -69 + 3 \cdot polar + 1.3 \cdot \Delta Gw + \sqrt[3]{-6.5 \cdot polar} + \sqrt[3]{-34.9 \cdot \Delta Gw}$$

(5)

The values range permits calculation of the similiarity between two sequences in a BLOSUM fashion.

Table 2. The amino acids' similarity coefficients were calculated by their polarity and ΔG of solvation.

Amino Acid	Similarity Coefficient	Amino Acid	Similarity Coefficient
L	0	M	0.4
I	0	Y	0.6
F	0.1	C	0.6
W	0.2	A	0.7
V	0.4	T	2
G	2.5	H	−6
S	3	R	−6
Q	5	D	8
N	−8	K	−8
E	−9	P	12

Once the synthesis is complete, the program moves to the fragmentation, with a probability P that is proportional to the length of the polypeptide.

$$P = \frac{length}{threshold}; P \leq 1 \tag{6}$$

The threshold is the length beyond which the polypeptide is certainly fragmented. Once fragmentation is complete, the set of new molecules represents the starting pool of the new generation.

We used as the starting pool the six amino acids (A, V, G, E, D, and S) found in the Yamato meteorite [29], for which plausible prebiotic synthesis pathways are known. The initial pool can be modified during the experiment to simulate the appearance of new molecules in the amino acid pool.

The program generates text files containing the sequences and the number of occurrences in the sample. The similarity was calculated by multiple alignments with the ClustalW algorithm implemented in MEGA7 [30], with the Jones, Taylor, and Thornton (JTT) method [31]. Identical sequences have a distance of 0. A threshold for the distance values of 0.300 was set. Due to the large number of amino acid sequences obtained, they were clustered on the basis of their length and similarity in order to be able to analyze them. In this way, for two or more extremely similar sequences only one representative sequence was taken, increasing its number of occurrences. The clustering process is called epoch. The sequences obtained after clustering were submitted again to the program for a new epoch.

The whole experiment saw the creation of 20 epochs, equal to 8000 generations.

In the restart phase, equivalent to a second experiment, we wanted to verify that if a sequence was eliminated from one of the previously generated epochs, the one closely related to it would not have formed. We proceeded, therefore, with the elimination of a sequence from epoch 4, and we repeated the previously described protocol for the remaining epochs. The choice of the sequence to be eliminated is arbitrary.

3. Results

At the beginning of the simulations, peptides can catalyze only two processes: the lengthening of similar peptides and the shortening of peptides at precise positions. As a result, peptide building blocks begin to form. Kauffman hypercycles can emerge spontaneously if the system continues to grow in complexity without the formation of an astronomical amount of peptides. In fact, the number of sequences produced by the Monte Carlo method increases extremely rapidly and exponentially. After a few hundred generations, the number of peptides generated is in the order of 10^5, and the number of those surviving the degradation is 10^3.

It is, therefore, reasonable to cluster similar sequences and let only the representatives of each cluster of similar peptides in the sequence pool. The number of occurrences of these sequences represents the size of the cluster. Each group of similar sequences tends to fade over the generations due to the degradation of the peptides. To be sure that the presence of oscillations in the number of occurrences is neither an artifact of the algorithm nor a stochastic fluctuation, starting from epoch 4 we deleted some sequences. The choice of the peptide to be removed depends on their condition of variable and periodic concentration. It is important to remember that, in the course of the analysis, we indicate as sequence a cluster of n sequences with reciprocal distance lower than the threshold value. If the sequence is part of a Kauffman cycle, variations in the concentrations of the other sequences of the hypercycle should be observed. For very complex systems, with different interacting hypercycles, not even the deletion of one sequence would lead to the complete suppression of the other peptides of the cycle.

Following the suggestion of Eigen, we could hypothesize that the peptide A could catalyze the formation of B, that B catalyzes the formation of X, and the latter could catalyze the formation of A. In this simple cycle, each peptide catalyzes the formation of other peptides and ultimately sustains its creation.

As we can see in Figure 1, the ability of the similarity to increase the frequency of polymerization has led to the presence of numerous sequences for which the concentration is oscillating over the generations. To highlight the presence of the hypercycle we have set equal to zero the number of occurrences of the sequence α (see Table 3) AAGGAGGGGGGGGGGGGGGGAGGGAAGGGAAVGGGGG. Figure 1b shows the sequences that disappear in subsequent generations once the α sequence has been deleted. This result cannot be considered conclusive proof of the presence of interconnected cycles, but it suggests the high level of interconnection of the considered peptide system.

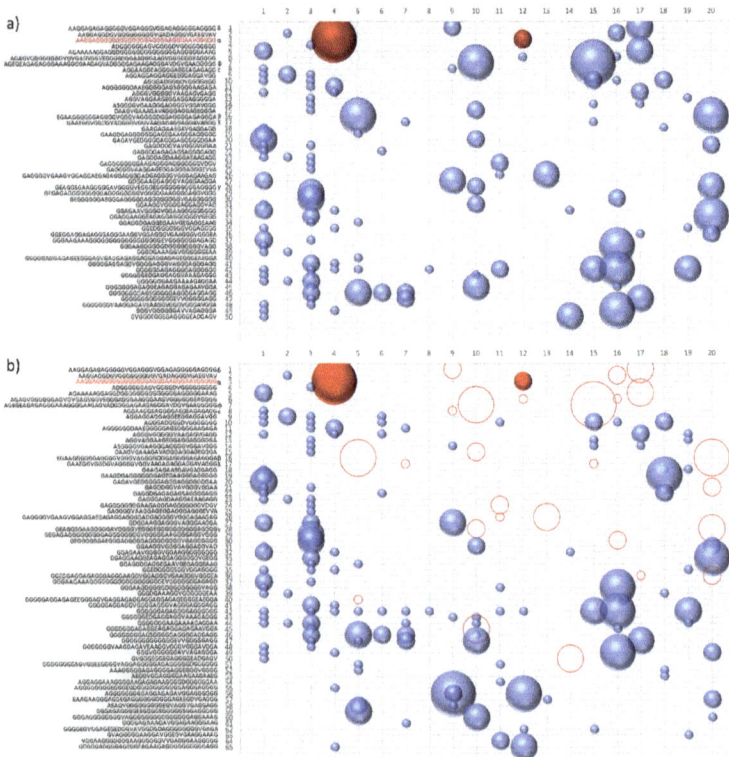

Figure 1. Clusters of similar sequences for each epoch. The number of copies of each group is shown as sphere size (**a**). Evolution of sequences after 8000 generations starting from a pool of six amino acids. (**b**) Evolution of sequences after the removal of the sequence in red from the pool at epoch 4 (corresponding to 1600 generations). Empty circles correspond to sequences that do not appear in the evolutive process upon the removal of the sequence in epoch 4.

Table 3. Sequences belonging to the same hypercycle.

Abbreviation	Sequences of the Cluster
α	AAGGAGGGGGGGGGGGGGGAGGGAAGGGAAVGGGGG
β	EGAAGGGGGGAGGGGVGGGVAGGGGDGGAGGGGAGAGGGA
χ	GAAEGGVGGDGVAGGGGVGGVAAGAGAGGAGGAVAGGG
δ	AAGGAGAGAGGGGGVGGAGGGVGGAGAGGGGGAGGGG
ε	AGGAAGGEAGGGGAEGEAGAGAGG
φ	GEGEAGAGAGGGAAAGGGGAAGAGVADGGGGAGAAGAGGGAVDGVGAAGGGGG
γ	GEAGGSGAAGGGGGAVGGGGVEGGGEGGGGGGGGGGAGGGG

The system of peptides represented in Figure 1a,b suggest how different sequences are associated with sequence α. To confirm the presence of an autocatalytic cycle, we have eliminated, again at epoch 4, another sequence chosen from among those that appear to interact with the sequence α.

Figure 2 shows the result of the new evolution.

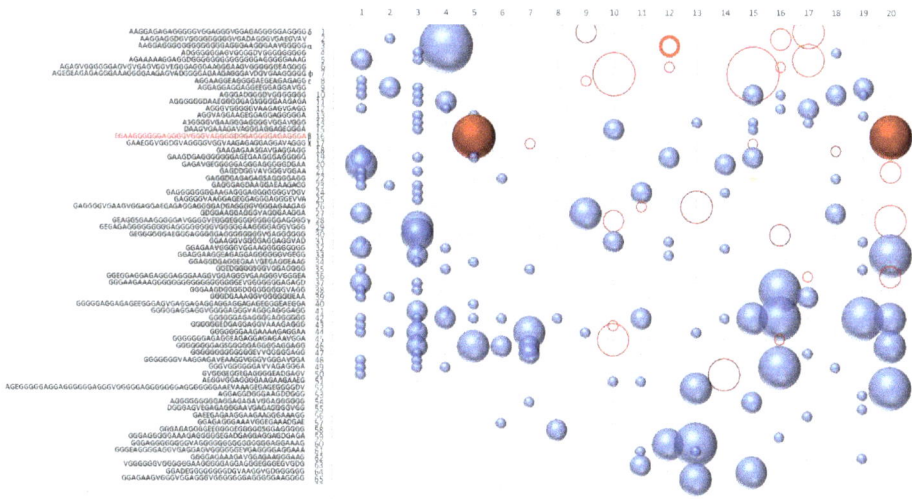

Figure 2. Clusters of similar sequences for each epoch. The number of copies of each cluster is shown as sphere size. Evolution of sequences after the removal of the sequence in red from the pool at epoch 4 (corresponding to 1600 generations). Empty circles correspond to sequences that do not appear in the evolutive process upon the removal of the sequence in epoch 4. For the sake of clarity, with the marked empty red circle, we indicate the sequence α.

The deletion of the sequence β (see Table 3) causes the extinction of the sequences χ, δ, and ε. The sequences belonging to the same hypercycle are listed in Table 3. The deletion of the same sequences by α or β confirms that α and β belong to the same hypercycle. Figure 3 shows some representative sequences obtained in the described experiments.

The presence of self-catalytic cycles is a necessary condition for the emergence of life, but not a sufficient one. We have observed that longer sequences can be formed and how sequences can catalyze the formation of similar sequences and therefore the establishment of self-catalytic cycles.

The obtained results are not accidental. We have repeated the same experiment from 50 slightly different initial conditions varying the concentration of glycine in the initial pool. The pool was constituted by the main six amino acids with constant concentration of 38 A, 4 D, 11 E, 100 G, 1 S, and 10 V. In the validation set we span the number of glycine from 75 to 124 and, in every case, we have observed the emergence of hypercycles. The sequences tend to be different in the exact composition but similar as calculated with Equation (4). In the Supplementary Information we have reported the list of the 83 most abundant sequences, and their mean concentrations with the standard deviation. The data support the result that the appearance of autocatalytic cycles is not accidental or dependent on a specific choice of initial conditions.

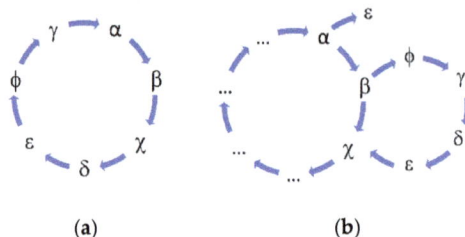

| (a) | (b) |

Figure 3. The hypercycle that emerged spontaneously imposing a similarity control to peptide replication. (**a**) An isolated hypercycle; (**b**) two interconnected hypercycles.

Along with the increase in sequence complexity, new properties can emerge—for example, amino acids bearing carboxylic groups in the side chain permit chelation of metal ions. Another example is the possibility of a disulfide bridge to increase the structural stability. It can be hypothesized that the role of some peptides may be the protection of other peptides from degradation. Finally, when a system is populated with enough copies of each protein, it is easy to imagine the encapsulation of materials in protovesicles. Vesicles can be formed by fatty acids when the concentration of water decreases. Upon water addition, this vesicle begins to swell and splits into two, incorporating a sufficient number of structures to continue the cycle.

It is also important to underline that different sequences correspond to different three-dimensional structures and different protein structures correspond to different functions. In this way, our model offers a framework to accommodate the emergence of new features, their diffusion in the population, the establishment of long-term memory and, above all, the possibility of evolution of the pool of peptides into new and more complex hypercycles. When a peptide becomes long enough, it starts to fold to reach a minimum of free energy. With a specific three-dimensional structure, certain catalytic activity may appear. For example, the appearance of sequences containing 2 or 3 residues of aspartic acid may allow the absorption of metallic cations and this new type of peptide may spread in the population using the mechanism described above.

4. Conclusions

Some of the earliest modelings of the origins of life came with the works of Eigen and Kauffman. The authors discussed the interlocking of reaction cycles as an explanation for the self-organization of prebiotic systems. The authors described the so-called hypercycle type of functional organization and stressed its possible relevance concerning the origin and evolution of life. We have shown how the introduction of similarity in the process of mating and degradation of nascent peptides can lead to a significant accumulation of some sequences.

More importantly, it was observed that the small number of sequences together with the concept of similarity allowed the emergence of a kind of dynamic memory. The presence of hypercycles is responsible for the conservation of information over time. The memory of the sequences is not preserved in other molecules such as DNA or RNA but it is an emerging property, and an obvious consequence of the similarity applied to the synthesis. It does not elude our attention that this type of memory can be present in every oscillating system, for example, in the brain. A prebiotic environment in which the presence of catalytic hypercycles may have preceded nucleic acid synthesis is plausible and even likely. It is worth noting that the described model does not explicitly take into account the chirality of amino acids. However, for the above considerations, the similarity may be responsible for a kind of template effect and lead to the accumulation of homochiral sequences.

This work has thermodynamic sense in an open system. Qualitatively, one can think of entropy S as a "fluid" that can be created or produced and which flows. Stationary states correspond to processes with no variation of entropy. The entropy change is the sum of the entropy changes generated inside the system with those of the entropy exchanged with its surroundings. An open system may undergo a decrease in its symmetry as in the presented work, thanks to the increase of the surrounding environmental entropy. Stability has its thermodynamic expression in Prigogine's theorem on minimum entropy production [32].

The entropy production reveals the rate of dissipation and the entropy exchange is the flow of entropy to or from the environment. The matter fluxes that drive entropy production are crucial for determining the thermodynamic stability of an open system. The entropy production is an important concept that is central to the thermodynamics of evolution [33]. Even though the investigation of entropy production is of extreme importance, it is not the goal of the present work. Finally, this sort of dynamic memory allowing the accumulation of a continuously expanding set of peptides is the prerequisite to proceeding with the slow optimization of the first proteins through a well-known model of Darwinian evolution. A prebiotic environment in which the presence of catalytic hypercycles may have preceded nucleic acid synthesis is plausible and even likely.

Supplementary Materials: The following are available online at http://www.mdpi.com/2075-1729/9/2/33/s1. The program Genesys is freely available upon request to the authors.

Author Contributions: Conceptualization, S.P.; methodology, S.P. and A.M.N.; software, A.P.; validation, S.P. and A.M.N.; writing—original draft preparation, S.P. and L.S.; writing—review and editing, S.C; visualization, A.M.N.

Funding: This research was funded by COST Action CM1304 Emergence and evolution of complex chemical systems.

Acknowledgments: We wish to express our gratitude to Bruno Maresca for the many long and passionate discussions of the onset of chemical and biological organization, and to Stuart Kauffman for his inspiring view of the emergence of life presented in COST meeting CM1304 in 2017 in Sopron (Hungary).

Conflicts of Interest: The authors declare no conflict of interest.

References

1. Pross, A. Toward a general theory of evolution: Extending Darwinian theory to inanimate matter. *J. Syst. Chem.* **2011**, *2*, 1. [CrossRef]
2. Luisi, P.L. About various definitions of life. *Orig. Life Evol. Biosph.* **1998**, *28*, 613–622. [CrossRef] [PubMed]
3. Perret, J. Biochemistry and bacteria. *New Biol.* **1952**, *12*, 68–69.
4. Eigen, M.; Schuster, P. A principle of natural self-organization. *Naturwissenschaften* **1977**, *64*, 541–565. [CrossRef] [PubMed]
5. Smith, J.M. Hypercycles and the origin of life. *Nature* **1979**, *280*, 445–446. [CrossRef] [PubMed]
6. Waldrop, M.M. Did life really start out in an RNA world? *Science* **1989**, *246*, 1248–1250. [CrossRef] [PubMed]
7. Kurland, C.G. The RNA dreamtime: Modern cells feature proteins that might have supported a prebiotic polypeptide world but nothing indicates that RNA world ever was. *Bioessays* **2010**, *32*, 866–871. [CrossRef]
8. Chen, X.; Li, N.; Ellington, A.D. Ribozyme catalysis of metabolism in the RNA world. *Chem. Biodivers.* **2007**, *4*, 633–655. [CrossRef] [PubMed]
9. Miller, S.L. A production of amino acids under possible primitive earth conditions. *Science* **1953**, *117*, 528–529. [CrossRef] [PubMed]
10. Pizzarello, S. The chemistry of life's origin: A carbonaceous meteorite perspective. *Acc. Chem. Res.* **2006**, *39*, 231–237. [CrossRef] [PubMed]
11. Longo, L.M.; Blaber, M. Protein design at the interface of the pre-biotic and biotic worlds. *Arch. Biochem. Biophys.* **2012**, *526*, 16–21. [CrossRef] [PubMed]
12. Lee, D.H.; Severin, K.; Ghadiri, M.R. Autocatalytic networks: The transition from molecular self-replication to molecular ecosystems. *Curr. Opin. Chem. Biol.* **1997**, *1*, 491–496. [CrossRef]

13. Kauffman, S.A. *The Origins of Order: Self-Organization and Selection in Evolution*; OUP USA: New York, NY, USA, 1993.

14. Eigen, M. Selforganization of matter and the evolution of biological macromolecules. *Naturwissenschaften* **1971**, *58*, 465–523. [CrossRef]

15. Klein, A.; Bock, M.; Alt, W. Simple mechanisms of early life—Simulation model on the origin of semi-cells. *Biosystems* **2017**, *151*, 34–42. [CrossRef] [PubMed]

16. Lancet, D.; Zidovetzki, R.; Markovitch, O. Systems protobiology: Origin of life in lipid catalytic networks. *J. R. Soc. Interface* **2018**, *15*, 20180159. [CrossRef]

17. Mann, S. The origins of life: Old problems, new chemistries. *Angew. Chem. Int. Ed.* **2013**, *52*, 155–162. [CrossRef] [PubMed]

18. Piotto, S. Lipid aggregates inducing symmetry breaking in prebiotic polymerisations. *Orig. Life Evol. Biosph.* **2004**, *34*, 123–132. [CrossRef]

19. Piotto, S.; Concilio, S.; Mavelli, F.; Iannelli, P. Computer simulations of natural and synthetic polymers in confined systems. In *Macromolecular Symposia*; Wiley-VCH: Weinheim, Germany, 2009; pp. 25–33.

20. Piotto, S.; Mavelli, F. Monte Carlo simulations of vesicles and fluid membranes transformations. *Orig. Life Evol. Biosph.* **2004**, *34*, 225–235. [CrossRef]

21. Rode, B.M. Peptides and the origin of life1. *Peptides* **1999**, *20*, 773–786. [CrossRef]

22. Miller, S.L.; Urey, H.C. Organic compound synthesis on the primitive earth. *Science* **1959**, *130*, 245–251. [CrossRef]

23. Swadling, J.B.; Coveney, P.V.; Greenwell, H.C. Clay minerals mediate folding and regioselective interactions of RNA: A large-scale atomistic simulation study. *J. Am. Chem. Soc.* **2010**, *132*, 13750–13764. [CrossRef] [PubMed]

24. Kun, Á.; Szilágyi, A.; Könnyű, B.; Boza, G.; Zachar, I.; Szathmáry, E. The dynamics of the RNA world: Insights and challenges. *Ann. N. Y. Acad. Sci.* **2015**, *1341*, 75–95. [CrossRef] [PubMed]

25. Cronin, J.; Gandy, W.; Pizzarello, S. Amino acids of the Murchison meteorite. *J. Mol. Evol.* **1981**, *17*, 265–272. [CrossRef] [PubMed]

26. Henikoff, S.; Henikoff, J.G. Amino acid substitution matrices from protein blocks. *Proc. Natl. Acad. Sci. USA* **1992**, *89*, 10915–10919. [CrossRef] [PubMed]

27. Available online: https://www.cgl.ucsf.edu/chimera/docs/UsersGuide/midas/hydrophob.html (accessed on 11 March 2019).

28. Gu, W.; Rahi, S.J.; Helms, V. Solvation free energies and transfer free energies for amino acids from hydrophobic solution to water solution from a very simple residue model. *J. Phys. Chem. B* **2004**, *108*, 5806–5814. [CrossRef]

29. Higgs, P.G.; Pudritz, R.E. A thermodynamic basis for prebiotic amino acid synthesis and the nature of the first genetic code. *Astrobiology* **2009**, *9*, 483–490. [CrossRef] [PubMed]

30. Kumar, S.; Stecher, G.; Tamura, K. MEGA7: Molecular evolutionary genetics analysis version 7.0 for bigger datasets. *Mol. Biol. Evol.* **2016**, *33*, 1870–1874. [CrossRef]

31. Jones, D.T.; Taylor, W.R.; Thornton, J.M. The rapid generation of mutation data matrices from protein sequences. *Bioinformatics* **1992**, *8*, 275–282. [CrossRef]

32. Kondepudi, D.; Prigogine, I. *Modern Thermodynamics: From Heat Engines to Dissipative Structures*; John Wiley & Sons: Hoboken, NJ, USA, 2014.

33. Hochberg, D.; Ribó, J.M. Stoichiometric network analysis of entropy production in chemical reactions. *Phys. Chem. Chem. Phys.* **2018**, *20*, 23726–23739. [CrossRef] [PubMed]

Article

Prebiotic Sugar Formation Under Nonaqueous Conditions and Mechanochemical Acceleration

Saskia Lamour [1,2,†], Sebastian Pallmann [1,†], Maren Haas [1,2] and Oliver Trapp [1,2,*]

[1] Department Chemie, Ludwig-Maximilians-Universität München, Butenandtstr. 5-13, 81377 München, Germany; saskia.lamour@cup.lmu.de (S.L.); sebpall@aol.com (S.P.); maren.haas@cup.lmu.de (M.H.)
[2] Max-Planck-Institut für Astronomie, Königstuhl 17, 69117 Heidelberg, Germany
* Correspondence: oliver.trapp@cup.uni-muenchen.de
† These authors contributed equally to this work.

Received: 2 June 2019; Accepted: 18 June 2019; Published: 20 June 2019

Abstract: Monosaccharides represent one of the major building blocks of life. One of the plausible prebiotic synthesis routes is the formose network, which generates sugars from C1 and C2 carbon sources in basic aqueous solution. We report on the feasibility of the formation of monosaccharides under physical forces simulated in a ball mill starting from formaldehyde, glycolaldehyde, DL-glyceraldehyde as prebiotically available substrates using catalytically active, basic minerals. We investigated the influence of the mechanic energy input on our model system using calcium hydroxide in an oscillatory ball mill. We show that the synthesis of monosaccharides is kinetically accelerated under mechanochemical conditions. The resulting sugar mixture contains monosaccharides with straight and branched carbon chains as well as decomposition products. In comparison to the sugar formation in water, the monosaccharides formed under mechanochemical conditions are more stable and selectively synthesized. Our results imply the possibility of a prebiotic monosaccharide origin in geochemical environments scant or devoid of water promoted by mechanochemical forces such as meteorite impacts or lithospheric activity.

Keywords: aldol reaction; mechanochemistry; minerals; monosaccharides; prebiotic chemistry

1. Introduction

The formose reaction is the classical route to carbohydrates based on the oligomerization of formaldehyde (C1) through a cascade of cross-, retro- and aldol reactions in aqueous solution in the presence of a basic catalyst, typically calcium hydroxide (Scheme 1) [1–4]. The resulting product mixture contains sugars of different length, constitution, and configuration [5–8]. The missing selectivity for specific monosaccharides is one of the major problems of the formose reaction in the context of the origin of life. Furthermore, isomerization and the instability of sugars under basic conditions are obstructive to the sugar formation in aqueous solutions [9,10]. The subsequent degradation reactions produce, for example, lactic acid, saccharic acid and α-dicarbonylic acids [11] and result in a dark tar-like substance. A water-independent reaction pathway to monosaccharides offers a consequential approach.

Contemplated scenarios for the origin(s) of life are as ample as the variety of building blocks, and molecules life is made of [12]. Examples of those are the warm little pond [13], hydrothermal vents [14,15], volcanic environments [16,17], drying lagoons [18–21], the primordial soup [22,23], eutectic solutions [24,25] or comet ponds [26–28]. Whereas each of them addresses distinct open questions in the context of the emergence of life, certain issues remain unresolved. A particular one lies with the contradictory assumption that life was formed in water when several chemical and biologically relevant transformations are condensation reactions that are disfavored in aqueous environments [29,30]. In this respect, only a scarce number of mechanochemical approaches [31,32] have been identified so far that plausibly demonstrate the formation of biologically relevant molecules when only grinding or milling of dry substrates is used to trigger reactivity. Such examples include the synthesis of α-amino acid derivatives [33,34] and modification of nucleosides and nucleotides [35]. Sources for mechanochemical energy considered are lithospheric activities such as weathering, erosion, diagenetic processes and meteoritic impacts on Earth. They can also be accounted for in asteroids tectonics [36]. Based on our interest in the formose reaction network [37] and on the feasibility of aldol reactions in mechanochemical setups [38,39], we investigated the potential formation of carbohydrates under nonaqueous and mechanochemical reaction conditions. We used an oscillatory and a planetary ball mill, which have a high mechanic input leading to shorter and therefore more practicable reaction times.

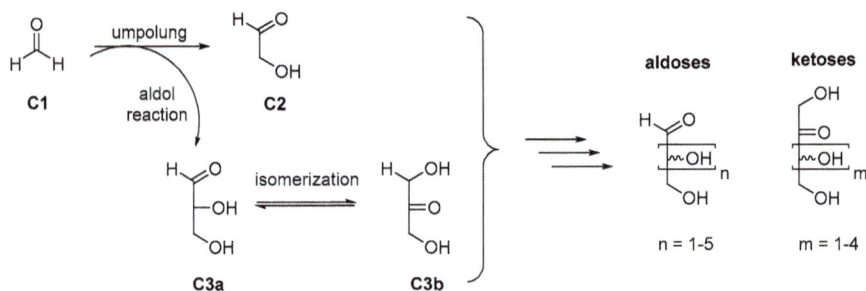

Scheme 1. Monosaccharide synthesis in the formose reaction network; **C1**: formaldehyde, **C2**: glycolaldehyde, **C3a**: glyceraldehyde, **C3b**: dihydroxyacetone.

2. Materials and Methods

All chemicals were used as received. O-ethylhydroxylamine hydrochloride (99%) N,O-bis(trimethylsilyl)trifluoroacetamide (BSTFA) (99%), pyridine (99%), phenyl-β-D-glucopyranoside (97%), paraformaldehyde (95%), glycolaldehyde (as dimer) (mixture of stereoisomers) and DL-glyceraldehyde (as dimer) (95%) were purchased from Sigma-Aldrich or TCI Germany GmbH. Calcium hydroxide (p.a.) was supplied by the chemical store of the Faculty for Chemistry and Pharmacy of the Ludwig-Maximilian University Munich, Germany. Portlandite and brucite were purchased from Seltene Mineralien, Gunnar Färber, Samswegen. Sodium montmorillonite clay was received from ABCR. Molecular sieves (4 Å, Type 514, pearl form) were acquired from Carl Roth GmbH & Co., KG. Water was deionized (DI) by a VWR Puranity PU 15 (VWR, Leuven, Belgium).

Adsorbed formaldehyde was prepared as follows: Anhydrous paraformaldehyde was heated under nitrogen flow to 150 °C and the gaseous monomer led through a column filled with dry molecular sieves (0.4 nm). The resulting molecular sieves contained approximately 12.5 wt% adsorbed formaldehyde, determined by weight increase.

For sugar formation under nonaqueous conditions, C2 (1.25 g, 20.8 mmol, 1.00 eq) and $Ca(OH)_2$ (310 mg, 4.18 mmol, 0.20 eq) were mixed using a vortex mixer in a 10 mL glass vial. In 2 mL glass vials, reaction mixtures of 156 mg were placed in an air-conditioned room at 23 °C.

The mechanochemically promoted synthesis of carbohydrates was conducted following these procedures: A 5 mL stainless steel ball mill jar was either charged with a) 125 mg C2 (2.08 mmol, 1.00 eq) and 31 mg $Ca(OH)_2$ (0.42 mmol, 0.20 eq) or b) 52 mg C2 (0.86 mmol, 0.50 eq), 78 mg C3a (0.86 mmol, 0.50 eq) and 26 mg $Ca(OH)_2$ (0.35 mmol, 0.20 eq). Reactions with mineral catalysts were also performed using 20 mol% catalyst and the same total mass, approximately 155 mg. The reaction mixtures were immediately ground using a single 7 mm stainless steel ball in the oscillatory ball mill CryoMill (Retsch GmbH, Haan, Germany) at a frequency of 30 Hz. Reactions with formaldehyde were conducted in 20 mL stainless steel grinding bowls with ten 10 mm balls. 1.0 g formaldehyde-loaded molecular sieves (125 mg formaldehyde, 4.16 mol, 0.50 eq.), 250 mg C2 (4.16 mmol, 0.50 eq.) and 123 mg $Ca(OH)_2$ (1.67 mmol, 0.20 eq.) were added, the jar flushed with nitrogen gas and immediately grinded in the Pulverisette 7 (Fritsch GmbH, Idar-Oberstein, Germany) at 400 rpm for 90 min.

For the reaction in aqueous solution, C2 (140 mM) was dissolved in degassed water at 40 °C, and 20 mol% Ca(OH)$_2$ was added.

If not analyzed immediately, samples were stored at −196 °C in liquid nitrogen to prevent further reaction. Aqueous solutions were lyophilized prior to derivatization.

GC-MS detection of carbohydrates followed a published protocol [40]. In short: About 2 to 5 mg of the sample was dissolved in 200 μL pyridine, mixed with 200 μL of a 40 mg/mL O-ethylhydroxylamine hydrochloride solution with 50 mM phenyl-β-D-glucopyranoside as internal standard and heated for 30 min at 70 °C on a rocking shaker. To this mixture, 120 μL BSTFA was added and the resulting solution was heated again for 30 min at 70 °C. The derivatized sugars were separated by GC-MS on a TraceGC Ultra system coupled to a PolarisQ MS (quadrupole-ion trap mass spectrometer [MS]) operated by Xcalibur software. Injections were performed using a split/splitless injector in split mode at 250 °C. Flame ionization detection was co-recorded with MS data and operated under carbon-correction at 250 °C. A SE-52 column (14 m length, ID 250 nm, 250 nm film thickness) with 80 kPa helium was used with the following temperature program: Beginning at 50 °C for 2 min and increasing temperature by 10 K/min to 140 °C and then 5 K/min to 240 °C and keeping that temperature for 2 min. Due to E/Z-isomerism of oximes, carbohydrate analytes can show two signals.

FID peak areas of respective carbohydrates (C2–C7) were corrected by their effective carbon number to account for different response factors. This number has been calculated for each derivatized sugar and the internal standard from literature values for individual functional groups [41]. For relative compositions, each sugar value was put in proportion with all relevant sugar groups. For kinetic studies, the absolute concentration of the derivatized sample was determined with the help of calibration lines and area ratios in relation to the internal standard. This composition was transferred to the molecular distribution of the mechanochemical approach. Errors of relative frequencies are standard deviations of duplicate experiments and duplicate or triplicate derivatization procedures.

3. Results and Discussion

3.1. Mineral-Catalyzed Mechanochemical Monosaccharide Synthesis

In classic organic chemistry, the formose reaction starts with formaldehyde (plus glycolaldehyde) and a base, most often calcium hydroxide, in aqueous solution. Since formaldehyde (C1) itself is gaseous and solid only below −92.15 °C, it is not easily deployable for solid phase reactions. Alternatively, the polymerized forms paraformaldehyde and trioxane could be used. However, neither of them depolymerized under the chosen conditions and were, therefore, unreactive. As a result, we employed glycolaldehyde (C2) and DL-glyceraldehyde (C3a) as smallest carbohydrate building blocks. Both have been shown to be prebiotically relevant and are connected to terrestrial and extra-terrestrial origins [42–47]. A plethora of minerals was already abundant on the early Earth during the Hadean Eon, the era when life and its building blocks is believed to have formed [48]. These represent possible catalysts for prebiotic reactions and have been shown to be active in aqueous formose settings conducted under the aspect of the origin(s) of life [49,50]. As basic catalysts are most active and the common catalyst is calcium hydroxide, three hydroxyl minerals were chosen: portlandite, brucite, and montmorillonite. Knowing about mechanochemical approaches in the context of aldol reactions [31,38,39,51–53], we tested ball milling for its ability to promote and direct the sugar formation. Upon ball milling glycolaldehyde with 20 mol% mineral additive (30 Hz, 90 min) and analysis of the product mixture after derivatization by GC-MS, we found the formation of tetroses and hexoses in all three cases (Figure 1). The reactions with brucite and montmorillonite only showed traces of hexoses, but (6.79 ± 0.93)% and (14.19 ± 2.66)% tetroses, respectively. The highest conversion was observed with portlandite (73%) yielding (42.91 ± 0.99)% tetroses and (12.16 ± 0.35)% hexoses. In comparison with pure calcium hydroxide, the product contribution is similar, but the conversion is slightly lower. After establishing the ability to form higher monosaccharides with minerals, further investigations were conducted with pure calcium hydroxide as a model catalyst.

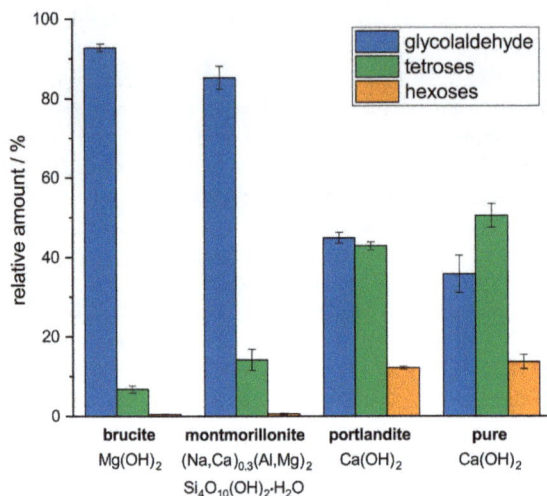

Figure 1. Comparison of different mineral catalysts with pure calcium hydroxide used in mechanochemical sugar formation starting from glycolaldehyde. Oscillatory ball mill, 30 Hz, 90 min, 20 mol% catalyst.

3.2. Model System: Nonaqueous Reaction

In a first instance, we investigated the reactivity of glycolaldehyde in the presence of 20 mol% calcium hydroxide at room temperature over the course of a month without any mechanical impact. The mixed solids were let to rest in a glass vial for a defined time interval before immediate derivatization and analysis. Already after 24 hours, we observed the formation of tetroses, namely erythrose, threose, and erythrulose, making up (1.12 ± 0.05)% of the reaction mixture. At further reaction progress, the consumption of glycolaldehyde increases while tetroses and hexoses form.

After 28 days, the initially colorless, powdery mixture turns into a yellow, viscous liquid with hexoses making up (66.36 ± 0.37)%. The change of appearance is typical for formose-type reaction of carbohydrates and is known as the yellowing point. This is when carbohydrates of higher order and complex branching occur. Further reaction results in an only partly solvable tar. Right before this point, we observe heptoses—the highest order carbohydrates resolvable by the here employed GC-MS method. The generation of heptoses is an indication of reversible reaction pathways since formal additions of C2 and multiples thereof can only result in even number carbon skeletons. Therefore, a growing proportion of retro-aldol reactions of long-chained carbohydrates can be considered.

We also investigated the temperature dependence for the initial sugar formation and found that elevated temperatures of (a) 50 °C and (b) 60 °C promote a significant monosaccharide formation within half an hour. Relative frequency for tetroses and hexoses were (a) (7.14 ± 0.23)% and (0.88 ± 0.03)% and (b) (38.38 ± 0.90)% and (5.10 ± 0.25)%, respectively. On the other hand, low temperatures (<0 °C) prevented reactivity and were, therefore, used for storage. Higher temperatures than those mentioned were not investigated due to the low melting point of glycolaldehyde. Further data on the temperature dependence are part of the Supplementary Materials.

3.3. Model System: Kinetic Investigations of Mechanochemical Reaction

After substantiating the feasibility of carbohydrate synthesis under dry conditions without any energy source, we proceeded to examine the influence of a mechanic energy source instead of temperature. In our experiments, we used an oscillatory ball mill charged with I) **C2** or II) **C2** and C3a together with Ca(OH)₂ (Figures 2 and 3). We made sure that for comparability the employed mixtures were of the same total mass. Thus, momentum transfer is equal. Grinding was performed at 30 Hz and did not cause a significant increase in temperature. On average, one hour of grinding resulted in a temperature rise of 1 K. Catalyst loading was 20 mol%. We also tested the feasibility of lower catalyst loading. It was found that it slowed down the reaction rate. For further details, see Supplementary Materials.

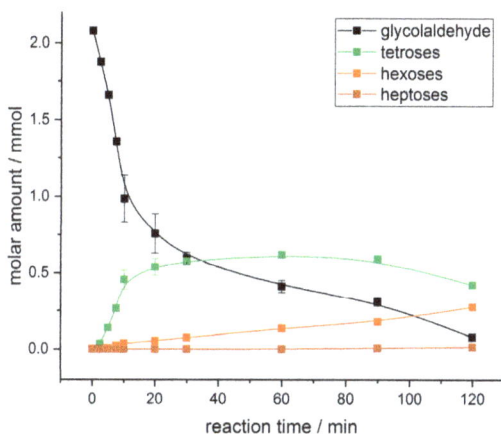

Figure 2. Reaction progress of glycolaldehyde and 20 mol% calcium hydroxide in an oscillatory ball mill at 30 Hz up to 120 min.

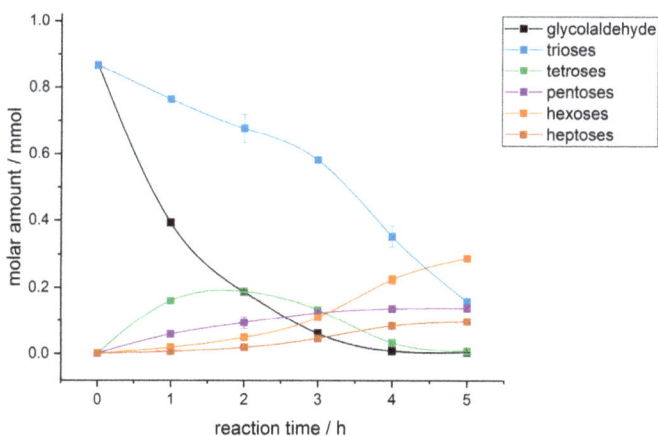

Figure 3. Reaction progress of glycolaldehyde and glyceraldehyde (1:1) plus 20 mol% calcium hydroxide in an oscillatory ball mill at 30 Hz up to 5 h.

In both cases I) and II), we observed that carbohydrate synthesis is achievable and that the reaction rate is accelerated. Results are depicted in Figures 2 and 3. In order to verify the hypothesis that grinding promotes the sugar formation through energy transfer rather than by efficient blending of the reactants, we milled the glycolaldehyde samples at 2.5, 5, 7.5 and 10 min and derivatized them all simultaneously after 10 minutes in total, so that, for instance, the sample of 2.5 min was kept for 7.5 min etc. If efficient intermixing was the sole contributor for an enhanced reactivity, all samples should show the same product distribution as within a minute of milling the substrates are already mixed effectively. However, we observed an exponential consumption of glycolaldehyde instead and, thus, conclude a kinetic acceleration by the mechanochemical energy. In fact, for longer reaction times, no significant dependence between reaction progress and milling was found. We explain this observation with a changing aggregate state of the mixture. With proceeding milling time, the reaction mixtures turn, firstly, from colorless powders into yellow viscous pastes before, secondly, they eventually become dark yellow solids that are not completely dissolved in our derivatization solutions. The kinetic data were only recorded up to this point (I) 2 h, II) 5 h). For longer reaction times the derivatized samples still contained the described monosaccharides, undissolved tar was not further investigated. During the first transition of texture, milling becomes ineffective as the sticky paste decelerates the ball within the milling jar and thus reduces energy transfer. At this time, the reaction progress is mainly diffusion controlled and is not accompanied by effective blending. As the reactants are already thoroughly mixed and in close contact in the viscous solution, the reaction rate is still sufficiently high. With further reaction progress, the mixture solidifies and becomes responsive to grinding again.

For the reaction I) starting from glycolaldehyde the course of the reaction develops as follows: Glycolaldehyde was consumed rapidly leading to the bisection of its amount after 10 min. Meanwhile tetroses were formed with a slightly slower rate. In the next ten minutes, the rate of glycolaldehyde consumption slowed down to 1/16 of the starting rate, the tetrose amount stabilized and hexoses increased steadily. After 90 min, due to less abundance of glycolaldehyde as feedstock for tetroses, their amount started to decrease in favor of hexoses. Conversion of glycolaldehyde was near to complete after 120 min. Tetroses and hexoses were present in similar amounts by that time and heptoses occurred in minor amounts.

The product distributions for milled and not milled reactions are similar but not equal. For the final measured data points (28 d in a vial vs. 120 min milling), the mechanochemical reaction sample consists of about 3.7 times more tetroses while having the same glycolaldehyde, half the hexoses and one-fifth of the heptoses abundance in relative ratios. This approach, thus, exhibits a different selectivity in favor of shorter carbohydrates. Interestingly, also heptoses are formed probably due to similar reasons as already explained for the non-mechanochemical solid phase reaction of glycolaldehyde.

For increasing the complexity of the reaction network, we further investigated the mechanochemically accelerated reaction of glycolaldehyde with calcium hydroxide in the presence of glyceraldehyde. This way, also pentoses and heptoses are accessible by direct aldol condensation reactions. As it can be inferred from Figure 3, glycolaldehyde consumption is considerably faster (3.6 times) than the one of the trioses. Glyceraldehyde is only being significantly used up after glycolaldehyde is almost 10 times less available. At this time point, the reaction rate of hexoses increased more than trifold, whereas tetroses were abundant in the experiments discussed previously, in the presence of trioses they are effectively consumed after their maximum abundance at 2 h for the sake of heptoses. Moreover, dihydroxyacetone is formed during the cause of the reaction, which likely is both the result of enediolization of glyceraldehyde and retro-aldol reactions (Supplementary Materials). In contrast to glyceraldehyde being consumed, dihydroxyacetone is enriched over the cause of the reaction.

In the final product mixture, glycolaldehyde and tetroses were nearly fully converted into higher monosaccharides. Up to the extent of our kinetic studies, triose content decreased down to a fifth of the starting amount. Similar amounts were found for pentoses and heptoses. The major components of this final product distribution were hexoses as a statistic result of having two possible reaction paths generating them: C2 + C2 + C2 and C3 + C3.

The reaction mixture consists of unbranched aldoses and ketoses and a minor amount of branched monosaccharides, but only few decomposition products. In comparison with commercially purchased samples, several monosaccharides were identified from the mixture, namely xylose, lyxose, arabinose, xylulose, ribulose, ribose, apiose, tagatose, psicose as well as sorbose and/or fructose (latter two not separated by GC-MS). As a fundamental monosaccharide for life, ribose was synthesized during the mechanochemical reaction. Ribose was formed in 2% yield and in comparison, the relative ratio of ribose to all pentoses increases over the course of reaction up to 12% after 5 h (Figure 4).

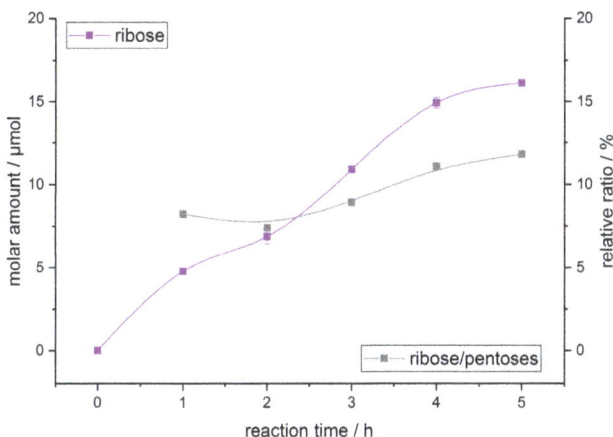

Figure 4. Synthesis of ribose and the relative ratio of ribose relating to all pentoses during the reaction of glycolaldehyde and glyceraldehyde (1:1) with 20 mol% calcium hydroxide in an oscillatory ball mill at 30 Hz up to 5 h.

3.4. Comparison with Aqueous Reaction (C2+C2)

The mechanochemical reaction was compared to its counterpart reaction in water starting with glycolaldehyde (140 mM) and calcium hydroxide at 40 °C (Figure 5). After 30 min glycolaldehyde was almost completely consumed for the higher monosaccharide synthesis. Hexoses represent the major proportion of the monosaccharides. Although the conversion was accelerated, the selectivity towards aldoses und ketoses was decreased and reached only 50%–60% with decomposition products like lactic acid taking up a substantial part of the product mixture. In the mechanochemical set-up, the reaction was much slower. Due to the deceleration, the decomposition is also delayed and selectivities around 95% are generated. In the context of the origin of life, the nonaqueous reaction supplies a variety of sugar building blocks and feedstock molecules for subsequent reactions towards biologically relevant molecules. As a result of the selectivity and decelerated formation, higher monosaccharides remain available to a higher degree for further reaction steps.

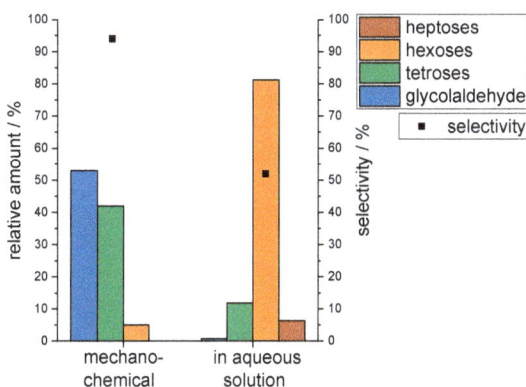

Figure 5. Composition of monosaccharides and selectivity of the aldol reaction using glycolaldehyde, catalyzed by 20 mol% calcium hydroxide, after 30 min under two different reaction conditions. Mechanochemical: oscillatory ball mill, 30 Hz. In aqueous solution: 140 mM glycolaldehyde, 40 °C.

3.5. Built-In of Formaldehyde

To investigate the reactivity of the C1 building block formaldehyde, the gaseous reactant needs to be made accessible for ball-milling. Depolymerization of paraformaldehyde or trioxane in the ball mill was not feasible in this reaction setup. Instead, availability of monomeric formaldehyde was achieved by thermal depolymerization of paraformaldehyde and adsorption to molecular sieves prior to the mechanochemical reaction and using the adsorbed formaldehyde as starting material. This relates to mineral adsorption of formaldehyde on the early Earth, which has been discussed as a prebiotic "sink" for formaldehyde [43]. These reactions were performed in a planetary ball mill in 20 mL stainless steel grinding bowls equipped with an air-tight lid under an oxygen-free atmosphere.

When the reaction was carried out solely with adsorbed formaldehyde and 20 mol% calcium hydroxide, no conversion was observed. However, when glycolaldehyde was added (1:1), monosaccharides were formed (Figure 6). In comparison, we added dry molecular sieves to the reaction of glycolaldehyde and calcium hydroxide where we found only the even-numbered carbon chain length showing that the changing product distribution is not due to the molecular sieves. In the reaction with both glycol- and formaldehyde additionally trioses, pentoses and heptoses were formed in significantly higher amounts, as has been observed previously in the reaction with only glycolaldehyde due to retroaldol reactions (Supplementary Materials). Therefore, we conclude that formaldehyde as the C1 building block is incorporated in this reaction. When the reaction was carried out not in equimolar amounts of formaldehyde and glycolaldehyde, but with catalytic amounts of glycolaldehyde (5 mol%), monosaccharide formation occurred. However, in the same reaction time only traces were detected.

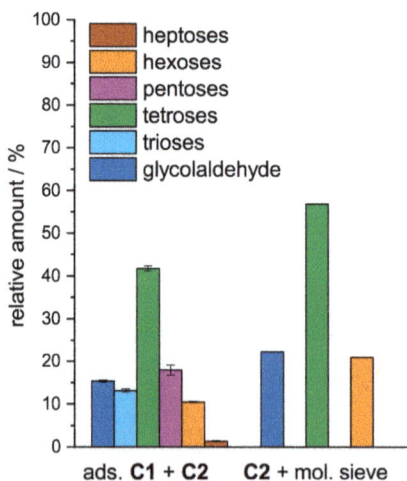

Figure 6. Product distribution of the mechanochemical aldol reaction in a planetary ball mill at 400 rpm after 90 min starting from glycolaldehyde, 20 mol% calcium hydroxide and (a) molecular sieves with adsorbed formaldehyde or (b) dry molecular sieves.

This indicates that the first step of the formose reaction, the dimerization of formaldehyde via an umpolung reaction [54,55], does not proceed under these conditions. Formaldehyde is incorporated as part of the aldol reaction network not only when equimolar amounts of glycolaldehyde are used, but also if minor amounts are already present. These can be accounted for by several suggested sources for the prebiotic occurrence of glycolaldehyde [42,43,47]. It has to be noted, that the initial concentrations of glycolaldehyde and glyceraldehyde on the early Earth were lower than the concentrations experimentally used in this study, which influences the rate of product formation, but not the intrinsic reaction rate constants.

Life **2019**, *9*, 52

4. Conclusions

In summary, we showed that the formation of carbohydrates is not exclusively feasible under an aqueous environment but also establishes under dry conditions, even with a higher selectivity towards unbranched monosaccharides. When glycolaldehyde and glyceraldehyde or formaldehyde are in contact with a basic catalyst, such as minerals or the surrogate calcium hydroxide higher sugars of complex composition form, among them those of biological relevance, such as ribose. Further, we have presented proof that by using a mechanic energy source, the nonaqueous synthesis of sugars is kinetically accelerated and that its product distribution is altered in favor of aldoses and ketoses. Based on those results, we deem it is likely that scenarios of dry and wet cycles, meteoritic impacts and lithospheric activity contributed to the origin of a complex carbohydrate reaction network with the aid of mineral catalysts. The given example of a solid phase reaction, therefore, widens the narrow scope of identified reactions feasible under conditions devoid of water in the context of the origin(s) of life.

Supplementary Materials: Additional information is available online at http://www.mdpi.com/2075-1729/9/2/52/s1.

Author Contributions: Conceptualization, S.L., S.P. and O.T.; formal analysis, S.L., S.P. and M.H.; funding acquisition, O.T.; investigation, S.L. and M.H.; methodology, S.L.; supervision, O.T.; writing—review & editing, S.L., S.P., M.H., and O.T.

Funding: This research was funded by the Max-Planck-Society and the German Research Foundation (DFG) by CRC/SFB 235 Project 05. S.P. gratefully acknowledges the Fonds der Chemischen Industrie for a Ph.D. fellowship.

Conflicts of Interest: The authors declare no conflict of interest.

References

1. Butlerow, A. Bildung einer zuckerartigen Substanz durch Synthese. *Justus Liebigs Ann. Chem.* **1861**, *120*, 295–298. [CrossRef]
2. Breslow, R. On the mechanism of the formose reaction. *Tetrahedron Lett.* **1959**, *1*, 22–26. [CrossRef]
3. Delidovich, I.V.; Simonov, A.N.; Taran, O.P.; Parmon, V.N. ChemInform Abstract: Catalytic Formation of Monosaccharides: From the Formose Reaction Towards Selective Synthesis. *ChemInform* **2014**, *45*. [CrossRef]
4. Zafar, I.; Senad, N. The Formose Reaction: A Tool to Produce Synthetic Carbohydrates Within a Regenerative Life Support System. *Curr. Org. Chem.* **2012**, *16*, 769–788. [CrossRef]
5. Socha, R.; Weiss, A.; Sakharov, M. Homogeneously catalyzed condensation of formaldehyde to carbohydrates: VII. An overall formose reaction model. *J. Catal.* **1981**, *67*, 207–217. [CrossRef]
6. Tambawala, H.; Weiss, A.H. Homogeneously catalyzed formaldehyde condensation to carbohydrates: II. Instabilities and Cannizzaro effects. *J. Catal.* **1972**, *26*, 388–400. [CrossRef]
7. Huskey, W.P.; Epstein, I.R. Autocatalysis and apparent bistability in the formose reaction. *J. Am. Chem. Soc.* **1989**, *111*, 3157–3163. [CrossRef]
8. Simonov, A.N.; Pestunova, O.P.; Matvienko, L.G.; Parmon, V.N. The nature of autocatalysis in the Butlerov reaction. *Kinet. Catal.* **2007**, *48*, 245–254. [CrossRef]
9. Evans, W.L. Some Less Familiar Aspects of Carbohydrate Chemistry. *Chem. Rev.* **1942**, *31*, 537–560. [CrossRef]
10. El Khadem, H.S.; Ennifar, S.; Isbell, H.S. Evidence of stable hydrogen-bonded ions during isomerization of hexoses in alkali. *Carbohydr. Res.* **1989**, *185*, 51–59. [CrossRef]
11. De Bruijn, J.M.; Kieboom, A.P.G.; Bekkiun, H.V. Alkaline Degradation of Monosaccharides VI1: The Fhucto-Fobmose Reaction of Mixtures of D-Fructose and Formaldehyde. *J. Carbohydr. Chem.* **1986**, *5*, 561–569. [CrossRef]
12. Cleaves, H. Prebiotic Chemistry: Geochemical Context and Reaction Screening. *Life* **2013**, *3*, 331. [CrossRef]
13. Darwin, C. *Letter to J.D. Hooker*; Hooker, J.D., Down, United Kingdom, 1871, published in: Darwin Correspondence Project, "Letter no. 7471". Available online: http://www.darwinproject.ac.uk/DCP-LETT-7471 (accessed on 19 June 2019).
14. Martin, W.; Baross, J.; Kelley, D.; Russell, M.J. Hydrothermal vents and the origin of life. *Nat. Rev. Microbiol.* **2008**, *6*, 805–814. [CrossRef] [PubMed]

15. Kitadai, N.; Nakamura, R.; Yamamoto, M.; Takai, K.; Li, Y.; Yamaguchi, A.; Gilbert, A.; Ueno, Y.; Yoshida, N.; Oono, Y. Geoelectrochemical CO production: Implications for the autotrophic origin of life. *Sci. Adv.* **2018**, *4*, eaao7265. [CrossRef] [PubMed]

16. Huber, C.; Eisenreich, W.; Wächtershäuser, G. Synthesis of α-amino and α-hydroxy acids under volcanic conditions: Implications for the origin of life. *Tetrahedron Lett.* **2010**, *51*, 1069–1071. [CrossRef]

17. Yamagata, Y.; Watanabe, H.; Saitoh, M.; Namba, T. Volcanic production of polyphosphates and its relevance to prebiotic evolution. *Nature* **1991**, *352*, 516–519. [CrossRef] [PubMed]

18. Robertson, M.P.; Miller, S.L. An efficient prebiotic synthesis of cytosine and uracil. *Nature* **1995**, *375*, 772–774. [CrossRef]

19. Shapiro, R. Prebiotic cytosine synthesis: A critical analysis and implications for the origin of life. *Proc. Natl. Acad. Sci. USA* **1999**, *96*, 4396–4401. [CrossRef]

20. Shapiro, R. Comments on 'concentration by evaporation and the prebiotic synthesis of cytosine'. *Orig. Life Evol. Biosph.* **2002**, *32*, 275–278. [CrossRef]

21. Ross, D.S.; Deamer, D. Dry/Wet Cycling and the Thermodynamics and Kinetics of Prebiotic Polymer Synthesis. *Life* **2016**, *6*, 28. [CrossRef]

22. Rode, B.M.; Fitz, D.; Jakschitz, T. The first steps of chemical evolution towards the origin of life. *Chem. Biodivers.* **2007**, *4*, 2674–2702. [CrossRef] [PubMed]

23. Bada, J.L. How life began on Earth: A status report. *Earth Planet. Sci. Lett.* **2004**, *226*, 1–15. [CrossRef]

24. Gull, M.; Zhou, M.; Fernandez, F.M.; Pasek, M.A. Prebiotic phosphate ester syntheses in a deep eutectic solvent. *J. Mol. Evol.* **2014**, *78*, 109–117. [CrossRef] [PubMed]

25. Menor-Salvan, C.; Marin-Yaseli, M.R. Prebiotic chemistry in eutectic solutions at the water-ice matrix. *Chem. Soc. Rev.* **2012**, *41*, 5404–5415. [CrossRef]

26. Clark, B.; Kolb, V. Comet Pond II: Synergistic Intersection of Concentrated Extraterrestrial Materials and Planetary Environments to Form Procreative Darwinian Ponds. *Life* **2018**, *8*, 12. [CrossRef] [PubMed]

27. Clark, B.C. Primeval procreative comet pond. *Orig. Life Evol. Biosph.* **1988**, *18*, 209–238. [CrossRef] [PubMed]

28. Oberbeck, V.R.; Aggarwal, H. Comet impacts and chemical evolution on the bombarded Earth. *Orig. Life Evol. Biosph.* **1991**, *21*, 317–338. [CrossRef]

29. Akouche, M.; Jaber, M.; Maurel, M.-C.; Lambert, J.-F.; Georgelin, T. Phosphoribosyl Pyrophosphate: A Molecular Vestige of the Origin of Life on Minerals. *Angew. Chem. Int. Ed.* **2017**, *56*, 7920–7923. [CrossRef]

30. Coveney, P.V.; Swadling, J.B.; Wattis, J.A.; Greenwell, H.C. Theory, modelling and simulation in origins of life studies. *Chem. Soc. Rev.* **2012**, *41*, 5430–5446. [CrossRef]

31. Wang, G.-W. Mechanochemical organic synthesis. *Chem. Soc. Rev.* **2013**, *42*, 7668–7700. [CrossRef]

32. Boldyreva, E. Mechanochemistry of inorganic and organic systems: What is similar, what is different? *Chem. Soc. Rev.* **2013**, *42*, 7719–7738. [CrossRef] [PubMed]

33. Bolm, C.; Mocci, R.; Schumacher, C.; Turberg, M.; Puccetti, F.; Hernández, J.G. Mechanochemical Activation of Iron Cyano Complexes: A Prebiotic Impact Scenario for the Synthesis of α-Amino Acid Derivatives. *Angew. Chem. Int. Ed.* **2018**, *57*, 2423–2426. [CrossRef] [PubMed]

34. Bolm, C.; Hernández, J.G. From Synthesis of Amino Acids and Peptides to Enzymatic Catalysis: A Bottom-Up Approach in Mechanochemistry. *ChemSusChem* **2018**, *11*, 1410–1420. [CrossRef] [PubMed]

35. Eguaogie, O.; Vyle, J.S.; Conlon, P.F.; Gîlea, M.A.; Liang, Y. Mechanochemistry of nucleosides, nucleotides and related materials. *Beilstein J. Org. Chem.* **2018**, *14*, 955–970. [CrossRef] [PubMed]

36. Buczkowski, D.; Wyrick, D. Tectonism and Magmatism on Asteroids. In Proceedings of the European Planetary Science Congress, Nantes, France, 27 September–2 October 2015.

37. Pallmann, S.; Šteflová, J.; Haas, M.; Lamour, S.; Henß, A.; Trapp, O. Schreibersite: An effective catalyst in the formose reaction network. *New J. Phys.* **2018**, *20*, 055003. [CrossRef]

38. Belén, R.; Angelika, B.; Carsten, B. A Highly Efficient Asymmetric Organocatalytic Aldol Reaction in a Ball Mill. *Chem. Eur. J.* **2007**, *13*, 4710–4722.

39. Heintz, A.S.; Gonzales, J.E.; Fink, M.J.; Mitchell, B.S. Catalyzed self-aldol reaction of valeraldehyde via a mechanochemical method. *J. Mol. Catal. A Chem.* **2009**, *304*, 117–120. [CrossRef]

40. Haas, M.; Lamour, S.; Trapp, O. Development of an advanced derivatization protocol for the unambiguous identification of monosaccharides in complex mixtures by gas and liquid chromatography. *J. Chromatogr. A* **2018**, *1568*, 160–167. [CrossRef] [PubMed]

41. Willis, D.E.; Scanlon, J.T. Calculation of Flame Ionization Detector Relative Response Factors Using the Effective Carbon Number Concept. *J. Chromatogr. Sci.* **1985**, *23*, 333–340.
42. Ritson, D.; Sutherland, J.D. Prebiotic synthesis of simple sugars by photoredox systems chemistry. *Nat. Chem.* **2012**, *4*, 895–899. [CrossRef]
43. Cleaves Ii, H.J. The prebiotic geochemistry of formaldehyde. *Precambrian Res.* **2008**, *164*, 111–118. [CrossRef]
44. Meinert, C.; Myrgorodska, I.; de Marcellus, P.; Buhse, T.; Nahon, L.; Hoffmann, S.V.; d'Hendecourt Lle, S.; Meierhenrich, U.J. Ribose and related sugars from ultraviolet irradiation of interstellar ice analogs. *Science* **2016**, *352*, 208–212. [CrossRef]
45. de Marcellus, P.; Meinert, C.; Myrgorodska, I.; Nahon, L.; Buhse, T.; d'Hendecourt Lle, S.; Meierhenrich, U.J. Aldehydes and sugars from evolved precometary ice analogs: Importance of ices in astrochemical and prebiotic evolution. *Proc. Natl. Acad. Sci. USA* **2015**, *112*, 965–970. [CrossRef] [PubMed]
46. Goesmann, F.; Rosenbauer, H.; Bredehoft, J.H.; Cabane, M.; Ehrenfreund, P.; Gautier, T.; Giri, C.; Kruger, H.; Le Roy, L.; MacDermott, A.J.; et al. COMETARY SCIENCE. Organic compounds on comet 67P/Churyumov-Gerasimenko revealed by COSAC mass spectrometry. *Science* **2015**, *349*, aab0689. [CrossRef] [PubMed]
47. McCaffrey, V.P.; Zellner, N.E.; Waun, C.M.; Bennett, E.R.; Earl, E.K. Reactivity and survivability of glycolaldehyde in simulated meteorite impact experiments. *Orig. Life Evol. Biosph.* **2014**, *44*, 29–42. [CrossRef]
48. Hazen, R.M. Paleomineralogy of the Hadean Eon: A preliminary species list. *Am. J. Sci.* **2013**, *313*, 807–843. [CrossRef]
49. Cairns-Smith, A.G.; Ingram, P.; Walker, G.L. Formose production by minerals: Possible relevance to the origin of life. *J. Theor. Biol.* **1972**, *35*, 601–604. [CrossRef]
50. Schwartz, A.W.; de Graaf, R.M. The prebiotic synthesis of carbohydrates: A reassessment. *J. Mol. Evol.* **1993**, *36*, 101–106. [CrossRef]
51. Chauhan, P.; Chimni, S.S. Mechanochemistry assisted asymmetric organocatalysis: A sustainable approach. *Beilstein J. Org. Chem.* **2012**, *8*, 2132–2141. [CrossRef]
52. Rodríguez, B.; Rantanen, T.; Bolm, C. Solvent-Free Asymmetric Organocatalysis in a Ball Mill. *Angew. Chem.* **2006**, *118*, 7078–7080. [CrossRef]
53. Kulla, H.; Haferkamp, S.; Akhmetova, I.; Röllig, M.; Maierhofer, C.; Rademann, K.; Emmerling, F. In Situ Investigations of Mechanochemical One-Pot Syntheses. *Angew. Chem. Int. Ed.* **2018**, *57*, 5930–5933. [CrossRef] [PubMed]
54. Castells, J.; Geijo, E.; López-Calahorra, F. The "formoin reaction": A promising entry to carbohydrates from formaldehyde. *Tetrahedron Lett.* **1980**, *21*, 4517–4520. [CrossRef]
55. Eckhardt, A.K.; Linden, M.M.; Wende, R.C.; Bernhardt, B.; Schreiner, P.R. Gas-phase sugar formation using hydroxymethylene as the reactive formaldehyde isomer. *Nat. Chem.* **2018**, *10*, 1141–1147. [CrossRef] [PubMed]

Article

Formation of Abasic Oligomers in Nonenzymatic Polymerization of Canonical Nucleotides

Chaitanya V. Mungi [1], Niraja V. Bapat [1], Yayoi Hongo [2,3] and Sudha Rajamani [1,*]

[1]   Indian Institute of Science Education and Research (IISER), Pune 411008, Maharashtra, India
[2]   ELSI, Tokyo-Tech (Earth-Life Science Institute, Tokyo Institute of Technology), 2-12-1, Ookayama, Meguro-ku, Tokyo 152-8550, Japan
[3]   OIST, Okinawa Institute of Science and Technology Graduate University, 1919-1 Tancha, Okinawa 904-0412, Japan
*   Correspondence: srajamani@iiserpune.ac.in; Tel.: +91-020-25908061; Fax: +91-020-25908186

Received: 31 May 2019; Accepted: 1 July 2019; Published: 4 July 2019

Abstract: Polymerization of nucleotides under prebiotically plausible conditions has been a focus of several origins of life studies. Non-activated nucleotides have been shown to undergo polymerization under geothermal conditions when subjected to dry-wet cycles. They do so by a mechanism similar to acid-catalyzed ester-bond formation. However, one study showed that the low pH of these reactions resulted in predominantly depurination, thereby resulting in the formation of abasic sites in the oligomers. In this study, we aimed to systematically characterize the nature of the oligomers that resulted in reactions that involved one or more of the canonical ribonucleotides. All the reactions analyzed showed the presence of abasic oligomers, with purine nucleotides being affected the most due to deglycosylation. Even in the reactions that contained nucleotide mixtures, the presence of abasic oligomers was detected, which suggested that information transfer would be severely hampered due to losing the capacity to base pair via H-bonds. Importantly, the stability of the *N*-glycosidic linkage, under conditions used for dry-wet cycling, was also determined. Results from this study further strengthen the hypothesis that chemical evolution in a pre-RNA World would have been vital for the evolution of informational molecules of an RNA World. This is evident in the high degree of instability displayed by *N*-glycosidic bonds of canonical purine ribonucleotides under the same geothermal conditions that otherwise readily favors polymerization. Significantly, the resultant product characterization in the reactions concerned underscores the difficulty associated with analyzing complex prebiotically relevant reactions due to inherent limitation of current analytical methods.

Keywords: prebiotic polymerization; nucleotide oligomerization; abasic oligomers; nucleotide stability; stability as a selection pressure; dry-wet cycles

1. Introduction

Polymerization of monomers under prebiotically plausible conditions would have been an essential event during the origin of life on Earth. Aligning with the RNA World hypothesis, most efforts in the past were targeted towards making RNA molecules using nonenzymatic polymerization methods. Several studies have looked at polymerization of imidazole 'activated' nucleotides. However, both the availability of these monomers in concentrated amounts and their polymerization process under early Earth conditions remains uncertain [1]. A few studies have looked at the polymerization of non-activated nucleotides and have reported the formation of RNA-like polymers by subjecting these monomers to dehydration-rehydration (DH-RH) cycles in the presence of lipids [2,3]. These DH-RH reactions were carried out under simulated volcanic geothermal conditions, a niche thought to have been prevalent on the prebiotic Earth. The chemical and thermal fluxes associated with this niche are considered to have facilitated pertinent prebiotic processes relevant to the emergence of life [4–6]. Detailed biochemical characterization of the reaction products from lipid-assisted polymerization of adenosine 5′ monophosphate (5′-AMP), suggested the presence of abasic sites in the resultant oligomers [7]. Loss of base resulted from the cleavage of the *N*-glycosidic bond, as the reactions were carried out at low pH. Depurination has been previously studied in the biological context in which abasic RNA was found to be significantly more stable than abasic DNA [8]. This study looked at the rate of strand cleavage by β,δ-elimination and 2′,3′-cyclophosphate formation and reported a 15-fold reduction in cleavage when abasic sites were present in RNA as compared to in DNA. In particular, the mechanism for loss of base has been predominantly studied in DNA-based systems due to their biological relevance. Loss of base was found to be favored at low pH, especially below pH 2.4, which corresponded to the pKa of the nucleobases that were studied [9]. Another study reported a steep increase in depurination with increased temperatures [10]. The sigmoidal curve obtained for the rate of depurination vs. temperature showed a dramatic increase in slope due to the loss of base above 85 °C.

Thereafter, a systematic kinetic analysis of the deglycosylation reaction was carried out for modified and canonical nucleosides [11]. Acidic conditions lowered the enthalpic activation parameter (ΔH) of deglycosylation, thus enhancing the ability of the leaving group and resulting in the loss of base, especially in the case of purines. The pKa of the nucleobases was found to be correlated with the stability of the glycosidic bond; bases with higher pKa (more basicity) also had higher ΔH, which resulted in lower rate constants for deglycosylation. Specifically, purine deoxyribonucleosides had half-lives ($t_{1/2}$) close to 15 min, whereas their ribonucleoside counterparts showed half-lives of about 7–8 days under low pH (0.1 M HCl) and physiological temperatures (37 °C). The study also reported half-lives at higher temperatures of 50 °C, and the $t_{1/2}$ for deoxyadenosine decreased further to only 3.2 min. In particular, the main claim of this study was that the canonical bases have the lowest rates of deglycosylation at physiological pH when compared to other modified bases. It has been argued that the stability of *N*-glycosidic linkages would have been one of the pertinent selection pressures that allowed for the refining of the genetic alphabet during transition from the RNA World(s) to a DNA-based system, allowing for efficient encoding of information [12,13].

Given the aforementioned data and our observation of depurination in polymerization reactions involving 5′-AMP [7], we decided to systematically characterize the products resulting from other contemporary RNA nucleotides when used as starting monomers. Towards this extent, we have already reported the preliminary results from reactions involving 5′-GMP, 5′-CMP, and 5′-UMP (and combinations thereof). These monomer-based reactions also resulted in the formation of oligomers (Supplementary Figure S3 in Supplementary Materials File 2) [7], but their exact biochemical nature was not ascertained. In a relevant study, native pyrimidines were observed to have long half-lives ($t_{1/2}$ of over 400 days) at pH 1 and 37 °C [14]. This therefore makes pyrimidine-based monomers better candidates for studying acid-catalyzed polymerization due to their higher glycosidic bond stability. Furthermore, the possibility of base pairing has been previously demonstrated in oligomers formed in DH-RH reactions [15], which was argued to positively impact the yield of 'intact' oligomers. To study this in greater detail, mixtures of nucleotides capable of base pairing (i.e., AMP + UMP and GMP + CMP), as well as a mixture of all four nucleotides, were subjected to DH-RH reaction conditions at low pH. All the resultant products were analyzed using mass spectroscopy to delineate the oligomers formed in various reactions. It is pertinent to consider the complexity that could emerge in the resultant products, as such 'mixed' nucleotide reactions would result in heteropolymers containing more than one type of nucleobase. This complexity is reflective of how processes would have progressed in a heterogeneous prebiotic soup. In this kind of a scenario, multiple reactants and reactions would have been affected concomitantly, with the product distributions reflecting the varying reactivity and availability of the reactants involved. Similar trends were also observed in our reactions, emphasizing the importance of factoring in molecular complexity while analyzing prebiotically pertinent reactions.

2. Materials and Methods

2.1. Materials

Adenosine 5′-monophosphate (AMP), uridine 5′-monophosphate (UMP), guanosine 5′-monophosphate (GMP), cytidine 5′-monophosphate (CMP), and ribose 5′-monophosphate (rMP) were purchased as disodium salts from Sigma-Aldrich (Bangalore, India) and used without further purification. The phospholipid, 1-palmitoyl-2-oleoyl-sn-glycero-3-phosphocholine (POPC), was purchased from Avanti Polar Lipids Inc. (Alabaster, AL, USA). All other reagents used were of analytical grade and purchased from Sigma-Aldrich (Bangalore, India).

2.2. Methods

2.2.1. Dehydration-Rehydration Cycles

Oligomerization reactions were carried out in the same set up described previously [7]. The parameters used for DH-RH cycles were selected based on previous experiments. Seven DH-RH cycles were carried out at 90 °C with 1 h of drying time. The pH of the reaction was lowered using H_2SO_4 and Milli-Q water was used as rehydrating agent for subsequent cycles. In order to check the polymerization of nucleotides, reactions were carried out with a ratio of 1:5 of lipid:nucleotide. Additionally, binary mixtures of nucleotides capable of base pairing, i.e., AMP + UMP and GMP + CMP, were also subjected to DH-RH cycles using a 1:1 ratio of both nucleotides (e.g., 2.5 mM AMP + 2.5 mM UMP). Finally, a reaction with all four nucleotides in equal ratio (1.25 mM each) was also carried out under the same reaction conditions.

2.2.2. Analysis of Deglycosylation

Deglycosylation reaction analysis was carried out mainly for AMP, as it has been reported to have the highest N-glycosidic bond stability amongst the canonical purines [11]. The AMP solution was maintained at pH 2 using H_2SO_4 and dried at 90 °C. Separate reactions vials were used for individual time points (as detailed in Section 3), and the reactions were analyzed for the loss of base. Deglycosylation reactions were also carried out without dehydration of the starting reaction mixture by heating the solutions in closed tubes. This was done in order to prevent the oligomerization that also takes place under these conditions. These samples were then analyzed by HPLC for evaluating the breakdown in each sample, and the percentage depurination was estimated as follows: (area of breakdown peak/area of monomer peak) × 100. The time required for the breakdown of N-glycosidic linkage was calculated by plotting the percentage depurination against time.

2.2.3. HPLC Analysis

Chromatography was performed using an Agilent 1260 chromatography system (Agilent Technologies, Santa Clara, CA, USA) and DNAPac PA200 column (4 × 250 mm) from Dionex (now Thermo Scientific, Sunnyvale, CA, USA). Samples were analyzed with a linear gradient of $NaClO_4$ in 2 mM Tris buffer at pH 8 using a flow rate of 1 mL/min. All solvents, purchased from Sigma-Aldrich (Bangalore, India), were of HPLC-grade and used after filtering through a 0.22 μm nylon filter followed by degassing. Samples were detected using a high-sensitivity flow cell (60 mm path length) in a diode array detector. As some reactions contained more than one type of monomer, each nucleotide was injected separately as a control, and the retention time of the respective peaks was noted. In some cases, more than one monomer peak eluted together (AMP and GMP) due to the specificity of column for phosphate groups and not necessarily for the nitrogenous bases. This aspect of the column chemistry did not allow for further optimization of peak elution profiles. Since this column offered the best single-nucleotide resolution, we performed qualitative analysis using this column despite the aforementioned limitation of this technique.

2.2.4. Mass Analysis

Detailed mass analysis of nucleotide mixtures was carried out at the Earth Life Science Institute (ELSI), Tokyo, Japan. Samples were lyophilized and shipped to ELSI at Tokyo Institute of Technology, and further mass analysis was carried out there using the Acquity UPLC+ system from Waters Corp. (Milford, MA, USA) with a CORTCES UPLC C18 column (1.6 μm, 2.1 × 50 mm) using a water/acetonitrile gradient containing 0.1% trifluoroacetic acid. Mass determination was carried out in the positive ion mode with XEVO G2-XS QTof Mass Spectrometry using MassLynx ver. 4.1 (Waters Corp.)

3. Results and Discussion

3.1. Polymerization of Canonical Nucleotides

Initially, DH-RH reactions were performed with individual nucleotides. The reaction mechanism relies on the protonation of the phosphate group and subsequent nucleophilic attack of the 2′/3′ OH of a neighboring monomer, which results in a phosphodiester bond. This proposed mechanism is independent of the nucleobase. Given this, in principle, all four nucleotides should effectively polymerize by the aforementioned acid-catalyzed esterification mechanism. We replicated the previously reported results and observed similar HPLC chromatograms for all nucleotides (Supplementary Figure S1A–F in Supplementary Materials File 2). Purine reactions showed greater polymerization; however, peaks observed in the dead volume indicated loss of nucleobases and formation of abasic oligomers.

This was further confirmed by performing mass analysis of these samples. Mass spectrometry (MS) was performed on individual nucleotide controls and on the reaction samples. Controls for all nucleotides showed the expected mass for the monomer. Additionally, mass numbers corresponding to stacked oligomers, which could result from association of molecules due to ionization, were also observed (Supplementary Figure S2 in Supplementary Materials File 2). Fragmentation of the monomer was also observed during MS as the control nucleotide samples showed mass numbers corresponding to free bases, ribose 5′-phosphate (rMP), and, in some cases, phosphate (iP) as well. Due to the possibility of such fragmentation during MS acquisition, loss of base was considered to have taken place during the reaction only if it was observed in both HPLC as well as the MS analysis.

Analysis of the reaction samples showed the presence of oligomers in all the reactions. The mass numbers obtained from this analysis are summarized in Table 1. Purine nucleotide-based reactions showed the presence of mass numbers corresponding to the respective free bases, monomers, abasic dimers, and abasic trimers. For the reactions containing pyrimidine nucleotides (UMP and CMP), mass numbers corresponding to the monomer and intact dimers were observed in both the cases. Notably, these were the only two reactions that showed the presence of intact dimers, indicating greater stability of the glycosidic bond in the pyrimidine nucleotides. However, mass numbers corresponding to free bases and abasic dimers were also observed in the pyrimidine-based reactions. Nonetheless, this did not corroborate with the observations from the corresponding HPLC analysis wherein no breakdown peaks were observed for these reactions. The presence of free base in the mass spectrum can thus be attributed to fragmentation during the ionization, especially in the case of reactions involving only pyrimidine nucleotides. Mass spectra of pyrimidine control samples also showed the presence of free bases, confirming that they potentially resulted from fragmentation during MS data acquisition. Based on the combined evidence from HPLC and MS, we can infer that the glycosidic bond cleavage in pyrimidine reactions might not have occurred during the DH-RH cycles. As stated in the introduction section, the rate of deglycosylation is variable amongst nucleobases, with purines being the most prone. Our results corroborate this observation as deglycosylation was predominantly observed in the purine reactions.

Table 1. Mass numbers observed in the reactions containing individual nucleotides. Detailed mass spectra and peaks obtained for the individual nucleotide polymerization reactions are included in the Supplementary Materials File 1.

Chemical Species	Expected Mass	Observed Mass	ppm Error
AMP Reaction			
Adenine	136.0617	136.0635	13.2290
AMP monomer	348.0703	348.0691	3.4475
Abasic Dimer	560.0778	560.0764	2.4996
Abasic Trimer	772.0874	772.0899	3.2379
GMP Reaction			
Guanine	152.0566	152.057	2.6305
GMP Monomer	364.0652	364.0627	6.8669
Abasic Dimer	576.0737	576.0715	3.8189
Abasic Trimer	788.0824	788.0803	2.6646
UMP Reaction			
Uracil	113.0345	113.0346	0.8846
UMP Monomer	325.0431	325.0435	1.2306
Intact Dimer	631.0684	631.0682	0.3169
Abasic Dimer	537.0517	537.0533	2.9792
CMP Reaction			
Cytosine	112.0505	112.0494	9.8170
CMP Monomer	324.0591	324.0596	1.5429
Intact Dimer	629.1004	629.1017	2.0664
Abasic Dimer	536.0677	536.0677	0.0000

3.2. DH-RH Reactions for Nucleotide Mixtures Capable of Hydrogen Bonding

Subsequently, reactions were also carried out under aforementioned conditions with nucleotide mixtures capable of hydrogen bonding (H-bonding). In a previous study, it was shown that the oligomers that resulted during the DH-RH reactions of AMP and UMP could hydrogen bond as assessed by the hyperchromicity analysis of the reaction products [15]. However, results obtained from our experiments suggest that the formation of abasic oligomers from purine nucleotides would potentially decrease the number of H-bonds that can be formed for a given length of oligomer. Importantly, it is well known that the canonical nucleotide monomers do not form H-bonded pairs by themselves in solution [16]. Nonetheless, in order to check whether starting with a mixture of base-pairing nucleotides might indeed positively affect the outcome of oligomer formation, reactions were carried out with binary mixtures of nucleotides (AMP + UMP and GMP + CMP) and the samples were analyzed. Base-specific separation of monomers and/or oligomers could not always be achieved using the HPLC as in previous reports [7] (Supplementary Figure S1 in Supplementary Materials File 2). Both reactions showed the presence of dead volume peaks, which typically correspond to free bases. However, it was difficult to determine whether purines or pyrimidines were being predominantly lost. However, results from the individual nucleotide reactions suggested that the loss of purines might be the major contributor to the observed HPLC breakdown peaks in these binary reaction mixtures. Subsequent characterization was, therefore, carried out by mass analysis of the resultant oligomers.

Table 2 summarizes the mass numbers observed in the mass spectrometric analysis of A + U and G + C reactions. Though exact abundances of the peaks were different in these reactions, all these reaction mixtures were comprised of abasic oligomers with some purines and pyrimidines still left intact. Oligomers found in these reactions were similar to those observed in purine-only reactions that contained multiple abasic sites with a single intact base (as depicted in Supplementary Figure S3A–D in Supplementary Materials File 2). Intact dimers were observed only in the G + C reaction but with relatively poor abundance. Since quantitative MS analysis was not performed, it is beyond the scope of this work to comment on the yields of oligomers based just on the corresponding mass abundances. Peaks observed in the dimer and trimer populations mainly consisted of abasic oligomers, often with only one intact base remaining. Completely abasic oligomers (such as rMP-rMP dimers, Supplementary Figure S3E in Supplementary Materials File 2) were not observed in these reactions possibly due to ligation of such products with other species, resulting in higher oligomers (e.g., formation of a species like AMP-rMP-rMP, etc., Supplementary Figure S3F in Supplementary Materials File 2). The loss of base could be predominantly occurring during the DH-RH reaction, as the breakdown peak was indeed observed in the HPLC analyses of these reaction mixtures.

Table 2. Peaks observed in MS analysis of mixed nucleotide reactions. Potential structures for some of the chemical species is depicted in Supplementary Figure S3 in Supplementary Materials File 2. Detailed mass spectra and peaks obtained for reactions containing base pairing nucleotides are included in the Supplementary Materials File 1.

Chemical Species	Expected Mass	Observed Mass	ppm Error
AMP + UMP Reaction			
Adenine	136.0617	136.0635	13.2290
AMP Monomer	348.0703	348.0691	3.4475
Uracil	113.0345	113.0346	0.8846
UMP Monomer	325.0431	325.0435	1.2306
Abasic A Dimer	560.0778	560.0812	6.0705
Abasic U Dimer	537.0517	537.0533	2.9792
Abasic A Trimer	772.0874	772.0899	3.2379
Abasic U Trimer	749.0603	749.0637	4.5390
GMP + CMP Reaction			
Guanine	152.0566	152.057	2.6305
GMP Monomer	364.0652	364.0627	6.8669
Cytosine	112.0505	112.0494	9.817
CMP Monomer	324.0591	324.0596	1.5429
Intact CC Dimer	629.1004	629.1017	2.0664
Intact CG Dimer	669.1065	669.1042	3.4374
Abasic G Dimer	576.0737	576.0715	3.8189
Abasic C Dimer	536.0677	536.0677	0.0000
Abasic G Trimer	788.0824	788.0803	2.6646
Abasic C Trimer	748.0763	748.0787	3.2082

The exact chemical structure of these oligomers could not be resolved due to the lack of purification methods that might have allowed for their evaluation by further analytical methods like MS-MS or NMR. Mass analysis of even the purified dimers was non-trivial due to the presence of excessive salt in the purified fraction, which resulted from the use of ion-exchange chromatography. Other chromatographic methods (such as ion-paired reverse phase chromatography) did not yield sufficient resolution for efficient purification. Nonetheless, these results strongly suggest that even in the reaction that contains base pairing nucleotides, loss of base continued to persist under our reaction conditions.

The reactions containing a mixture of all four nucleotides would, in principle, result in a complex mixture of oligomers that might be difficult to analyze. HPLC analysis of this reaction indicated the formation of oligomeric products similar to the ones seen in the binary combination of nucleotides, albeit with reduced yields [7]. Interestingly, the yields of the resultant oligomers seemed somewhat reduced in this reaction, as indicated by a lower peak intensity for the oligomers. This could potentially stem from the competition occurring between the monomers in these reactions. A breakdown peak was observed in this reaction as well, which most likely corresponded to purine bases that might have been lost due to cycling at low pH. The dimer peak resolved into multiple peaks, which could be attributed to high complexity in the dimer population resulting from the multiple combinations of interactions that are plausible between the four nucleotides. For example, there are at least 10 types of intact dimers that can form in the reaction independent of the order of the bases in them (viz. AA, AU, AG, AC, UU, UG, UC, GG, GC, CC). Apart from these, there would also be dimers that have one intact base and one abasic site (viz. Ar, Ur, Gr, Cr). MS characterization of this reaction, therefore, was found to be very challenging even at the level of the dimer population. This underscores the severe analytical constraints when working with multiple monomers under conditions that result in the preferential formation of specific products. Peaks observed in the spectrum had very low intensity/abundance, resulting in poor signal-to-noise ratios. This can partially also be attributed to poor ionization of the complex reaction mixture, thereby making the determination of the exact chemical species very difficult. Preliminary analysis indicated the formation of abasic oligomers similar to the ones that were observed in the A + U and G + C reactions. Nonetheless, further analysis could not be conducted due to the high ppm errors that were associated with the oligomers.

The difficulty in analyzing the products formed in the all four nucleotide reactions highlights the complexity that would have been intrinsic to a prebiotic reaction. Although polymerization of nucleotides ideally should take place independent of the nucleobase, this was not necessarily observed in any of our reactions. The polymerization of pyrimidines might have been slower due to the lack of efficient base stacking, and thus the formation of pyrimidine homopolymers might have been difficult in the presence of nucleotides containing other bases. Formation of abasic oligomers, which were predominantly detected in purine-only reactions, were also found to be the major products in this reaction. Importantly, the observed abasic oligomers would not be able to efficiently store information or transfer it, which diminishes their significance as prebiotically relevant informational polymers.

3.3. Deglycosylation Reactions during DH-RH Cycles

As both polymerization and deglycosylation reactions were being favored under similar conditions, we decided to further analyze the aspect of degradation of nucleotides under DH-RH conditions. The breakdown peak was predominantly observed in reactions involving purine nucleotides, either when used as monomers or even when present as mixtures. Since the AMP-based reactions always showed the most prominent breakdown, we studied the depurination aspect of this reaction in greater detail, i.e., at small time windows. The half-life ($t_{1/2}$) of a glycosidic bond has been shown to decrease by about 5 times for dAMP with a concurrent increase in temperature from 37 °C to 50 °C [11]. Given this, it was estimated that for AMP, this might correspond to a decrease from a few days to a few hours or minutes when the deglycosylation reactions were to be conducted at 90 °C. Previous experiments have shown that DH-RH cycling was necessary for and facilitated the formation of RNA-like oligomers [3]. Therefore, to minimize the oligomerization in our reactions and focus on deglycosylation, for specific time intervals post-dehydration, the samples were analyzed without invoking rehydration. The samples were analyzed at 5, 10, 15, 30, 45, and 60 min post-dehydration (Figure 1). HPLC analysis showed that the breakdown peak was observed in as early as the 5 min sample, which indicated that loss of base took place very early on. However, unlike previous reports (e.g., Ref. [3]), a small amount of oligomerization was also observed in these samples even in the absence of DH-RH cycling.

Such oligomer peaks would interfere with the quantitation of deglycosylation, as some of the monomers (intact and otherwise) would also be utilized in the formation of oligomers. Furthermore, accurate quantification of oligomeric peaks could not be conducted due to the generation of abasic site(s) in the oligomers. Given these, deglycosylation was studied in the absence of polymerization by conducting the reactions in solution.

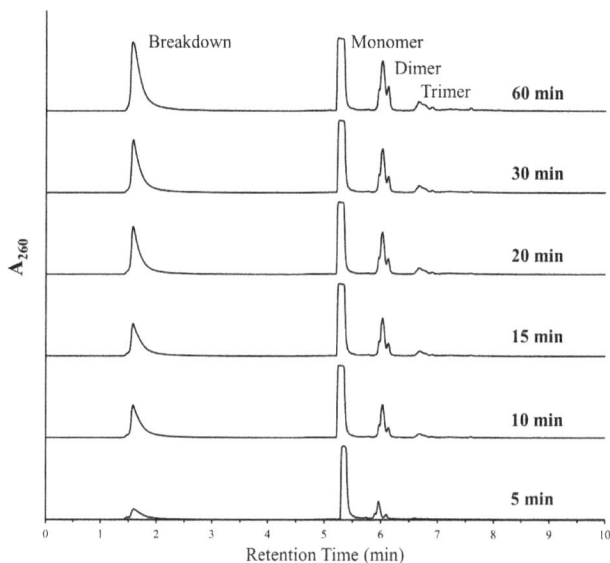

Figure 1. Breakdown and marginal polymerization was observed in reactions where AMP was heated at 90 °C, pH 2, without rehydration. HPLC analysis showed increasing peaks for free nucleobases (breakdown) and higher oligomers (predominantly dimers and trimers) with increased duration. This suggests that the cleavage of the glycosidic bond and oligomerization both occur under simultaneous experimental conditions.

Oligomerization generally requires complete dehydration, as loss of water is not feasible under aqueous conditions. We, therefore, heated the AMP reaction mix at pH 2 and 90 °C under aqueous conditions in closed vials. Initially, time points as mentioned above were taken for this reaction as well; however, only about 6–7% depurination was observed in 1 h under aqueous conditions. This reaction was then followed for 7 h, which is equivalent to the duration of seven DH-RH cycles. In the pilot reaction, almost 50% depurination was observed in the 7 h sample and this was confirmed by repeating the reaction in triplicate. The average of percentage depurination (from three reaction replicates) was plotted versus time (Figure 2), wherein the time taken for degradation of 5'-AMP to half of the starting concentration was found to be ~6.35 h. This was about a 30-fold reduction in the $t_{1/2}$ of AMP when compared to the previously reported $t_{1/2}$ of 8.2 days that was obtained for reactions analyzed at pH 1 and 37 °C [11]. This rapid degradation of AMP at high temperatures poses a serious challenge to the feasibility of undertaking long-term reactions, as a large amount of AMP would be lost in a fairly short period of time. Similar experiments were also carried out with other canonical nucleotides; only 5'-GMP showed a high rate of breakdown that was similar to AMP. Both the pyrimidines viz. UMP and CMP did not show significant deglycosylation under similar reaction conditions, even after 7 h (Supplementary Figure S4 in Supplementary Materials File 2). This was in line with previously reported results, which showed that pyrimidines have greater *N*-glycosyl bond stability against purines [17,18]. Significantly, pyrimidines did not yield oligomers with an efficiency that was comparable to purines, which rather undermines their glycosyl bond stability in the context of the formation of informational molecules under prebiotic conditions.

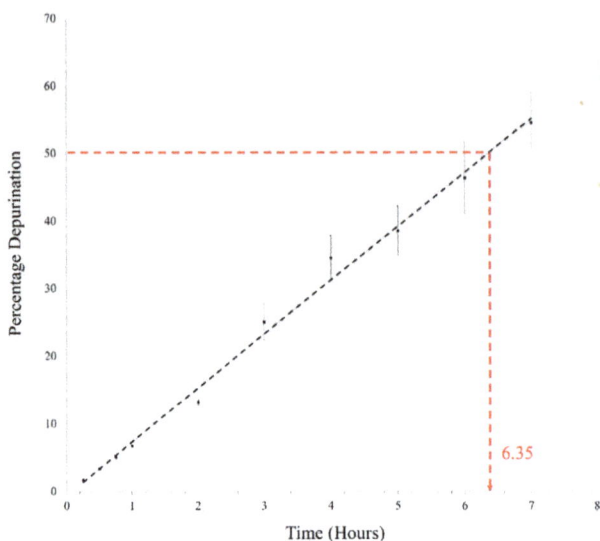

Figure 2. Percentage depurination is plotted against time for estimating the degradation of AMP under reaction conditions used for dehydration-rehydration (DH-RH) reactions. AMP gets degraded to about half of the starting concentration in ~6.35 h at pH 2 and 90 °C. (n = 3, R^2 = 0.9941).

4. Conclusions

Lipid-assisted polymerization was carried out with canonical nucleotides, as either individual monomers or binary mixtures capable of base pairing, or as a mixture of all four nucleotides. Oligomerization was observed in all the reactions, but the associated efficiency varied depending on the nature of the nucleobase. The results from MS analysis indicated the formation of abasic oligomers in almost all the reactions. Along with oligomers possessing abasic sites, intact dimers were observed only in pyrimidine-containing reactions. Purines underwent significant deglycosylation under the DH-RH conditions in all reactions studied. The extent of deglycosylation was also evaluated by characterizing the stability for purine ribonucleotides under the conditions used for DH-RH cycling. Specifically, under our DH-RH conditions, the apparent half-life of AMP was found to be only ~6.35 h, which was much lower than what has been previously reported, albeit under 37 °C [11]. Such high propensity for deglycosylation in purines raises imminent concerns about the stability of the glycosidic linkages in both monomers and the oligomers, with severe implications for storing and transferring information.

Results from this study, combined with previous relevant findings pertaining to a variety of issues surrounding canonical nucleosides [19–22], strongly advocate for the idea that modern nucleobases might have been preceded by different chemical ancestors. Specifically, low glycosidic bond stability in purines, under the same conditions that promote the formation of the phosphodiester bonds, indicates that the presence of both these moieties in the same monomer might have been unlikely under prebiotically pertinent conditions. Even though oligomerization involving non-activated nucleotides containing canonical bases may have been facilitated by other means [23,24], acid-catalyzed oligomerization would have not likely resulted in 'intact' informational polymers under prebiotic conditions.

Finally, mass analysis of complex reaction mixtures, such as the reaction containing all four nucleotides, is challenging at the very least. A study that reported MALDI analysis of oligomerization products using montmorillonite clay, had demonstrated the presence of up to 30-mer oligomers [25]. However, a common criticism has been the possibility of generating false positives during MS data acquisition, and hence there is a need for accurate calibration and careful sample preparation to minimize complications [26]. Few other studies have also resulted in a complex mix of products like the famous Formose reaction that presented major analytical challenges [27]. Nonetheless, despite the intrinsic difficulty associated with discerning such complex and heterogeneous mixtures, their analysis is very crucial for characterizing prebiotic reactions [28]. The contemplation of the inherently complex nature of prebiotic reactions and the efforts to simulate this and study them in detail has resulted in the recent emergence of the new field of 'messy chemistry' [29]. Nevertheless, it is also crucial to consider that the intrinsic heterogenic nature of the 'substrate chemical space' would potentially lead to an even more complex and diverse 'product chemical space'. More importantly, attempts to analyze the reactions with mixtures of starting material, which represent the complexity associated with prebiotic inventory, have recently begun [30]. Nonetheless, development of highly sensitive and robust analytical techniques, as well as bold approaches towards simulating and characterizing complex prebiotic reactions, is increasingly being acknowledged as being fundamental to solving the mystery of the origins of life on Earth.

Supplementary Materials: The following are available online at http://www.mdpi.com/2075-1729/9/3/57/s1, Supplementary Materials File 1: Detailed mass spectra and peaks obtained from individual nucleotide polymerization reactions as well as reactions containing base pairing nucleotides; Figure S1: HPLC chromatograms of oligomerization reactions. (Reproduced from Ref. [4]); Figure S2: Mass spectrums of nucleotide monomer controls showing presence of free bases due to fragmentation.; Figure S3: Potential chemical structures of dimers and trimer based on the masses observed; Figure S4: Analysis of deglycosylation of nucleotides when heated at 90 °C, pH 2, under aq. solution.

Author Contributions: This work was conceptualized by C.V.M. and S.R. C.V.M., Y.H., and S.R. contributed towards the methodology. C.V.M., N.V.B., Y.H., and S.R. analyzed and interpreted the data. C.V.M., N.V.B., and S.R. contributed towards the manuscript preparation, with inputs from Y.H. Funding for this work as acquired by S.R.

Funding: This research was funded by the Science and Engineering Research Board (SERB), DST, Government of India, vide grant No. EMR/2015/000434. The authors would also like to acknowledge that this publication was also made possible through the support of the ELSI Origins Network program (ELSI-EON), funded by a grant from the John Templeton Foundation, at the Earth-Life Science Institute of the Tokyo Institute of Technology. C.V.M. and N.V.B. acknowledge the research fellowship received from CSIR, Government of India.

Acknowledgments: The authors wish to express their gratitude to Kuhan Chandru, Tony Jia, Irena Mamajanov, Hanako Ricciardi, and the ELSI-EON Staff for their help and support. We also wish to express our sincere gratitude to IISER, Pune, for their constant support.

Conflicts of Interest: The authors declare no conflict of interest.

References

1. Huang, W.; Ferris, J.P. One-step, regioselective synthesis of up to 50-mers of RNA oligomers by montmorillonite catalysis. *J. Am. Chem. Soc.* **2006**, *128*, 8914–8919. [CrossRef] [PubMed]
2. Deamer, D. Liquid crystalline nanostructures: Organizing matrices for non-enzymatic nucleic acid polymerization. *Chem. Soc. Rev.* **2012**, *41*, 5375–5379. [CrossRef] [PubMed]
3. Rajamani, S.; Vlassov, A.; Benner, S.; Coombs, A.; Olasagasti, F.; Deamer, D. Lipid-assisted synthesis of RNA-like polymers from mononucleotides. *Orig. Life Evol. Biosph.* **2008**, *38*, 57–74. [CrossRef] [PubMed]
4. Van Kranendonk, M.J.; Deamer, D.W.; Djokic, T. Life springs. *Sci. Am.* **2017**, *317*, 28–35. [CrossRef] [PubMed]
5. Djokic, T.; Van Kranendonk, M.J.; Campbell, K.A.; Walter, M.R.; Ward, C.R. Earliest signs of life on land preserved in ca. 3.5 Ga hot spring deposits. *Nat. Commun.* **2017**, *8*, 15263. [CrossRef] [PubMed]
6. Forsythe, J.G.; Yu, S.; Mamajanov, I.; Grover, M.A.; Krishnamurthy, R.; Fernµndez, F.M.; Hud, N.V. Ester-Mediated Amide Bond Formation Driven by Wet–Dry Cycles: A Possible Path to Polypeptides on the Prebiotic Earth. *Angew. Chem. Int. Ed.* **2015**, *54*, 9871–9875. [CrossRef] [PubMed]
7. Mungi, C.V.; Rajamani, S. Characterization of RNA-Like Oligomers from Lipid-Assisted Nonenzymatic Synthesis: Implications for Origin of Informational Molecules on Early Earth. *Life* **2015**, *5*, 65–84. [CrossRef]
8. Leumann, C.J.; Küpfer, P.A. The chemical stability of abasic RNA compared to abasic DNA. *Nucleic Acids Res.* **2007**, *35*, 58–68.
9. Suzuki, T.; Ohsumi, S.; Makino, K. Mechanistic studies on depurination and apurinic site chain breakage in oligodeoxyribonucleotides. *Nucleic Acids Res.* **1994**, *22*, 4997–5003. [CrossRef]

10. Lindahl, T.; Nyberg, B. Rate of depurination of native deoxyribonucleic acid. *Biochemistry* **1972**, *11*, 3610–3618. [CrossRef]

11. Rios, A.C.; Yua, H.T.; Tor, Y. Hydrolytic fitness of N-glycosyl bonds: Comparing the deglycosylation kinetics of modified, alternative, and native nucleosides. *J. Phys. Org. Chem.* **2015**, *28*, 173–180. [CrossRef] [PubMed]

12. Rios, A.C.; Tor, Y. Refining the Genetic Alphabet: A Late-Period Selection Pressure? *Astrobiology* **2012**, *12*, 884–891. [CrossRef] [PubMed]

13. Rios, A.C.; Tor, Y. On the Origin of the Canonical Nucleobases: An Assessment of Selection Pressures across Chemical and Early Biological Evolution. *Isr. J. Chem.* **2013**, *53*, 469–483. [CrossRef] [PubMed]

14. Shapirof, R.; Melvyn, D.J. Acidic Hydrolysis of Deoxycytidine and Deoxyuridine Derivatives. The General Mechanism of Deoxyribonucleoside Hydrolysis. *Biochemistry* **1972**, *11*, 23–29. [CrossRef] [PubMed]

15. DeGuzman, V.; Vercoutere, W.; Shenasa, H.; Deamer, D. Generation of Oligonucleotides Under Hydrothermal Conditions by Non-enzymatic Polymerization. *J. Mol. Evol.* **2014**, *78*, 251–262. [CrossRef] [PubMed]

16. Ts'o, P.O.P.; Melvin, I.S.; Olson, A.C. Interaction and Association of Bases and Nucleosides in Aqueous Solutions. *J. Am. Chem. Soc.* **1963**, *85*, 1289–1296. [CrossRef]

17. Garrett, E.R.; Seydel, J.K.; Sharpen, A.J. The Acid-Catalyzed Solvolysis of Pyrimidine Nucleosides1. *J. Org. Chem.* **1966**, *31*, 2219–2227. [CrossRef]

18. Gates, K.S. An overview of chemical processes that damage cellular DNA: Spontaneous hydrolysis, alkylation, and reactions with radicals. *Chem. Res. Toxicol.* **2009**, *22*, 1747–1760. [CrossRef]

19. Orgel, L.E. Prebiotic chemistry and the origin of the RNA world. *Crit. Rev. Biochem. Mol. Biol.* **2004**, *39*, 99–123.

20. Hud, N.V.; Cafferty, B.J.; Krishnamurthy, R.; Williams, L.D. The origin of RNA and "my grandfather's axe". *Chem. Biol.* **2013**, *20*, 466–474. [CrossRef]

21. Kim, H.; Benner, S.A. Prebiotic Glycosylation of Uracil with Electron-Donating Substituents. *Astrobiology* **2015**, *15*, 301–306. [CrossRef] [PubMed]

22. Mungi, C.V.; Singh, S.K.; Chugh, J.; Rajamani, S. Synthesis of barbituric acid containing nucleotides and their implications for the origin of primitive informational polymers. *Phys. Chem. Chem. Phys.* **2016**, *18*, 20144–20152. [CrossRef] [PubMed]

23. Da Silva, L.; Maurel, M.-C.C.; Deamer, D. Salt-Promoted Synthesis of RNA-like Molecules in Simulated Hydrothermal Conditions. *J. Mol. Evol.* **2015**, *80*, 86–97. [CrossRef] [PubMed]

24. Himbert, S.; Chapman, M.; Deamer, D.W.; Rheinstädter, M.C. Organization of nucleotides in different environments and the formation of pre-polymers. *Sci. Rep.* **2016**, *6*, 31285. [CrossRef] [PubMed]

25. Zagorevskii, D.V.; Aldersley, M.F.; Ferris, J.P. MALDI analysis of oligonucleotides directly from montmorillonite. *J. Am. Soc. Mass Spectrom.* **2006**, *17*, 1265–1270. [CrossRef] [PubMed]

26. Burcar, B.T.; Cassidy, L.M.; Moriarty, E.M.; Joshi, P.C.; Coari, K.M.; McGown, L.B. Potential Pitfalls in MALDI-TOF MS Analysis of Abiotically Synthesized RNA Oligonucleotides. *Orig. Life Evol. Biosph.* **2013**, *43*, 247–261. [CrossRef]

27. Islam, S.; Powner, M.W. Prebiotic Systems Chemistry: Complexity Overcoming Clutter. *Chem* **2017**, *2*, 470–501. [CrossRef]

28. Cleaves, H.J. Prebiotic chemistry: Geochemical context and reaction screening. *Life* **2013**, *3*, 331–345. [CrossRef]

29. Nicholas, G.; Nathaniel, V.; Kuhan, C.; Caleb, S.; Irena, M. Bulk measurements of messy chemistries are needed for a theory of the origins of life. *Philos. Trans. R. Soc. A Math. Phys. Eng. Sci.* **2017**, *375*, 20160347.

30. Meringer, M.; Cleaves, H.J. Computational exploration of the chemical structure space of possible reverse tricarboxylic acid cycle constituents. *Sci. Rep.* **2017**, *7*, 17540. [CrossRef]

MDPI

St. Alban-Anlage 66

4052 Basel

Switzerland

Tel. +41 61 683 77 34

Fax +41 61 302 89 18

www.mdpi.com

Life Editorial Office

E-mail: life@mdpi.com

www.mdpi.com/journal/life